Leadership in the Construction Industry

This book presents a new framework for leadership in the construction industry which draws from the authentic leadership construct. The framework has three major themes: self-leadership, self-transcendent leadership, and sustainable leadership.

Despite its significance, leadership has not been given due importance in the construction industry as focus is placed on managerial functionalism. At the project level, even with the technological advances in the industry in recent years, construction is realized in the form of people undertaking distinct interdependent activities which require effective leadership. The industry faces many challenges including: demanding client requirements and project parameters; more stringent regulations, codes and systems; intense competition in the industry; and threats from disruptive enterprise. In such a complex environment, technology-driven and tool-based project and corporate management is insufficient. It must be complemented by a strategic, genuine, stakeholder-focused and ethical leadership.

Leadership in the Construction Industry is based on a study on authentic leadership and its development in Singapore. Leadership theories and concepts are reviewed; the importance of leadership in the construction industry is discussed; and the grounded theory approach which was applied in the study is explained. Many eminent construction professionals in Singapore were interviewed in the field study. Emerging from the experiences of the leaders documented in this book are three major themes (1) self-leadership: how leaders engage in various self-related processes such as self-awareness, self-regulation, and role modeling; (2) self-transcendent leadership: how leaders go beyond leading themselves to leading others through servant leadership, shared leadership, spiritual leadership, and socially-responsible leadership; and, finally, (3) sustainable leadership or the strategies leaders employ to make the impact of their leadership lasting. A synthesis of these themes and their implications for leadership development is presented before the book concludes with some recommendations for current and aspiring leaders about how they can engage with them. This book is essential reading for all construction practitioners from all backgrounds; and researchers on leadership and management in construction.

George Ofori is Professor and Dean of the School of the Built Environment and Architecture at London South Bank University, UK. He is joint co-ordinator of CIB TG95 on Professionalism and Ethics in Construction, and Deputy Chair of the Board of CoST, the Infrastructure Transparency Initiative and a Member of the Board of Trustees of Engineers Against Poverty. He has published over 110 refereed journal papers, 5 books and 29 chapters, mainly on construction industry development, leadership in construction, and sustainability in construction.

Shamas-ur-Rehman Toor received his PhD from the National University of Singapore. He is an international development practitioner and currently works at the Asian Infrastructure Investment Bank (AIIB), Beijing, China. Previously, he worked at the Islamic Development Bank (IsDB) Group, Jeddah, Saudi Arabia, and the University of New South Wales (UNSW), Sydney, Australia. He has published over 30 international refereed journal papers, 3 book chapters and 38 international peer-reviewed conference papers in the areas of leadership development, organizational behavior, and project management.

Leadership in the Construction Industry

Developing Authentic Leaders in a Dynamic World

George Ofori and Shamas-ur-Rehman Toor

Routledge
Taylor & Francis Group

LONDON AND NEW YORK

First published 2021
by Routledge
2 Park Square, Milton Park, Abingdon, Oxon OX14 4RN

and by Routledge
52 Vanderbilt Avenue, New York, NY 10017

Routledge is an imprint of the Taylor & Francis Group, an informa business

British Library Cataloguing-in-Publication Data
A catalogue record for this book is available from the British Library

Library of Congress Cataloging-in-Publication Data
Names: Ofori, George, author. | Toor, Shamas-ur-Rehman, author.
Title: Leadership in the construction industry : developing authentic leaders in a dynamic world / George Ofori and Shamas-ur-Rehman Toor.
Description: First edition. | Abingdon, Oxon ; New York : Routledge/Taylor & Francis Group, 2021. | Includes bibliographical references and index.
Identifiers: LCCN 2020041902 (print) | LCCN 2020041903 (ebook) | ISBN 9780367482312 (hardback) | ISBN 9780367482152 (paperback) | ISBN 9781003038757 (ebook)
Subjects: LCSH: Construction industry. | Leadership.
Classification: LCC HD9715.A2 O37 2021 (print) | LCC HD9715.A2 (ebook) | DDC 624.068/4--dc23
LC record available at https://lccn.loc.gov/2020041902
LC ebook record available at https://lccn.loc.gov/2020041903

ISBN: 978-0-367-48231-2 (hbk)
ISBN: 978-0-367-48215-2 (pbk)
ISBN: 978-1-003-03875-7 (ebk)

Typeset in Baskerville
by Taylor & Francis Books

George Ofori:
This book is dedicated Nana Ofori. She has been patient, generous and supportive.

Shamas-ur-Rehman Toor:
This book is dedicated to my father, Mohammad Zafar ul Islam Toor, and my mother, Aziz Akhtar, who shaped my personality and my leadership before I had heard about it.

Contents

Illustrations

Figures

Tables

Box

Preface

Leadership plays an important role in all walks of life. Leadership is critical in the construction industry. It is necessary at all levels of the construction industry – projects, firms, professional institutions, and trade associations.

Project leadership is more necessary in construction than in any other field of activity owing to the nature of the construction project. Construction projects are among the most complex of all human activities. On a construction project site, leadership is needed at the team and at the intra- and inter-trade levels to ensure the attainment of the project objectives. Construction companies are confronted with a market structure, the nature of the clients and the operating environment to which they must respond with an industry structure and working practices and procedures, which constitutes a recipe for less than optimal performance. Construction companies are also typically slow to apply new business practices, adopt technology, and embrace innovation. Good leadership can make a big difference in how construction companies conduct their business and affect their ability to provide lasting value to their clients. Leadership is key in the professional institutions and trade associations as the social image of construction is not attractive in most countries and the industry must do much to recover the loss of trust by the society. Finally, considering the vital role that the construction industry plays in socio-economic development, it is even more crucial that we deepen our understanding of leadership.

Leadership has been studied since antiquity. The field has gone through many stages, during each of which a particular paradigm has been popular, although at all stages, the approaches proposed up to that point co-exist, and there is a tendency for the earlier ideas to make a strong return, although in somewhat modified forms. One of the more recent trends in leadership studies is a focus on positive forms of leadership. The authentic leadership construct is one of these. Some of its adherents and advocates argue that it is a root construct underlying all positive forms of leadership. We focus on authentic leadership in this book.

Leadership development is possible and necessary. The writings of the ancient philosophers like Plato and Aristotle, Confucius, and Lao Tzu make this clear. We need better and more capable leaders to deal with the challenges that we collectively face. We deserve to be led by positive, ethical, and authentic leaders who are true to themselves and to others. Leaders who can think big and in the best interest of everyone around them.

This book is based on a study of "Authentic Leadership Development in the Construction Industry in Singapore." It explores this new construct which we considered to be particularly important to construction owing to the issues the industry faces in most countries. The study applied the grounded theory approach. As authors, we have produced many papers on leadership, focusing on leadership development in the construction industry. This book covers the entire study on which most of the papers were based. It is not a collection of the previously published papers. We believe that it provides fresh insights into the subject of leadership in construction.

We are grateful to the many eminent professionals and proven authentic leaders in Singapore who gave so generously of their time to make the study possible. The interview sessions for the study were major learning experiences for us. The journey of each leader we spoke to helped us uncover various dimensions of authentic leadership and eventually write this book. In the book, we let the leaders speak; we use extracts from the interviews to present the analysis of the field study. We believe this lends authenticity to the message of the book.

We are trying to accomplish many things in this book. First, we set out to add to the understanding of leadership and its importance to construction. We start with a comprehensive review of the knowledge on leadership in general, and focus on the new construct of authentic leadership. We tackle some of the unsettled debates in the field, in particular, the "born or made" debate, and that on "chance or choice."

Second, we want existing leaders and those aspiring to be leaders to realize that it is possible for them to make efforts, on their own, to continuously develop their leadership capabilities. The discussion of the antecedents of leadership development, of reflection, and of self-realization is important to help leaders and aspiring leaders understand what their leadership really stands for.

Third, we offer insights from the lives of leaders "who have been there and done that." Extensive interviews with leaders who had gained a reputation of being authentic provide examples of how leaders develop, and how they exercise leadership in real life.

Fourth, we want to provide some guidance on appropriate research approaches in studies on leadership. The discussion of research methodology is deliberately detailed, especially for the benefit of new researchers. We share a model for authentic leadership development which we have built, from the findings in the study. In the Conclusion, more discussion on areas for further research is presented.

Fifth, we believe our research can help develop interventions to accelerate authentic leadership development. While reading this book can bring several insights to the existing and aspiring leaders and help them discover their own leadership, we believe that targeted interventions to develop authentic leadership need to be carefully designed, tested, and implemented. There is already some progress in this area but much more needs to be done and we invite our fellow researchers, practitioners, and trainers further advance this work.

We hope this book will be useful to everyone in our five audiences: leadership researchers; leaders in the construction industry; construction practitioners aspiring to be leaders; educational institutions and training providers; and students and trainees on leadership courses, either as part of degree programmes, or standalone training programmes. We hope that readers of this book will read it not only from the academic perspective but also from the perspective of discovering, practicing, and sustaining their own authentic leadership.

The structure of the book

Chapter 1 The leadership crisis

The unethical managerial practices and bad leadership of a number of high-profile corporate executives have resulted in a loss of trust of leaders by society. Examples of these include the Enron and Worldcom failures. Many consider the severe economic recession of 2008 not only as a collapse of the economy but also a meltdown of leadership. There have been many corporate scandals in which leaders have deceived and breached the trust not only of the shareholders and direct stakeholders in their organizations, but also, indirectly, of many others around the world. There is a need for a fresh look at how ethical and authentic leadership is practiced in organizations.

The construction industry does not have a good social image in most countries. This widely held perception has been formed from instances of corruption; exposure of mal-practice and mismanagement; poor quality of work which sometimes, leads to fatalities; and the adverse environmental and social impact of some projects. High profile corporate collapses, tragedies in built items and fraud and corruption scandals have been recent occurrences in the construction industries in a number of countries.

Chapter 1 begins by highlighting the crucial role that leadership plays in commercial orga-nizations in the challenging operating environment today. It is argued that a new approach to leadership that is grounded in the authenticity of leaders must be adopted in order to enable construction firms to face the challenges of their businesses as well as reap the benefits that factors such as advanced technologies and today's knowledge workers have to offer to firms.

Chapter 2 The role of leadership in the construction industry

Organizations in all sectors of the economy are under increasing pressure to offer value-added services, and learn to survive and grow in the face of increased competition, rapid change, and disruption. Construction deals with the strategic agendas and commercial needs of businesses and other parties involved in it, and the economic and social aspirations of the beneficiaries of its projects and the wider society in an era of technological advances and an increasing role of governments as a regulator. In order to achieve the project objec-tives, the customary control-driven management approach has less potential to succeed. It should be catalyzed with leadership.

The changing landscape of the construction industry also demands that the practitioners should be able to anticipate and drive the change instead of responding to, and coping with, developments. The construction industry needs its managers to develop leadership capabilities to successfully achieve the strategic objectives of the projects which involve several stakeholders with different vested interests, who also come from many socio-cultural backgrounds. Chapter 2 focuses on the importance of leadership for the construction industry, construction companies, and construction projects.

Chapter 3 Understanding leadership

Leadership has attracted the attention of researchers in a wide range of fields, from history to sociology, from military studies to political science, and from business to education. The study of leadership dates back many centuries, and a huge volume of literature has been formed. However, there are still many gaps in the knowledge on leadership and there continues to be a debate about the best strategies for developing and exercising leadership.

It is suggested in the literature that the development of leadership research can be divided into various distinct stages. The most evident progression is that from inherent traits of special people most suited to be leaders; and from determinism and certainty in the practice of lea-dership to flexibility and context-specificity. Some of the earlier concepts remain current while the field continues to develop. In recent years, biological scientists have begun to show interest in leadership, and some highly significant work is underway, which might lead to scientific confirmation or determination of some of the leadership concepts and relationships.

Some of the ways in which leadership has been conceptualized over many centuries are discussed and compared in Chapter 3. An agenda for future research on leadership is presented.

Chapter 4 Authentic leadership

There has been a surge of interest in authentic leadership over the past decade. Researchers on leadership have recognized that leadership is more than style, traits, or behaviors. The rising complexity of business, uncertainty of market conditions, increasing competition, and the greater demands of knowledge workers call for a focus on restoring confidence and optimism among employees and society as a whole; displaying resilience; fostering a new self-awareness; and the establishment of genuine relationships with all stakeholders. There is growing recognition among scholars and practitioners that a more authentic leadership development strategy has become relevant, and is needed in order for organizations to attain desirable outcomes in their present operating environments.

The first part of Chapter 4 discusses the authentic leadership construct and reviews the research on the subject. The chapter discusses how authentic leadership differs from other forms of leadership. It also presents an integrated model for authentic leadership development and considers sub-themes of the model which are taken up in the subsequent chapters of the book.

"Effective leadership" is one of the main answers to the many problems of the construction industry which hinder the performance of this important segment of the economy. Thus, attention should be paid to leadership development, at least for the sake of construction. The debate on whether leaders are born or can be developed continues. However, the current consensus appears to be that, whereas certain traits which are desirable in leaders may be naturally endowed, one can develop some, if not most, of the attributes and capabilities of leaders through appropriate structured interventions.

The second part of the chapter covers leadership development. It considers the suggestions which have been made on how leaders can be developed, and the quality of leadership in individuals enhanced, with appropriate interventions.

Chapter 5 Research design and the grounded theory approach

For the benefit of researchers who will read this book, it is necessary to discuss the methodology suitable for leadership research, and the method and approach adopted for the study on which the book is based. Leadership research has changed in many ways in recent decades. There is greater methodological diversity in such research as the complexity of the subject has been recognized. There has been an increase in the use of qualitative research approaches to study leadership. These are deemed to provide greater richness and comprehensiveness in the exploration of the subject. The study for this book used the infrequently adopted research approach: grounded theory. It is pertinent to discuss the nature of this approach, why it was adopted for the study, and lessons following its adoption and application.

Chapter 5 explains the methodology used to accomplish the research objectives. The research focuses on "authentic leadership development and influence in the construction industry" and the key output of this research is to develop a "grounded theory framework of authentic leadership development and influence" in the construction industry. The discussion also covers the data collection approach and the instrument used, the target population, the sampling method and other related issues. The data analysis and validation approach and, finally, the issue of validity and reliability of the application of the resulting grounded theory are considered. As the field study was undertaken in Singapore, a note on Singapore's economy and its construction industry is included to provide a background. It includes the structure and practices of the construction industry; and measures for developing the construction industry and enhancing its performance.

Chapter 6 Leadership development

The field study was undertaken to explore the subject of authentic leadership development and the related one of leadership influence. The field study had two components: a survey of postgraduate students in project management; and interviews with eminent construction professionals and practitioners. The results of the simple quantitative analysis of the exploratory survey formed the basis of the development of the interview questions for the larger study. The information from the interviews was analyzed using the guidelines of the grounded theory approach.

Chapter 6 presents the results of the field study. It begins with consideration of the exploratory survey of project management. The summary results of the in-depth field interviews of authentic leaders is presented; these reflect the process of authentic leadership development over a person's life span. The concepts and categories that emerged as a result of open and axial coding of the interviews with authentic leaders are presented in summary form. A framework of authentic leadership development that emerged through the integration of the main categories is presented.

Chapter 7 Self-leadership

Leadership begins from the self. An authentic leader is a person who is deeply aware of his or her values and life story and is able to align these with his or her purpose of leadership. An authentic leader also engages in self-regulation through the leader's behavior and the relationships the leader establishes. Such a leader engages in authentic role modeling and makes efforts to develop his or her professional knowledge and technical competence, and to hone his or her leadership approach.

Chapter 7 considers the processes that make up self-leadership. Authentic leaders are inner-focused (self-aware) and able to align their values, purpose, goals, and ambitions to each other. Such an individual is able to connect to his or her core-self, and are then able to move to the next step, that is, self-transcendent leadership, discussed in Chapter 8.

Chapter 8 Self-transcendent leadership

Another major theme of leadership that emerged during the interviews with authentic leaders in Singapore's construction industry was "self-transcendence." This was evident in the interviewees' motives, goals, leadership philosophy, decisions, and actions. They were not only keen on attaining their own goals or achievements, but also were concerned about the welfare and well-being of their followers, colleagues, organizations, and society. Six sub-themes emerged under the category of self-transcendent leadership. These are: soulful leadership, servant leadership, spiritual leadership, shared leadership, service-oriented leadership, and socially responsible leadership.

Authentic leaders are able to translate their self-awareness into self-transcendence or the desire to contribute to the lives of others. Self-transcendent leaders give priority to the needs of their followers, organizations, and other stakeholders. Chapter 8 explores this theme, and its relevance to construction projects and companies.

Chapter 9 Sustainable leadership

The third category that emerged in the study under "authentic leadership influence" was "sustainable leadership." Authentic leaders strive to sustain their leadership in order to ensure that the good aspects of it do not wane over time. They take steps to establish systems and a culture within their organizations that encourage their subordinates to grow as authentic followers.

Categories that emerged from the lower-level concepts in Chapter 9 include: creative problem-solving, sustainable human capital development, sustainable social capital development, building high-performance and sustainable teams, building sustainable organizations, and ensuring sustained succession planning. Higher-order abstraction of these categories resulted in the broader category of "sustainable leadership" which refers to strategies, actions, and decisions of authentic leaders to ensure the sustenance of their leadership. Chapter 9 focuses on how leaders engage in various strategic social processes to sustain their leadership in their organizations.

Chapter 10 Integration: reconciliation of self with social realities

A major step in the guidelines of the grounded theory approach to research is the integration of categories which emerge in the analysis of the results, and the findings. The step is selective coding, the final coding phase in the grounded theory approach. Selective coding integrates all the interpretive work of the analysis.

In Chapter 10, the categories presented in the preliminary model of the study are further developed and integrated. The core category of the study was "dynamic and creative reconciliation of self and social realities." The story line to this core category and the other categories is developed. The validation of the final model of the study and key findings is also presented. It is compared with the conceptual model based on a review of the literature, and the differences highlighted as contributions to the body of knowledge on authentic leadership.

Chapter 11 Further development and application of the root construct

Chapter 11 presents a summary of the study and the findings of the research. Authentic leadership is compared with other forms of leadership, from the new insights derived from the study. It is inferred that the findings confirmed that authentic leadership is a root construct of positive forms of leadership. The research propositions outlined in the earlier chapters of the book are discussed. The contributions the study has made to knowledge are outlined. The implications for practice to current leaders and aspiring leaders; and human resource management are considered. A critical evaluation of the grounded theory research methodology is presented. The limitations of the research are also discussed. Recommendations for future research are presented. The conclusion to the whole study ends the chapter.

Chapter 12 A final reflection

A final reflection on authentic leadership in construction is presented in Chapter 12. A follow-up study in 2020 which sought the current views of some of the leaders interviewed in the main study is analysed. The questions related to how their leadership has changed, and what their leadership in the current era involves, and their views on leadership development in construction in the future. Finally, the authors make personal comments on the research and on the subject of leadership development, consider the place of authentic leadership in current events, and look forward in anticipation of significant developments in the field of leadership research.

Abbreviations

ALI	Authentic Leadership Inventory
ALQ	Authentic Leadership Questionnaire
ANOVA	Analysis of Variance
ASCE	American Society of Civil Engineers
BCA	Building and Construction Authority
BIM	building information modeling
BRICS	Brazil, Russia, India, China, and South Africa
CAQDAS	computer-aided qualitative data analysis software
CEO	chief executive officer
CIB	International Council for Research and Innovation in Building and Construction
CIDB	Construction Industry Development Board
CIJC	Construction Industry Joint Council
CoST	Infrastructure Transparency Initiative
COVID-19	Coronavirus-19
ELS	Ethical Leadership Scale
FIDIC	Fédération Internationale des Ingénieurs-Conseils
GDP	gross domestic product
GMALD	grounded model of authentic leadership development
IAMLD	Integrated Antecedental Model of Leadership Development
ICONDA	international construction database
ICT	information and communication technology
IDD	integrated digital delivery
IoT	Internet-of-Things
ISO	International Organization for Standardization
LBDQ	Leadership Behavior Description Questionnaire
LMX	leader-member-exchange
LPC	least preferred co-worker
LPI	Leadership Practices Inventory
MCLQ	Multicultural Leadership Questionnaire
MD	managing director
MLQ	Multi-factor Leadership Questionnaire
NYAA	National Youth Achievement Award

SE	significant experiences
SI	significant individuals
SMWT	self-managing work teams
VDL	vertical dyad linkage
VP	vice president

1 The leadership crisis

Introduction

The world faces several complex challenges, many of which are unprecedented. Demographic change, a rapidly worsening climate which is increasing the frequency and severity of natural disasters, rapid urbanization leading to crises in the management of cities, social and political instability, repeated cycles of economic recessions, and deadly pandemics are threatening the world. Many view the less than adequate response to these challenges as a collective failure of leadership at many levels of society.

Leadership has never been more critical, and yet more difficult. A UK government report notes that the challenges of leadership and management in the twenty-first century include: working in an environment of constant change; having to deal with a difficult economic climate and growing international competition; learning about and effectively using the new technologies, the knowledge economy, and the rise of social media (Department for Business, Innovation and Skills Leadership and Management Network Group, 2012). There is also a need for greater transparency; an increase in consumer demands; concern with environmental resources and climate challenges underlined by evidence of destructive weather events. The World Economic Forum's (2016a) Global Risks Report identified these risks for 2016: failure of climate change mitigation and adaptation; weapons of mass destruction; water crises; large-scale involuntary migration; and severe energy price shocks. With good leadership, these risks can be turned into opportunities, which may prevent the risks.

Corporate scandals in many countries have exposed leaders who not only have deceived the internal stakeholders of their organizations but also have breached the trust of a large number of other legitimate stakeholders, often across national boundaries. Over recent decades, high amounts of executive compensation, evidence of unethical managerial practices, and poor corporate governance have left scars on the world's history. Despite the efforts which have been made to restore confidence in corporate leadership, including regulations, the advocacy of pressure groups and consumer action, some leaders have continued to break the basic tenets of ethics. The public has had good cause to lose confidence in corporate leadership. The repeated episodes of corporate leadership failure make a fresh look at ethical leadership practices in organizations necessary.

This chapter highlights the crucial role that leadership plays in commercial organizations. Organizations need to have, to deploy and to further develop appropriate forms of leadership at all levels. It is argued here that a new approach to leadership that is grounded in the authenticity of leaders must be adopted in order for corporations to reap the benefits that the talented people employed in today's business organizations have to offer.

Leadership crises in business

> In the best of times, we tend to forget how urgent the study of leadership is. But leadership
> always matters, and it has never mattered more than it does now … the subject is vast,
> amorphous, slippery, and, above all, desperately important.
>
> (Bennis, 2007, p. 2)

In the early 2000s, the corporate scandals in the telecommunication organizations, One. Tel, WorldCom, and Adelphia Communications; the energy firm Enron; and the accounting firm Arthur Anderson all made the headlines. The frequency of the occurrence of these incidents of corporate malpractice, the magnitude of the losses, the geographical spread of the negative impact, and the number of other organizations and persons affected during this era were unprecedented. Scholars and commentators started to point the finger at the type of leaders who were at the head of these organizations. The number of papers and books on leadership grew substantially in the aftermath of these scandals, and new models of leadership were proposed.

After that wave of scandals, the world might have thought that appropriate lessons had been learned, regulatory measures had been introduced and improvements in corporate governance and oversight had been effected. The optimism was misplaced. In mid-2008, the shockwaves of an approaching financial crisis, perhaps the worst economic crisis in history, began to be felt. The wave of what many now call the 2008 Global Financial Crisis, which started in the United States, had an impact on almost every country. The fall of the housing finance institutions in the United States, Fannie Mae and Freddie Mac, was the beginning of the episode. When world giants of the finance sector such as Merrill Lynch and Lehman Brothers faced major difficulties, the whole world felt the jolts of the severe economic earthquake. Companies in various countries such as Bradford and Bingley (the United Kingdom), Wachovia (the United States), Lloyds TSB (the United Kingdom), and Hypo Real Estate (Germany) followed in the series of financial failures.

Thus, not only did many venerable financial institutions in many countries go bankrupt and many major stock markets lose large proportions of their values, but also, seemingly unrelated sectors such as manufacturing, information and communications technology (ICT) and services were severely affected. The ordinary person around the world was also facing high levels of uncertainty. Several companies were forced to lay off workers, and some reduced salaries, did not pay other compensations, and slashed workers' benefits. It was estimated that some 50 million people around the world lost their jobs.

In the global financial crisis of 2008, it became evident that poor and/or dishonest leadership and bad managerial practices had resulted in the loss of the savings and livelihoods of large numbers of people. During the period, there were renewed calls for authentic forms of leadership. Many conferences were held to debate the importance of ethics, authenticity, and the service of leaders. Books on authentic leadership topped the bestseller lists. The momentum for substantive change built up, only to subside again as the business world seemed to get back to normal.

A few years later, the world faced another series of corporate leadership crises. Nissan, Volkswagen, and Tesla (in the car manufacturing sector); Facebook and Google (in the information and technology sector); WeWork (commercial real estate), Theranos (health technology); Commonwealth Bank (in Australia), Wells Fargo (in the United States) and Abraaj (based in the Middle East) (in the finance sector) were in the news. For example, it was revealed that the privacy of Facebook users might have been compromised; users' private information and data might have been sold to another firm without their knowledge. People's confidence in the security of their personal information on social media platforms dipped.

Research shows that ineffective management costs businesses in the United Kingdom over £19 billion per year in lost working hours; 43 percent of UK managers rate their line managers as ineffective; nearly three-quarters of organizations in England reported a deficit of management and leadership skills in 2012, and this deficit was considered to be contributing to the productivity gap with countries, such as the USA, Germany, and Japan; and incompetence or bad management of company directors was found to be causing 56 percent of corporate failures (Department for Business, Innovation and Skills Leadership and Management Network Group, 2012). Studies show that employees are not satisfied with the leadership of their organizations. The 2019 Edelman Trust Barometer (Expectations for CEOs) Report shows that "CEOs are falling short of expectations among employees, underperforming in their societal leadership and in areas of ethics and character" (Edelman, 2019). Some 79 percent of employees expect their management to "tell the truth" but only 55 percent believe they do. Similarly, 74 percent of employees expect chief executive officers (CEOs) to embody their organizations' values but only 60 percent agree that CEOs are doing so.

On the other hand, good leadership and management have a significant impact on organizational performance (Department for Business, Innovation and Skills Leadership and Management Network Group, 2012): best-practice management development can result in a 23 percent increase in organizational performance; and a single point improvement in management practices (on a five-point scale) is associated with the same increase in output as a 25 percent increase in the labor force or a 65 percent increase in invested capital. The report concludes that: "Improving our leadership and management capability … helping managers at all levels to develop the right skills and behaviours will ensure organisations have the ability to adapt, innovate and evolve, and seize the growth opportunities that lie ahead" (ibid., p. 7).

Leadership at all levels has seen perhaps the most difficult challenge of recent times in the form of the Coronavirus-19 (COVID-19) pandemic. Leaders at the national level and in public and private organizations are having to make some difficult decisions. Their leadership values, competence, ability to deal with crises, and emotional intelligence are being tested. The leaders' ability to engage with their teams, communicate with effectiveness, and show empathy toward their employees is more critical than ever. As companies go under and large numbers of people lose their jobs, leaders at all levels of society have to demonstrate that their leadership is relevant and needed in these times of crisis.

It is pertinent to note that most corporate leaders receive high monetary compensations and many other perks. According to the American Federation of Labor and Congress of Industrial Organizations (Executive Paywatch, 2018), there is a massive imbalance between the remunerations of CEOs and those of the workers. The average total compensation of CEOs of S&P 500 companies was US$14.5 million per annum, resulting in a CEO-to-worker pay ratio of 287 to 1.

A critical reality that these crises have exposed is the lack of leadership qualities on the part of executives, including those at the helm of global giants. Public polls also suggest widespread disappointment with leadership. For example, a Gallup Poll revealed that public confidence in the business leadership in the USA has not changed over a period of 30 years. On a scale from 0–100, the confidence in the business leadership averages below 30 whereas the average of confidence in the military leadership is 80 (see Avolio and Luthans, 2006). The world is facing a leadership deficit:

Around the globe, humanity currently faces three extraordinary threats: the threat of annihilation as a result of nuclear accident or war, the threat of a worldwide plague or ecological catastrophe, and a deepening leadership crisis in most of our institutions. Unlike the possibility of plague or nuclear holocaust, the leadership crisis will probably not become the basis for a best-seller or a blockbuster movie, but in many ways it is the most urgent and dangerous of the threats we face today, if only because it is insufficiently recognized and little understood.

(Bennis, 1996)

The transgressions among executives revealed in the many corporate scandals (see Mehta, 2003; Revell, 2003; Brown and Treviño, 2006; Manz et al., 2008) showed that there were unethical and even toxic leaders who were able to override governance procedures and management systems to take actions to the detriment of their organizations, their employees, and society (see Padilla et al., 2007; Schaubroeck et al., 2007). Rock (2006) observed that many business organizations lacked the right people and adequate talent to take up the leadership positions. The ethics and authenticity of leaders have been questioned (see Avolio and Gardner, 2005).

Leadership crises in government

The leadership crisis is not limited to the corporate world. Nearly all institutions of society are experiencing it. The way the populace have been expressing their unhappiness with their governments in many parts of the world in the past decade is unprecedented. The Arab Spring, which started in 2011, engulfed several countries in the Middle East, and led to the toppling of many governments, began with protests in Tunisia and quickly spread to several other nations in the region. The protests were led by the educated youth over high levels of youth unemployment, lack of freedom, and perceived lack of future prospects. More recently, in 2018–2019, another wave of civil unrest occurred in several countries in Latin America: in Bolivia, Chile, Ecuador, Honduras, Nicaragua, and Venezuela for different reasons including high prices and shortages of basic items, unemployment, political corruption, and lack of personal safety owing to the high levels of crime, and these problems still persist in some of these countries. The mass protests are an indicator that people have lost trust in their leaders and in the political system.

The global score on the Democracy Index of The Economist Intelligence Unit (2019) declined between 2014 and 2017, and stabilized in 2018. With several countries around the world facing social unrest marked by massive demonstrations (Chile, France, Hong Kong, Iran, Iraq, Lebanon, and so on) and instability (Democratic Republic of Congo, Libya, and Yemen), the democracy index for 2019 declined to the worst average global score since 2006.

A poll conducted by WorldPublicOpinion.org for the University of Maryland's Program on International Policy Attitudes, in 2008, showed that no major international political leader enjoys a high rating of public trust and confidence abroad.[1] Reporting on a more recent survey, the same organization noted: "It's … remarkable to see confidence ratings for major world figures cluster within such a narrow and low band," said the pollsters (see also, World Public Opinion, 2019). According to Freedom House, an independent watchdog organization which advocates the expansion of freedom and democracy around the world, freedom (political rights and civil liberties) declined for 13 years consecutively from 2005 to 2018. Countries for which the freedom score declined were consistently greater in number than those whose scores improved over these years.

Where have all the leaders gone?

Where have all the leaders gone? Bill George (2003) raises a simple, but in reality, a very complicated question in his bestselling book on authentic leadership. Describing the challenges facing modern business enterprises, George argues that capitalism has become a victim of its own success. To him, measures of business development such as growth, cash flow, and return on investment are fading away and the new criterion of success is "meeting the expectations of security analysts." To reach the earnings targets, investments are cut back, negatively affecting the long-term growth potential of the company. To achieve the targets of profit maximization and to satisfy shareholders, CEOs forget the need to nurture the human and social capitals which earn long-term success for the organizations.

In their book on positive organizational scholarship, Cameron et al. (2003) present a picture of organizations which are characterized by greed, selfishness, manipulation, secrecy, and a single-minded focus on winning. Employees suffer from anxiety, fear, burn-out, and feelings of abuse. Mistrust is common in social interactions. Such leadership illness is found in several organizations today. Sensing the so-called "leadership crisis," scholars have realized that leadership is not merely a style, a matter of charisma, motivation, inspiration, or strategy. It should be considered as character, positive behavior, and authenticity (see, for example, Luthans and Avolio, 2003; Avolio and Gardner, 2005).

The increasing unethical practices among leaders evident in the many corporate scandals, coupled with an increase in societal challenges, have together necessitated a more positive leadership that is genuine, reliable, trustworthy, ethical, self-disciplined, and selfless. George (2003) believes that these leaders, whom he calls "authentic leaders," possess the highest level of integrity, a deep sense of purpose, courage, genuine passion, and skillfulness in leadership.

Is a new type of leadership needed?

The current world situation is posing difficult scenarios for governments and business organizations. Globalization is an important challenge that leaders have faced; and it has now entered a new phase as supply chains are being disrupted and reshaped by global pandemics such as COVID-19, inter-government trade tensions and other dimensions of geopolitical competitions. The pace of technological advances and the need for adaptive capability is another major challenge for corporate leaders. Attracting, retaining, and developing the appropriate talent is yet another issue. Other challenges to which leadership should find solutions include environmental degradation, depleting stocks of vital natural resources and climate change, rapid urbanization, waves of epidemics and pandemics, and threats of possible major conflict over territorial disputes or sovereignty matters.

The changing business world calls for a renewed vision for leadership and demands fresh attention on restoring positive leadership in organizations. Researchers on authentic leadership argue that the reduction in ethical leadership, reduced trust of people in organizational leadership and the many socio-economic and cross-cultural challenges that today's businesses face call for positive leadership. There is growing recognition among scholars (Luthans and Avolio, 2003; Seligman, 2002) and practitioners (George, 2003) that a development strategy for a more authentic leadership is needed. They also argue that existing frameworks are not sufficient to develop the leaders of the future (Luthans and Avolio, 2003; May et al., 2003; Avolio and Gardner, 2005; Gardner et al., 2005).

The workplace is changing and in many ways. As the world enters the fifth industrial revolution and embraces a new generation of technologies (such as 5G, the Internet-of-Things, and advanced automation), the workplace behaviors and expectations of knowledge workers are being transformed. To save costs, companies are encouraging their employees to work from anywhere. Increasingly, more organizations want their employees to focus on delivery of output and not on completing an eight-hour day in the office; several organizations have been promoting flexible working hours and allowing their employees to manage their time according to their personal priorities. The outbreak of the COVID-19 pandemic has also changed the workplace in many ways. Organizations, both public and private, have had to use digital tools to enable employees to work from home. Applications such as Zoom, Webex, Microsoft Teams and Skype are now household names as people are using them to collaborate on assignments, and attend meetings virtually.

Another aspect of the changing workplace is progress toward gender parity, especially at the professional and managerial levels. More companies are trying to achieve gender parity in their workforce. More women are being promoted to managerial positions. Multinational organizations are promoting inclusion and diversity among their workforces, in terms of nationality, ethnic background, and cultures. As discussed in Chapter 2, corporations are realizing that diverse teams are more innovative and productive if given a conducive environment. Leaders are finding it even more challenging to derive effective ways and means to engage their team members, motivate them, and get the desired outcomes from them.

As several countries transform from industrial production to service orientation and knowledge economies, they need a new type of worker to support the new sectors. With more education, connectivity, information, and knowledge, people are more aware of their options and rights. Their attitudes are also changing; their loyalty is not solely to their organizations but to themselves, and this has an impact on the passion they bring to work. They are always looking for better opportunities, recognition of their efforts, more engaging work in a more informal environment, and a better work-life balance. They like working in teams and are keen to challenge themselves to solve difficult problems. Organizations have to learn how they can help their talent to be engaged and to grow, retain them, and get the best out of them.

There is need for change in the approach to leadership. A traditional transactional relationship between employees and managers does not help much in today's knowledge economies. A leader has to be emotionally intelligent to connect with the team members. His or her focus cannot just be on the task at hand but on the vision that brings people around, motivates them to perform to achieve a common goal, and at the same time helps them to learn and grow.

Leadership and the construction industry: an introduction

In the past few decades, it has become evident that success in business is determined by the quality of the leadership. Considering the fundamental changes in society, technology and in other areas, the tasks of leaders in all spheres of life are increasingly more difficult. Thus, it is necessary to understand leadership and use this knowledge to develop the kind of leaders the world needs.

The nature of the construction industry, the construction process, and the items the industry builds show the need for leadership in the industry. Construction projects typically comprise teams coming from a diverse range of professional backgrounds, countries, cultures, and socio-economic circumstances to undertake a large number of discrete activities. Today, a project manager is dealing with a greater proportion of highly trained workers, and advanced technologies on the construction site. Much of the work is also done off-site, as modular construction is playing an increasingly large role in the industry. The virtual team with members in different countries has been reality on large construction projects since the 2000s. Projects are growing ever larger and more complex. A project manager in construction is also dealing with an increasingly more demanding external environment which comprises government regulations and policies that are increasingly more stringent, and suppliers and sub-contractors have competitive pressures of their own.

The construction firm is different from any other enterprise in any other field of endeavor because, first, it is a price taker, with little ability to influence the demand on most projects. Second, its projects are fixed where they are required, making it impossible for the firm to influence the location, and necessary for construction firms to ensure that they have good logistics. Third, there are many participants on each project and an even larger number of stakeholders, both on the projects and beyond. Finally, the construction process is subject to arguably the widest range of regulations which are constantly increasing in terms of stringency. As a result of these peculiar characteristics, construction activity is fraught with a great deal of risks. The intense competition and price focus of clients lead to low margins in the industry which then hinder the ability of the firms to invest in technology and in their people. These challenges make leadership necessary in the construction industry.

The construction industry also faces a challenging future owing to the increasing size, complexity and duration of projects, and the intense competition in the industry which is leading to the development of new models of business and services, and greater sophistication on the part of clients. There are increasingly more stringent requirements in government policies and regulations, and the range of stakeholders on each project is widening and their access to instruments for participation in providing comments on the project improving (Sheffield, 2020). In recent years, some governments have sought to act to improve the performance of the construction industry by launching strategies and initiatives, and by forming agencies to manage the development of the industry. There have also been government-industry initiatives to improve the performance of the industry. With this set of tasks and operating environment, which are in a state of flux, it is evident that the construction industry needs good leaders for its projects and companies, and at the level of the industry.

The study

This book is based on a research study of the construction industry undertaken in Singapore. The field study mainly comprised interviews with some eminent professionals and senior practitioners who had a reputation for being authentic leaders in the construction industry. The objectives of the book are:

1   to establish the importance of leadership in the construction industry and consider the current state of application of leadership in it, and the implications of the prevailing level of leadership
2   to review the state of leadership research and consider, in particular, the authentic leadership construct and its processes
3   to explore how authentic leaders develop and remain authentic over their life span, and the processes by which authentic leaders influence their followers and their organizations, and contribute to society
4   to propose and validate a grounded theory of authentic leadership development and influence for the construction industry of Singapore, and consider its implications for current practicing leaders and aspiring leaders.

The structure of the book

The book is primarily based on the experiences and accounts of specially identified leaders who have been recognized in the construction industry in Singapore as individuals who practice solid values, lead with the heart, and demonstrate self-discipline. These leaders have established themselves as examples of authentic leadership in their organizations and among their peers in their professions and the wider industry. They have also developed the organizations they lead as sustainable organizations which are, themselves, leading the industry, performing at a high level and providing a good service to their clients. Thus, these leaders are appropriate subjects to study, with the view to gaining an understanding of the authentic leadership development process.

The main purpose of this book is to share with the readers the life stories of many leaders, and how they developed their own leadership styles. How they learned, through a host of life experiences, to become who they are. A leader develops his or her own leadership style over time, through introspection, reflection, framing and reframing his or her own experiences. This knowledge should be useful to readers.

Confucius once said: "By three methods we may learn wisdom: first, by reflection, which is the noblest; second, by imitation, which is the easiest; and third, by experience, which is the bitterest." This book offers all three ways of learning about leadership. It encourages existing as well as aspiring leaders to reflect on their own stories and inspires them to explore their inner selves, their deeply held values and what they wish to leave behind as their leadership legacy.

The book shares many examples from the lives of leaders: how they practice their leadership, gain followership, build effective teams, achieve excellence, and sustain leadership in their organizations. The book also shares how leaders face challenges, deal with failure, learn from both their successes and mistakes, and continuously learn to improve and succeed. In the book, the leaders speak, and tell their own stories. Extracts from the interviews are used to present the analysis of the field study. This lends authenticity to the message of the book.

The book will help readers to think about their own stories and consider their own core values. Readers who are already leaders can integrate the aspects of leadership found in the study in their own work as leaders. The book will help them to consider a number of questions. What their leadership really stands for. How to build their leadership influences. How to form a network among peers and business partners to benefit their leadership. What they want to get out of their leadership. What they want to leave behind as leaders.

If the readers are aspiring to be leaders, the book will provide some guidance on how they can develop as leaders. What tipping points in their own lives they should reflect on as possible antecedents of their leadership journey. How they can leverage on their followership to grow as leaders. Possible traps they should avoid in their leadership development journey. The framework for leadership development along the main themes proposed in the book should also contribute to efforts to create a new generation of leaders.

Note

1 See https://drum.lib.umd.edu/handle/1903/10686
   https://drum.lib.umd.edu/bitstream/handle/1903/10686/WPO_Leaders_Jun08_art.pdf?seque
   nce=6&isAllowed=y

2 The role of leadership in the construction industry

Introduction

The chapter presents a review of research on leadership in the construction industry. It outlines the leadership challenges the industry is facing and discusses the nature of the construction industry, its products, processes, and practices, and considers the implications of the attempts to realize efficiency and effectiveness in the operations of the industry. The strategic importance of construction is highlighted. The problems and challenges it faces and the efforts which have been made to address them are outlined. The need for leadership at various levels of the industry is set out. Focus then turns to leadership at the level of the construction project as it is the milieu in which construction operates and plays its role in contributing to the economy and society. It is argued here that no single leadership approach is best suited to all situations in construction projects, owing to their complexity and dynamism. There is a need for authentic leadership development of construction project leaders as the work is now undertaken in an environment which is growing more demanding and challenging.

Construction in a dynamic world

Organizations in all sectors of the economy are under increasing pressure to innovate, offer value-added services, and learn to survive and grow in the face of intense competition and rapid change. In the interdependent and interconnected global economy, when a typical medium-to-large-scale construction project involves several stakeholders with various interests and possibly many different socio-cultural backgrounds, the project manager has to keep their interests aligned in order to achieve a successful project. Construction, a people-intensive industry, also has to deal with the commercial and social needs of a broad range of stakeholders, technological advances in its processes and products, and increasingly more stringent regulations. Moreover, decisions made on construction projects have an impact upon the lives of the citizenry in several ways.

In all countries, the role of the construction industry in delivering the required assets which provide essential services is even more crucial today. For example, most of the 17 Sustainable Development Goals of the 2030 Global Development Agenda (United Nations, 2020) are directly or indirectly related to infrastructure, and thus, to the construction industry (The Economist Intelligence Unit, 2017). A sustainable future will depend on how the construction industry transforms itself, and adopts the necessary technology and practices for its increasingly complex tasks. The industry should be constructing buildings and infrastructure that are sustainable in the sense that they use less energy and other non-renewable resources, and are of high quality, inclusive, and resilient (United Nations, 2017).

The well-documented problems of construction discussed in more detail later in this chapter may be summarized as including: (1) the constraints on the project with respect to the performance requirements (usually, cost, time, quality, health and safety, and, more recently, sustainability) deriving from the client's stipulations and the regulatory system; (2) the intense competition at both the local and global levels; and (3) the disruption, real or potential, from the emergence of e-businesses. Construction projects also consume significant public and private resources, owing to their high expense. At the same time, they are typically known for time delays and cost overruns. Construction is not best known for high quality output, occupational safety and health, sustainable use of materials, or preserving ecosystems.

The construction industry and its activities and products have the potential to make a significantly positive impact on all lives. To unleash the full potential of the industry, the construction leaders and professionals have an important role to play. However, the importance of leadership in construction is not well recognized and is not valued. Leadership is not systematically taught in construction programs. As discussed in Chapter 1, in order to achieve the project objectives especially in the complex projects of today's world, the customary control-driven approach to management has to be supplemented with leadership. This chapter considers the importance of leadership at various levels of the construction industry. It focuses on the need for leadership in the successful delivery of construction projects.

Construction: a strategic industry

Construction plays an important role in economic and human development. From houses and simple schools through hospitals and factories to large and complex hotels; and from rural feeder roads to highways, ports, airports, and other infrastructure projects, the construction industry, in many ways, shapes all aspects of life.

The construction industry is arguably the most strategic segment of all economies. It is a significant contributor to the economy (Lopes, 2012). Its supply chain reaches far and wide into numerous other industries (Tennant and Fernie, 2014). Construction drives demand in many sectors of the economy, and its output is a necessary input into the operations of all other sectors (Hillebrandt, 2000). Thus, construction activity has implications for production in the whole economy, and for employment generation by itself and in other sectors, and therefore, economic growth and long-term national development.

According to the World Economic Forum (WEF) (2016b), the construction industry accounts for nearly 6 percent of global gross domestic product (GDP) which translates into US$10 trillion in revenue and US$3.6 trillion of value addition. In the USA, nominal value added in construction was 4.1 percent of GDP, with the industry accounting for US$890.3 billion[1] and employing nearly 7.5 million workers in 2019.[2] Similarly, in 2018, the construction industry contributed £117 billion to the UK economy, 6 percent of the total (Rhodes, 2019). There were 2.4 million construction industry jobs in the UK in 2019, 6.6 percent of all jobs. There were 343,000 construction businesses in the UK, 13 percent of the total. The industry is unusual because of the high proportion of self-employment in it: 36 percent in 2019, compared to the average for the whole economy of 13 percent.

In some emerging economies, the share of construction in GDP is significant and the industry is a major employer of the workforce. For example, in China, the construction industry accounted for 6.8 percent of GDP and its total output was valued at US$894 billion in 2019, employing over 50 million workers.[3] Similarly, in India, construction accounted for 9 percent of GDP in 2017 and employed nearly 51 million workers.[4] Construction and its role in the GDP are potentially even more significant in the developing countries which still need large volumes of social, industrial, and economic infrastructure.

Global Construction Perspectives and Oxford Economics (2015) estimates that, as of 2015, the growth of the global construction industry remained on an upward trajectory and its overall size is set to grow to US$15 trillion by 2025 and US$17.5 trillion by 2030. Nearly 57 percent of this growth will come from three countries: China, the United States, and India. Other countries that will make significant contributions to global construction output include Indonesia, the United Kingdom, Mexico, Canada, and Nigeria.

Infrastructure development remains an important sector of construction, driving global economic growth and development. The Global Infrastructure Outlook of the Global Infrastructure Hub estimated global infrastructure investment needs to be US$94 trillion between 2016 and 2040.[5] Current trends show that infrastructure investment needs stand at US$79 trillion. This represents a US$15 trillion investment gap compared to what the world will require for infrastructure needs by 2040. The majority of the infrastructure demand will be in Asia, mostly in China and India. Other countries that will contribute to this include Japan, Indonesia, Turkey, Bangladesh, Pakistan, the Philippines, Thailand, and Vietnam. According to the report, the infrastructure investment gap is proportionately the largest for the Americas (47 percent) and Africa (39 percent).

Within infrastructure, energy and transport are the two key sectors where future investments will be needed. McKinsey estimates that global spending on transport infrastructure (roads, railways, ports, and airports) was nearly US$1 trillion in 2015 (Woetzel et al., 2017). To support economies, more investments are needed in transport, an average of US$1.5 trillion on an annual basis from 2017 to 2035, the McKinsey analysis shows. According to the International Renewable Energy Agency, in order to meet the objectives of the Paris Climate Agreement, significantly more investments are required for a global energy transformation. This would require nearly US$22.5 trillion in new renewable installed capacity through 2050 which will mean a doubling of current annual investments.[6]

Another important area in construction globally is housing. Due to the growing populations and rapid urbanization, the need for housing continues to rise while availability and affordability remain complex issues in urban centers in countries at all levels of development. According to a report by the McKinsey Global Institute,[7] the global housing affordability gap amounts to some US$650 billion per year, or 1 percent of global GDP. The housing affordability gap can reach double digits of local GDP in cities with a high cost of living.

From the above studies, it is clear that the construction industry will continue to grow in the coming years if countries want to meet their development aspirations, and the world is to achieve the targets on the global development agenda.

The unique challenges of the construction industry

The uniqueness of the construction industry and each of its projects and products is probably the most often mentioned feature in the literature on the industry (Hillebrandt, 2000; Ofori, 2019). This uniqueness also makes construction project management a distinct discipline as it faces a considerable number of challenges in various contexts. The World Economic Forum (2015) highlights a number of challenges that the construction industry faces, including:

1   lack of innovation and delayed adoption of new technologies;
2   informal process or insufficient rigor and consistency in process execution;
3   insufficient knowledge transfer from project to project;
4   weak project monitoring;
5   lack of cross-functional cooperation;
6   lack of collaboration with suppliers;
7   conservative company culture;
8   a shortage of young talent and people development.

Toor and Ofori (2008a) present a taxonomy of construction industry challenges which they categorize into: industry-specific challenges, general business challenges, and operating environment challenges (socio-cultural, economic, technological, legal and regulatory, and ethical). Toor and Ofori note that the socio-cultural challenges include political upheavals and cross-cultural issues. The economic challenges include funding difficulties and uncertain economic conditions (compared with the long period of gestation, and the long life of the built item). Technological challenges include the need for increased use of ICT, such as in Building Information Modelling (BIM), e-procurement and project management systems; and a growing need for innovation. Legal and regulatory challenges include the frequent changes in the law and the generally widening extent and growing stringency of the regulatory regime. Firms operating internationally also need to cope with different legal systems and regulations in the host countries, the litigation procedures, and arbitration methods within and across countries.

Construction activity is typically known as being demanding, dirty, dangerous, and low-paying (Construction 21 Steering Committee, 1999). Construction has a poor record of health and safety in most countries. Its record in terms of the occurrence and severity of accidents, as well as mortality, is among the worst of all sectors (International Labour Organization, 2001; Health and Safety Executive, 2020), and the construction industry is failing to attract and retain the talent it needs (Toor and Ofori, 2008b); thus, in many countries, it faces a lack of quality people and an aging workforce (Songer et al., 2006).

Within the industry, the relationships among the parties can be adversarial (Latham, 1994); conflicts leading to contractual disputes and litigation are common (Winch, 2014). The construction industry does not have a good social image in most countries; it is well known for having unethical practices, and a reputation for corruption, as discussed below.

The growing volume of construction demand and activity in many countries is resulting in severe pressure on resources and is having a negative impact on ecosystems and habitats (Ofori, 2003). It is estimated that the construction industry is the biggest consumer of the world's raw materials (World Green Building Council Europe Regional Network, 2019). Thus, construction activity is seen as one of the industries with the greatest potential to contribute to the path toward a circular economy (UK Green Building Council, 2019). Governments have instituted legislation, policies and programs, and set requirements for attainment on sustainable construction (Building and Construction Authority, 2014). The professional institutions have launched technical guides and manifestos (UK Green Building Council, 2016) for attaining sustainability in construction.

Clients are becoming more knowledgeable and better informed and hence more demanding and selective in what they want from consultants and in deciding to which contractors they should award the work (Watermeyer, 2018). There is also greater global competition for construction projects (World Economic Forum, 2015). The operating models in construction are changing. Projects are growing larger and more technologically complex, with higher numbers of stakeholders who may have multi-ethnic and multi-cultural backgrounds (Ofori and Toor, 2009). There is also increased private-sector participation in infrastructure projects (Akintoye et al., 2015). At the business level, vertical integration in the packaging of projects, strategic alliances, such as joint ventures and consortia, and partnering relationships; mergers and acquisitions, some resulting in multi-disciplinary corporations, are increasingly common, and make management tasks more complicated.

Consequently, there is a need for changes in many aspects of the operations of the construction industry (Ofori, 2003). There should be greater focus on the following issues:

- how the project procurement processes can be improved (Kumaraswamy and Dulaimi, 2002; Watermeyer, 2018);
- how the construction value chain can be made more efficient (Atkin, 1998);
- how the concerns of health and safety in project delivery and sustainable development can best be addressed (Lingard and Rowlinson, 2006);
- how the level of professionalism in the industry can be enhanced (Vee and Skitmore, 2003; Morrell, 2015; Foxell, 2018);
- how the level of adversarialism in the industry can be reduced and a collaborative and partnering approach promoted (Li et al., 2001);
- how corruption in the industry can be significantly reduced, if not eradicated (Stansbury, 2005; Transparency International, 2011; Stansbury and Stansbury, 2018);
- how the social image of construction can be enhanced (Rameezdeen, 2007).

Future trends in construction

Peiffer (2017) predicted ten key trends in US construction in 2017. They included:

1 Collaborative project delivery methods would become more popular.
2 The labor shortage would continue to plague the industry.
3 Offsite/modular construction will gain a stronger foothold in the market.
4 Construction firms would be cautiously optimistic for an infrastructure spending boost.
5 The Internet-of-Things (IoT) has the potential to revolutionize the construction site.
6 Construction costs would continue to rise owing to persistent increase in materials prices and labor costs.
7 The adoption and application of Virtual Reality and Augmented Reality technologies would increase.
8 The industry would consider changing its message on sustainable construction to build on the success it has attained, for example, by publicizing the value of smart, high-performing buildings and infrastructure in such a way that more people can understand it.
9 Construction firms would face increased scrutiny and prosecution of safety and fraud incidents.

In Singapore, the Building and Construction Authority (BCA) (2017) identifies three global trends which will be shaping construction: (1) the digital revolution – the advent of smart buildings, new construction technologies and digitalized work processes; (2) rapid urbanization – the need for advanced technologies to build faster and better; and (3) climate change – strong demand for green building expertise.

W55 on Building Economics, a working commission of the global network of researchers for the International Council for Research and Innovation in Building and Construction (CIB), notes that the big issues for the construction industry are (Newton et al., 2016):

- business analytics and performance metrics;
- demographic change and the aging population
- digitization and smart cities;
- ethics and regulatory controls;
- globalization and integration;
- green building and retrofit;
- infrastructure needs and public finance;
- labor and material supply chains;
- natural disasters and risk management;
- new construction methods and markets;
- productivity and innovation;
- urbanization and housing affordability.

Similarly, after analyzing the contemporary situation and potential future challenges, the research roadmap of another CIB commission, W065 on Construction Management and Organization, indicates the need for new business models for construction enterprises (Wamelink et al., 2016):

1   How to align their organization to continuously changing contexts.
2   Changing roles – changing division of work among the professionals.
3   Changing expectations from society or other stakeholders in the network – collaborative attitude; ethical issues; administrative changes; transparency; sustainability, and the circular economy.

On the WEF's Future of Construction platform, set up in 2014 for the partners of the WEF to shape the agenda of the construction industry and to find solutions, the following challenges are outlined, on each of which the WEF has a working group, and presents a series of blogs:

- *affordability* – creating high-quality, affordable infrastructure and housing;
- *disaster resilience* – making infrastructure and buildings resilient against climate change and natural disasters;
- *flexibility / liveability / well-being* – creating infrastructure and buildings that improve the well-being of end-users;
- *lifecycle performance* – reducing the lifecycle costs of assets and designing for reuse;
- *project delivery* – creating certainty to deliver on time and on budget, and improving the productivity of the industry;
- *sustainability* – achieving carbon-neutral assets and reducing waste during construction.

Construction's social image: ethical challenges

The construction industry is perceived in many countries as one of the most undesirable sectors to work in (Toor and Ofori, 2007). Corruption and fraudulent and unethical practices are prevalent in both developing and industrialized countries (Toor and Ofori, 2008e; PwC, 2014). The levels of professionalism in construction are deemed to be poor in most countries. Transparency International (2011)[8] found construction to be the most corrupt industrial sector. Studies show that the construction industry does not enjoy the trust of society (Morrell, 2015). The perception of social undesirability has been built up from:

exposures of malpractice and mismanagement on projects; poor quality of work which sometimes leads to fatalities as buildings collapse; and the adverse environmental and social impact of construction projects. High profile corporate failures in construction such as that of Carilion plc in the UK, and fraud and corruption scandals, such as those involving Oderbrecht S.A. in Brazil (which implicated a sitting President and a former one) (Watts, 2017), and several companies in South Africa (over collusion in bidding for World Cup venues), have been recent occurrences in the construction industries in a number of countries.

The social undesirability of the construction industry and its practices has repercussions for the quality of social and economic infrastructure and eventually on economic growth. It has the greatest impact on the poor. In construction, the lack of transparency is commonly seen in procurement procedures when large projects are awarded. Although governments, multilateral development banks, and bilateral donors have sought to ensure transparency on public projects, reports of the occurrence of bribes, kickbacks, and other forms of corruption to secure major construction contracts are not uncommon (Hawkins and Prado, 2020).

There is interest among researchers in studying the reasons for unethical behavior in construction. Fan et al. (2001) suggest that the increasing ethical problems in the construction professions may be due to the perception gap about ethical issues between senior and relatively younger professionals. The quantity surveyors responding to their survey ranked the employer, self, and client as more important when they faced any ethical dilemma. On the contrary, the interest of the general public was ranked relatively low. This indicates where the solutions to the lack of ethics can be found. Various initiatives are in place to enhance transparency or reduce corruption in construction. These include:

1   The Infrastructure Transparency Initiative (CoST), a 19-member multi-stakeholder program for promoting transparency on public projects which involves disclosure of specified project data by procuring entities; independent assurance of the data; and social accountability (Infrastructure Transparency Initiative, 2020).
2   The use of Integrity Pacts by project parties, developed and supported by Transparency International and Fédération Internationale des Ingénieurs-Conseils (FIDIC) (International Federation of Consulting Engineers).
3   The work of non-governmental organizations such as the Global Infrastructure Anti-Corruption Centre, UK (Stansbury and Stansbury, 2018).
4   Efforts by professional institutions to promote adherence to ethics, such as the production of the Engineers Charter by the American Society of Civil Engineers (ASCE) (2007), which it has been encouraging its members around the world to sign, promoting it as a way for the engineering community to help improve the welfare and quality of life for all.
5   The formation of interest groups among researchers such as the CIB Task Group 95 on Professionalism and Ethics in Construction which formulated a research road map and organized a major conference on the subject in 2018 (Egbu and Ofori, 2018).
6   The publication, in 2016, of the International Organization for Standardization's (ISO) anti-bribery management standard for organizations, ISO 37001. The standard focuses on bribery but can be expanded to include other corruption offences. It is applicable to small, medium and large organizations in the public, private and voluntary sectors. Organizations can get certified to ISO 37001.
7   The establishment of anti-corruption agencies in many countries.
8   The promulgation of laws against corruption such as the UK's Bribery Act 2010 (which came into force in 2011) and the well-known United States Foreign Corrupt

Practices Act of 1977, which are increasingly being widened to include providing for prosecution in the country where the bribery took place and/or in the person or organization's home country; and for the prosecution of both organizations and individuals for corruption.

9   The signing of international treaties which require member states to implement anti-corruption laws and procedures, such as the United Nations Convention against Corruption (2003), and the Organization for Economic Cooperation and Development Convention on Combatting Bribery (1999).

Stansbury and Stansbury (2018) suggest that preventing corruption requires five core elements: leadership; awareness raising; transparency; controls; and law and enforcement. They note that: "Leadership requires leaders of governments, public sector bodies and private sector organizations to refuse to participate in any corruption personally, and to ensure that the government, organization, department or people that they lead also do not participate" (ibid., p. 33).

The war for talent

A study by McKinsey and Company highlighted the "war for talent" as a strategic business challenge, and proposed success in it as a critical driver of corporate performance (see Michaels et al., 2001). The authors found that the secret of high-performing companies, as compared to the average performers, was a pervasive talent mindset. The high-performing companies believed in the retention and appropriate management of superior talent as their competitive advantage. On the other hand, most average- and low-performing companies were either incapable of attracting the right talent or failed to cherish and leverage it when they incidentally had it.

In the highly competitive construction industry, organizations must develop their own people (Dulaimi, 2005). However, construction firms do not appear to be responding to their long-term needs for human resources with a systematic approach. Many authors have noted that human resource development is not accorded priority in the construction industry (Regumyamheto and Batatia, 1994). Studies have also shown that many reputable construction firms have no core strategies for these purposes (Maloney, 1997). Many do not have any human resource management (HRM) department to deal with the training, deployment, management, and development of professionals and other workers (Mezher and Tawil, 1998). At best, the HRM departments have a limited role in corporate decision-making in many construction firms. While the current workforce is aging, the industry is also struggling to retain its existing talent and to attract high quality personnel to join the construction professions (Chinowsky and Taylor, 2007). As a result, many studies have found that there are shortages of construction professionals in many countries.

In their analysis of key trends in the construction industry, Young and Bernstein (2006, p. 6) considered "A continuous shortage of labor" to be among the top five challenges which the US construction industry would face. They identified the reasons for this shortage as: shifting demographics, with fewer people to replace retiring workers; a lack of interest in the construction industry among the younger generation; and "natural disasters causing increased demand and bidding wars for good workers" (ibid., p. 18). Brandenburg et al. (2006) reported similar findings in another study on the US construction industry. They concluded that the shortage of human resources was likely to continue unless adequate measures were put in place to address the problem at all levels.

A cross-industry research report published in the UK revealed that there is a significant shortage of workers in construction-related occupations and that the situation would get worse when the UK leaves the European Union. The skills with greatest shortages included quantity surveyors, project managers, civil engineers, trades supervisors, bricklayers, carpenters, and general laborers (Association for Consultancy and Engineering et al., 2019). Similar trends are seen in the United States where an article in *Forbes* (Cilia, 2019) noted that more than 434,000 construction jobs were vacant in April 2019. According to the article, workers do not prefer construction jobs primarily due to the perceived dangers and difficulties in construction. This trend may well be contributing to the rising cost of construction in the US, the article claimed. The situation is not much different in India where an estimated 45 million additional skilled construction workers will be needed over the next 10 years, according to the Confederation of Real Estate Developers Association of India (The Economic Times, 2019).

In summary, many factors are highlighted as reasons for the human resource shortages in the construction industry. First, in many countries, construction is not perceived as an attractive sector due to its negative social image and the characterization and perception of it as being risky and unsafe, and offering only low-paid and uncertain employment opportunities (Campbell, 2006; Rameezdeen, 2007). Second, the project-based nature of construction activity, casualization of employment in many countries and the reliance on labor-only subcontracting (Debrah and Ofori, 1997; Loosemore et al., 2003), leading to what is known as "hollowed out firms" have resulted in a perception among many that the job opportunities in the industry are insecure (Storey et al., 2002). Thus, the construction industry is not seen as offering opportunities for its personnel to develop their careers. Third, as mentioned above, the construction workforce is aging and the younger generation shows little interest in careers in construction (Young and Bernstein, 2006). This phenomenon is evident in many countries.

It has also been suggested in the literature that it is inappropriate or impossible to apply the established aspects of strategic HRM in construction firms, owing to the fluctuations in demand which characterize construction activity (Debrah and Ofori, 1997; Dainty et al., 2000). The constant economic pressure on construction firms largely due to the irregular flow of work has made it difficult for the developmental side of strategic HRM to be applied (Drucker, 1996; Debrah and Ofori, 1997; Dainty et al., 2000). Strategic HRM programs receive little formalized consideration in most construction firms, and continue to be emergent rather than deliberate (Brandenburg et al., 2006; Dulaimi, 2005).

The World Economic Forum and Boston Consulting Group (2018) have made proposals to solve the talent gap in the construction industry. Companies should strategically plan talent supply and demand; promote and support employee training and development; and adopt new technologies to improve productivity and job satisfaction of their workers. They should also update their work culture to appeal to young and talented workers and more women. The professional institutions and trade associations should run image campaigns to promote careers in the industry; focus attention on new pools of talent; publicize career paths and opportunities in the industry; and collaborate with educational and training institutions. Government should use its role as client and regulator to formulate regulations and launch initiatives to foster innovation and, in this way, increase the attractiveness of the industry. It should harmonize building regulations and standards to reduce complexity and labor intensity in projects, and maintain the currency and rigor of public apprenticeships, academic, and training programs.

Gender imbalance

Various surveys and statistical analyses show that women continue to be underrepresented in the top echelons of leadership in businesses. Only 33 companies among the Fortune 500 companies were led by women in 2019; thus, only 6.6 percent of the group as a whole (Zillman, 2019).[9] The situation is even less encouraging when it comes to construction which is largely considered an "unattractive" sector for women. Consequently, there is a smaller proportion of women in the construction workforce, and an even smaller number of women executives. Data from the US Bureau of Labor Statistics (2020b) has put the proportion of women in the construction workforce at about 9 percent over the past 25 years but this includes administrative and office positions which traditionally has more women. Only 3.4 percent of workers in field production in the construction and extraction sectors in 2018 were women.[10] Again, in the United States, in 2018, women made up the following proportions of the construction industry workforce (Zitzman, 2020): construction managers, 7.5 percent; staff executives, 14 percent; line executives, 7 percent. Some 13 percent of construction firms are owned by women, with 9 percent of female-owned firms achieving revenues of more than $500,000; 4 percent of new construction firms in 2017 were launched by women. Similarly, in the UK, whereas 11–13 percent of workers in construction are women, the breakdown is as follows: 3 percent in manual trades; 5 percent in engineering; and 12 percent in professional roles (Brown, 2019). This, in an age when social equity is considered to be an important pillar of sustainable development (Ofori and Toor, 2008a).

Research shows that organizations benefit from diversity. McKinsey & Company (Hunt et al., 2015) found that gender-diverse companies are 15 percent more likely to outperform their peers and ethnically-diverse companies are 35 percent more likely to do the same. There have also been studies that have derived models of leadership, based on the traits of diverse leaders (The Diversity Practice, 2007). In a follow-up study two years later with a wider and more global base (Hunt et al., 2018), McKinsey & Company confirmed the earlier findings. It found a statistically significant correlation between a more diverse leadership team and financial outperformance demonstrated in the global data set. Companies in the top-quartile for gender diversity among executive teams were 21 percent more likely to outperform on profitability and 27 percent more likely to have superior value creation. It appears that it goes beyond gender. Companies in the top-quartile for ethnic/cultural diversity on executive teams were 33 percent more likely to have industry-leading profitability.

Several studies have shown that the construction workplace is typically competitive and discriminates against women (Gale, 1994). Researchers note that the construction industry has been unsuccessful in appealing to women as it is perceived as being physically demanding, combative, and male-dominated (Fielden et al., 2000). Zitzman (2020) suggests that the factors which explain the gender gap include: unconscious gender bias; the pay gap; the lack of career advancement; the lack of adequate training; inadequate facilities on site; a lack of role models; overall perceptions of women working in construction, which is traditionally a male-dominated career. Dainty et al. (2004) in their grounded theory of women career underachievement in large UK companies observe that the construction industry does not have a good track record in terms of equal opportunities and that it continues to suffer from the issues of discrimination, disadvantage, and underrepresentation. Dainty et al. note that "women are overtly and covertly discriminated against by men, who use structural systems to undermine their participation" (ibid., p. 239). Women face several barriers which usually include: "differential effect of structural organizational factors which had restricted opportunities, an incompatibility of construction work with women's expected societal roles, and a culture which militated against women's participation through the maintenance of an exclusionary and discriminatory work environment" (ibid., pp. 246–247).

However, there are signs of positive change in many areas as the response to careers in construction has been improving in some countries; Brown (2019) and Zitzman, 2020 give examples of many women in higher executive positions, and highlight indications of possible change in the overall situation. Many proposals have been made for addressing the gender imbalance in construction including (Brown, 2019; Zitzman, 2020):

- implementing initiatives at the national level to change the image of construction;
- addressing the gender pay gap;
- formation of groups focused on attracting women into the industry;
- encouraging more women into training programs by convincing them it is a viable career;
- changing company practices to remove gender bias from the work culture,
- providing flexible working hours to address the needs of women;
- developing local training programs and mentorship schemes specific to the needs of women;
- ensuring cleaner working spaces;
- taking such obvious actions as providing toilet facilities on site suitable for both males and females;
- offering construction site open days for women and girls;
- encouraging women in construction to become role models for other women.

Transforming the construction industry

The construction industry has remained traditional in most countries. Construction has not seen major technological breakthroughs, and thus it has not realized the efficiency gains evident in other sectors. Until now, technology uptake has been slow and gradual. Given the current scale and future growth pattern of the construction industry, even small improvements in its long and complex value chains can bring major benefits to economies. A WEF (2015) report on how to reshape the future of the construction industry, presents 30 measures across eight areas. The first four areas are where individual companies in the construction industry can take action on their own; five and six are the areas where more collaboration is required across the industry and the last two areas are where government can play a major role as the regulator as well as owner of major infrastructure projects. WEF suggests that the construction industry can be transformed, but players within the industry will have to take charge of this transformation (ibid.).

Similarly, Barbosa et al. (2017) of the McKinsey Global Institute make proposals for reforming the construction industry to improve its productivity. They note that examples of innovative firms and regions suggest that acting in seven areas simultaneously could boost productivity by 50–60 percent:

- reshaping regulation;
- rewiring the contractual framework to reshape industry dynamics;
- rethinking design and engineering processes;
- improving procurement and supply-chain management;

- improving onsite execution;
- infusing digital technology, new materials, and advanced automation;
- reskilling the workforce.

Many countries have launched programs to improve the performance of their construction industries. The UK has had a number of these strategies. The most recent of these were prepared by the government and the Construction Industry Council. The first made proposals for the industry up to 2025 (HM Government, 2013). The second one, the Construction Sector Deal (HM Government, 2018), was prepared as the construction industry's program for attaining the objectives of the country's Industrial Strategy. The Deal's main policies include:

- the Construction Leadership Council, comprising government officials, academics, construction companies, and clients which co-ordinates policy and highlights future challenges;
- the "Transforming Construction Programme" to lead innovation in construction methods, training and recruitment;
- support for skills development and retention;
- greater and improved use of modern construction methods;
- improved business practices, including prompt payment of small firms (Rhodes, 2019).

The construction industry featured prominently in the new economic strategy for Singapore (Economic Strategies Committee, 2010). Another example is the program of Malaysia (Construction Industry Development Board, 2015). Singapore's Construction Industry Transformation Map, launched in 2017, is the collective effort of the industry, the institutes of higher learning, unions, and government (Building and Construction Authority, 2017). The vision is: "An advanced and integrated sector with widespread adoption of leading technologies, led by progressive and collaborative firms well-poised to capture business opportunities, and supported by a skilled and competent workforce offering good jobs for Singaporeans" (ibid., p. 3). The three areas to transform the industry are:

1   *Design for Manufacturing and Assembly ("Build Efficiently")*: Design upfront for ease of manufacturing and assembly; highly automated offsite production facilities; efficient and clean on-site installation process.
2   *Green Buildings ("Build Green")*: Design for green buildings; sustainable practices in operations and maintenance.
3   *Integrated Digital Delivery (IDD) ("Build Efficiently")*: Enabled by building information modelling, IDD fully integrates processes and stakeholders along the value chain through advanced ICT and smart technologies.

In a report prepared for the UK Department for International Development, Arup (2018) proposed a framework for assessing the capacity and capability of any construction industry, to form a basis for creating a development program for the industry. Green (2011) is among those authors who are skeptical of the likelihood of success of national level construction industry strategies. He suggests that the government-commissioned strategies for construction industry reform in the UK are not firmly based on the context of the industry and reality of the prevailing trends.

Leadership in construction

In construction, leadership is required at the strategic (national) level of the industry; in business and other organizations; and on projects, and within their teams. This book is mainly on leadership of business organizations. However, it is pertinent to cover leadership at the industry and project levels, the former briefly, and the latter in greater detail owing to its relative necessity and most common occurrence.

At the industry level, the construction industry, in all countries, has professional institutions (for architects, different types of engineers, planners, quantity surveyors, and project managers); and trade associations (such as different types of building and construction contractors). These organizations represent the interests of different sectors and this has served to entrench the fragmentation of the construction industry.

In a few countries, national agencies have been established to regulate the industry and manage its continuous development. They include the Building and Construction Authority of Singapore (first set up in 1984); the Construction Industry Development Board in South Africa (established in 2001), and the Construction Industry Development Board in Malaysia (set up in 1994); and the National Construction Council in Tanzania (established in 1981). The BCA of Singapore, arguably the most successful of such national agencies, describes itself as: "BCA leads the transformation of the built environment sector through developing a highly competent professional workforce, adopting productive and game-changing technologies, and promoting the sector's niche expertise overseas" (Building and Construction Authority, 2020). Its vision is: "We transform the built environment sector and shape a liveable and smart Built Environment for Singapore." Its mission is: the most liveable and smart built environment, one that is safe, of high quality, sustainable, and inclusive; a strategic sector for Singapore's development; collaborative, transformative, and with professional personnel at all levels; and progressive, innovative, and globally successful firms.

Other examples of leadership at the industry level include industry umbrella organizations. The Construction Industry Joint Committee (CIJC) in Singapore comprises all the major professional institutions and trade associations in construction. It was established in 2000 to do the following:

- to be a platform to unite all key players in the construction industry;
- to co-ordinate the members' joint efforts to play a key role in the built environment sector and the economy;
- to provide quality feedback to the government on policies affecting the industry and problems facing the sector;
- to be a think-tank for the industry;
- to partner with the government to work out appropriate solutions to the problems affecting the industry (Construction Industry Joint Council, 2016).

Among the CIJC's current aims is to enhance the image of the construction industry. The CIJC comprises the presidents of nine professional institutions and trade associations. It meets bi-monthly; and also holds regular "dialogue meetings" with the BCA. The CIJC members are: the Association of Consulting Engineers Singapore; the Real Estate Developers Association of Singapore; Singapore Contractors Association Ltd; the Singapore Green Building Council; the Singapore Institute of Architects; the Singapore Institute of Building Ltd; the Singapore Institute of Surveyors and Valuers; the Society of Project Managers Singapore; and the Institution of Engineers Singapore.

The Construction Industry Council in the UK was set up in 1988 to provide "a single voice for professionals in all sectors of the built environment" (Designing Buildings Wiki, 2018); and Build UK (formed in 2015 as "a strong collective voice for the contracting supply chain in construction," after the merger of the UK Contractors' Group and the National Specialist Contractors' Council) (ibid.). In recent years, there have been government-industry entities such as the Construction Industry Leadership Council of the UK (established in 2013 and jointly chaired by the Secretary of State for Business, Innovation and Skills and an industry representative).

The importance of leadership of construction companies, owing to the unusual nature of the challenges they face, has already been considered in this chapter. Project leadership is discussed in the next few sections. This is an important aspect of the three levels in construction at which leadership is required, as the project creates the output of the construction industry; it involves participants from many backgrounds and many organizations and makes the leadership task very complex; it involves the formation, administration, and management of a temporary organization which must then be disbanded in an orderly manner; and it is subject to extensive regulation in all countries, with a range of agencies involved in aspects of it, which has an impact on efficiency on the projects.

The changing role of project managers in construction

Given the influence of so many factors on the construction industry, including the increasing complexity of projects, the role of project managers is changing. Quality standards are more stringent, and firms are facing more competition in their home countries and abroad. On large construction projects, an added complexity is the different interests of the various stakeholders, including those working on the project, and relevant ones from the construction industry and its related sectors, various levels of the government, the local population, civil society organizations and pressure groups.

The socio-economic and cultural changes during the last few decades have influenced the internationalization of construction activities, and, as a result, the players have faced several challenges (Raftery et al., 1998; Chan et al., 2001; Lewis, 2007). In particular, in the developing countries, where construction activity in the large-project segment is dominated by foreign players, cross-cultural interactions play a key role in negotiations, decision-making, problem-solving, and other aspects of business and technical operations.

Research supports the observation that cultural differences account for the varieties in the way management and leadership approaches are perceived and operationalized in different societies (Gerstner and Day, 1994; Loosemore and Lee, 2002; Wong et al., 2007). Studies show that failure to appreciate differences in culture among the participants in construction projects can lead to undesirable consequences (Norwood and Mansfield, 1999; Fellows et al., 2003). Using the example of doing business in China, Dahles and Wels (2002) note the importance of personal networks, and cultural norms and values. They observe that cultural issues are complex and can influence how a project leader deals with tensions and conflicts, uncertainties, and frustrations while negotiating and managing various national cultures, corporate identities, and business pressures and objectives.

Therefore, it is suggested that international firms should have managers who are culturally intelligent and have the "capacity to adapt to varying cultural settings based on the facets of cognitive and meta-cognitive processing, motivational mechanism and behavioral adaptation" (Early, 2006, p. 929). They must also have the reflexive capability (London and Chen, 2007) and the ability to communicate effectively with culturally diverse teams.

Selmer (2001) goes further to say that managers with these attributes can be used in a strategic way to develop business across borders.

Project managers operate with a significant degree of power and autonomy in return for increased responsibility for the operational performance of their projects (Dainty et al., 2006). They handle large budgets and manage large volumes of other resources. They are liable for everything happening on their projects and are accountable for the outcomes. They deal with several stakeholders of the project and should be able to satisfy the distinct demands of each of them. They resolve disputes among various parties involved in the project and they take critical decisions which have a long-term impact on the project's health. It is necessary to develop more sophisticated approaches to managing their performance in a way which supports both specific project objectives and the wider business goals of the organization (ibid.).

Project leadership: an overview

For three decades, there have been attempts to differentiate leadership from management (Zaleznik, 1977; Bennis and Nanus, 1985). As Toor and Ofori (2008a, 2008d) argue, leadership is long-term, visionary, and purpose-oriented, and seeks to attain innovation and change, while management is short-term, narrow, and task-focused, and aspires to achieve control and stability. Similarly, leaders and managers are different, as they apply different conceptualizations and approaches to work, exercise different problem-solving approaches, undertake different functions, and exhibit different behaviors, owing to their different intrinsic and extrinsic motivations. However, there have been very few efforts to differentiate project leadership from project management as well as organizational leadership. Norrie and Walker (2004) note that project management is the day-to-day implementation of a project plan to achieve pre-determined objectives within time and cost constraints. This limits the strategic thought process of project managers and impairs their vision for their people and organizations. Coleman and Bourne (2018) suggest that the skills required of a project leader are different from the skills developed for project management. Leadership is more about the future – setting direction, dealing with people, and working outside the project with stakeholders – whereas many aspects of project management are inward- and backward-looking.

Studies have found that the leader of a project is a highly critical success factor (Coleman and MacNicol, 2015). Smith (1999) argues that projects do not fail due to lack of project management, but rather from a lack of project leadership. Project managers give importance to managerial functionalism and focus on short-term project goals; they do not provide the needed leadership on the projects. In other words, project management alone does not work; project managers must be able to lead effectively. Thus, it is suggested that project management must progress toward project leadership (Bull, 2010; Madsen, 2015). Shenhar suggests that strategic project leadership shifts the focus in project management "from efficiency only to effectiveness and efficiency; from operational issues to strategic, operational, and human; and from getting the job done, to getting the business results and winning in the market place" (2004, p. 571).

As Gilbert (1983) puts it, project management must be complemented by project leadership that acts as a force of cohesion and catalyst, lifts management from routine bureaucracy, creates the environment for creativity and motivation, and helps achieve the strategic goals of projects. Shenhar (2004) highlights commercial considerations in proposing that, instead of the traditional focus on the project performance parameters of time, budget, quality, efficiency, and operational performance, companies should take a strategic look at their projects. It is only in this way that they can create value for their clients and other stakeholders and enhance their corporate competitiveness. Norrie and Walker (2004) propose a new perspective: projects should be completed on-time, on-budget, on-quality, and more importantly, on-strategy.

Owing to the many stakeholders involved, and the operating environments of many projects, and especially as projects become more complex, the project manager should be more than someone who administers the activities, delivers the technical output, but a facilitator, a co-ordinator, a motivator, and a politician. Thus, the project manager should also be a good project leader (Briner et al., 1996). Thite (2000) suggests that in the era of flatter organizational structures, outsourcing, and the emergence of knowledge workers (who need recognition, motivation, empowerment, and appreciation to perform well), boundaryless teams and virtual organizations, the task-oriented leadership style common on projects should be replaced by a participative, result-oriented style. Lewis (2003) presents the personal traits, principles, and examples of leaders and followers; and considers how to move from project management into project leadership.

The International Centre for Complex Project Management's Project Leadership (2018) noted that leadership is necessary to deal with complex projects of all types, including emergence, change management, maintaining strategic outcomes focus, and working with multiple stakeholders. Similarly, Remington (2011) reports on research on the successful leadership of complex projects.

Coleman and Bourne (2018) suggest that project leadership is different from general leadership. Project leaders have to deal with creating, organizing, developing, and dispersing the project team and the project organization. They also have to deal with change; they operate in an unstable and volatile environment and have to make fast judgment calls in ambiguous situations. Although many project leadership skills overlap with general leadership skills, simply developing general leadership skills is insufficient preparation for developing the leadership needed for major, complex projects.

Project leadership in construction

The term "project leadership" emerged only in the last few decades when scholars recognized that project leadership needs to be distinguished from organizational leadership. It first appeared in the *International Journal of Project Management* in the early 1980s (Gilbert, 1983). Slevin and Pinto (1991) used the term for the first time in the *Project Management Journal*. In construction research, the term first appeared in *Construction Management and Economics* (Liu et al., 2003) and in *Engineering, Construction and Architectural Management* (den Otter and Prins, 2003) in the early 2000s. Some books published in the 1990s carried "project leadership" in their titles (see, for example, Briner et al., 1990; 1996).

Toor and Ofori (2008a) reviewed works on leadership in the construction management literature and observed that empirical works on effectiveness and performance of construction leaders appeared in the 1980s (Bresnen, 1986; Bryman et al., 1987). In the 1990s, the focus moved to leadership style, attributes, and behaviors of project leaders (Dulaimi and Langford, 1999; Rowlinson et al., 1993). Toor and Ofori (2008a) argue that leadership research in construction has relied on old frameworks and empiricism embedded in the positivist approach.

There has been a surge of interest in project leadership since the early 2000s. Studies have addressed behaviors, traits, and styles (Giritli and Topcu-Oraz, 2004), leadership models (Chan and Chan, 2005), cross-cultural issues (Wong et al., 2007; Ofori and Toor, 2009); leadership development (Toor and Ofori, 2008a), the emotional intelligence of leaders (Songer et al., 2006), the negative sides of leadership (Toor and Ogunlana, 2009), ethical issues (Toor and Ofori, 2009a), and authenticity and psychological well-being (Toor and Ofori, 2009b). The establishment of the CIB Task Group 64 on Leadership ushered in a new era of research on construction project leadership (CIB Newsletter, 2006). Toor and Ofori (2008a) proposed a research agenda for leadership in construction. Observing this growing interest, McCaffer suggested that there should be a concerted effort by researchers around the globe, leading to the formation of a global research institute concentrating on leadership development in construction (1995, p. 306).

The Major Projects Leadership Academy was established in 2012 in the United Kingdom with the objective of transforming the implementation of government policy through world-class delivery of major projects. The academy seeks to do the following:

- develop project leaders to become world-class at successfully delivering major projects;
- create a cadre of world-class project leaders, formed into an expert support network;
- elevate the status of project leadership professionalism in central government.

The United Kingdom Project Leadership Programme was established in 2015 with the aim of developing a cadre of project leaders across government and improving project delivery. The program is designed to enhance project leadership skills and build excellent project leadership capacity at the level below major projects. It is based on three key leadership competences: leadership of self; commercial leadership; and technical project leadership.

Coleman and Bourne (2018) note that the literature on project leadership is limited, although many organizations are using projects to deliver products, services, and change, making projects critical to the future of organizations. The interest in the subject has not been uniformly on an upward trajectory. The ASCE journal, *Leadership and Management in Engineering* started in January 2001. However, it published its last issue in October 2013, although the editor acknowledged that there was much more to be done in the subject, and argued that there was the need to move the subjects of leadership and management in engineering into equity with the technical subjects of engineering (Hayden, 2013). He observed:

> Unless and until we change our perspective, we will continue to search for the quick and easy way to lead and manage others and the work itself. Until we see familiar scenarios in new ways, we will continue the search for "instant-oatmeal" solutions to management challenges.

The nature of project leadership

Cleland defines project leadership as "a presence and a process carried out within an organizational role that assumes responsibility for the needs and rights of those people who choose to follow the leader in accomplishing the project results" ((1995, p. 86). To Kloppenborg et al., project leadership is "the systemic application of leadership understanding and skills at each stage of a project life cycle" (2003, p. 15). Norrie and Walker define it as "the higher pursuit of the project team's creating purposeful, strategic action that will augment the organization's business strategy and achieve results within the norms and values of the organization" (2004, p. 48).

Coleman and Bourne (2018) discuss project leadership under five headings:

1   *The leader understanding themselves,* including their strengths, weaknesses, capabilities, and touchpoints, and how they will work with their team to deliver the project.
2   *Constructing the project organization,* from roles and responsibilities to the organizational culture, to enable it to perform well and deliver outcomes and benefits.
3   *Establishing, building, and developing the project team,* to deliver the project, and create project delivery capability for future projects.
4   *Delivering the outcomes and benefits of the project* by looking to the future, as opposed to managing the project schedule, resources, and immediate deliverable.
5   *Understanding the wider context in which the project is to be delivered,* and the challenges facing the team in the wider environment.

The features of project leadership make it different from organizational leadership. Some of these features are that project leadership:

- focuses on a single, unique project with a defined life-cycle;
- is constrained by fixed or predetermined performance parameters, such as limited time and budget;
- mostly deals with short-term goals but often has to find long-term solutions;
- co-ordinates a temporary, often loose, group formed from different organizations or across teams which are not accustomed to working with each other;
- deals with stakeholders with diverse backgrounds and often competing interests;
- deals with a loose and changing structure of the project;
- faces an environment of uncertainty;
- does not enjoy direct control or formal authority over teams or persons involved in the project;
- involves the performance of both internal and external roles related to the project;
- must establish a fine balance between leadership and management.

The complex and dynamic nature of projects poses distinct challenges to a leader, making his or her job multi-faceted, and demanding. Coleman and Bourne (2018) developed eight project leadership survival skills:

- *anticipating* – being prepared for what could knock the project off course next;
- *judgment and decision-making* – making timely decisions with incomplete information;
- *seeing it all* – feeling the totality of what is going on inside and outside the project;
- *building credibility and confidence* – belief in the leadership and the team;
- *being organizationally intelligent* – knowing when and how to engage with the organization;
- *learning* – being open-minded, and reflecting on and developing personal and team performance;
- *resolving conflicts and collaborating* – building a common purpose, despite the rules;
- *creating the project culture and environment* – deliberately defining and creating the culture and environment to succeed.

Effective project leaders

What makes an effective project leader? There have been many studies on the behavior and skills of project leaders. Smith (1999) observes that good project leaders are willing to contribute to the job even by dirtying their hands. Strang (2007) believes effective project leaders are aware of their behavior, traits, and skills and they try to reconcile their traits and behavior with the requirements of the situation and the context in which they operate. According to Toor and Ogunlana (2006) behaviors that are mentioned frequently in relation to project leadership include: visioning (Makilouko, 2004), inspiring (Dainty et al., 2005), goal-setting (Egbu, 1999), listening and empathizing (Odusami, 2002), considerate decision-making (Toor and Ogunlana, 2006), facilitating (Giritli and Topcu-Oraz, 2004), rewarding (Lee-Kelley and Leong, 2003), coaching/teaching (Giritli and Topcu-Oraz, 2004), people orientation (Fellows et al., 2003), and task orientation (Dulaimi and Langford, 1999). Behaviors that were considered important for successful project leadership on mega construction projects in Thailand included (Toor and Ogunlana, 2008): goal-setting, leadership by example, management of conflicts, facilitation of interaction, flexibility and accommodative attitude in decision-making, and result-orientation. The use of authority and punishment was among the lowest rated leadership behaviors.

Competencies that are covered in the discourse on project leadership include team skills, interpersonal skills, project management skills, communication skills, organizational skills, and technical skills. In a study of Thai and expatriate managers, Toor and Ogunlana (2008) found these competencies to be vital for leaders on cross-cultural projects: communication, teamwork, and personal and interpersonal skills. They also found that flexibility in decision-making, persistent performance, good listening skills, and a problem-solving style were also highlighted among the top-rated competencies of cross-cultural leaders.

Style in project leadership

Slevin and Pinto (1991) emphasize that the leadership style of the project leader is determined by the situation on the project. Hence, the leader may have different styles during different stages of the project. Several researchers also agree that leadership style should vary according to the situation and the leader must find a fit between his/her style, those who are being led, and situational factors (Kangis and Lee-Kelley, 2000). Rosenbaum (1991) maintains that effective leaders of technical projects are coaches and mentors who coach their followers to develop them and to expand individual productivity through teamwork. Mustapha and Naoum (1998) showed that high performing site managers employ a leadership style that is focused on team management. This is supported by Fraser (2000), who found that site managers scoring high on the effectiveness scale favored team-style leadership; those managers following a production style of leadership scored the lowest of all; and managers using a compromise leadership style had middle-range effectiveness scores. Ogunlana et al. (2002b) also found that the relationship-oriented leadership style was more appropriate than the task-oriented style for project managers in Thailand. Odusami et al. (2003) found that the most appropriate leadership style for Nigerian project leaders is that of the "consultative autocrat" (similar to the team leadership style).

Barber and Warn (2005) use the analogy of fire-fighting to describe a good project leader. They note that a proactive project leader pre-empts the problems and prevents the problems from occurring. Thus, an effective project leader is a "fire-lighter," who is able to inspire a shared commitment, encourage reflection before action, and coach others to achieve demanding objectives. Makilouko (2004) studied multicultural projects and found that the majority of project leaders adopted a task-oriented leadership style. They also showed cultural blindness, ethnocentrism, parochialism, or in-group favoritism. Leaders who indicated almost solely relationships-orientation, or both task and relationships-orientation, indicated also cultural sympathy and leadership strategies to maintain team cohesion and avoid cross-cultural problems.

To George (2003), authentic leaders have a sense of purpose, practice solid values, lead with the heart, and establish connected relationships. Such leaders empower others rather than concentrating power around them. They are consistent and self-disciplined individuals and never compromise on their principles. In addition to these characteristics, Toor and Ofori (2008a) note that authentic project leaders have good comprehension of cultural sensitivities and are highly motivated and self-aware. They understand the demands of the project from the perspectives of the key stakeholders, and give their best to make the project a success for all of them.

The leadership style dilemma in construction

Leadership disposition is a joint outcome of the leader's self-related cognitive information, personality traits, the underlying motives, and the leader's understanding of the operating situational variables (Toor and Ofori, 2006a; 2006b). Researchers on leadership have proposed a number of leadership styles. Each of these styles is on a continuum, such as: democratic and authoritarian (Tannenbaum and Schmidt 1958); task- and relationship-oriented (Fiedler, 1967); autocratic, consultative, and joint decision-making (Vroom and Yetton, 1973); servant leadership (Greenleaf, 1977); transactional, transformational, and *laissez-faire* (Avolio and Bass, 1991), charismatic leadership (House, 1977; Conger and Kanungo, 1987; Shamir et al., 1993), self-leadership (Manz and Sims, 1987), super leadership (Manz and Sims, 1989), and so on.

Researchers in the construction industry have also explored the leadership styles suitable for construction professionals. There are studies that favor task-orientation (see Monaghan, 1981) and a mix of "socially independent" and task-orientation (see Bresnen, 1986; Seymour and Elhaleem, 1991; Dulaimi and Langford, 1999) for better performance. On the other hand, there are also studies that support relationship-orientation as a more effective style (see: Toor and Ogunlana, 2005; Fellows et al., 2003). Other works note the importance of team-style (Fraser, 2000), supportive style (Fellows et al., 2003), and transformational style of leadership (Giritli and Topcu Oraz, 2004; Chan and Chan, 2005; Toor and Ogunlana, 2006). Other authors also note that the managers tend to vary their style during different stages of the project. They use a supportive style during the feasibility study and pre-contract stages of the works and a directive style as construction progresses (Rowlinson et al., 1993).

Several researchers (Makilouko, 2004; Odusami and Ameh, 2006) favor the argument for the varying leadership style across the project life and argue that a construction project follows a rapid lifecycle rather than being static, accentuating changes in project roles at various stages of the development. Because of this, the leader has to change roles from being a technical expert to a manager, then a leader and finally, at the end of the project, back to being a technical expert. Moreover, the project leaders usually have to lead several projects and teams simultaneously (Makilouko, 2004). The composition and dynamics of project teams differ; each requires a different leadership style. This suggests that a project leader may vary the leadership style over time, during different stages, under dissimilar situations, and across different teams and cultures as the project proceeds. Therefore, project leaders may need both task- and relationships-oriented leadership styles during different phases of the project to ensure order in the often chaotic situation of time and financial pressures, and simultaneous tasks.

There is no general agreement on what leadership style best suits construction professionals and project managers. However, an important question is whether general leadership characteristics (such as task- and/or relationship orientation, clarity of vision, intellectual stimulation, active or passive management, and so on) suffice in making up the character of an effective construction project leader. Furthermore, most of the leadership styles which are offered do not consider questions such as:

- how the leader develops a particular style;
- why the leader chooses to adopt a certain style;
- why some styles work within some teams but fail with others under similar conditions;
- whether there is an ideal set of qualities to constitute a best leadership style;
- how a leader switches from one style to another;
- whether it is possible to retain credibility by switching styles frequently;
- how a leader can have several styles at the same time if the leader is working on different projects.

More importantly, is leadership really about any specific style?

Although the literature shows that no one leadership style is good enough in all contexts, contemporary approaches to leadership fail to suggest how a leader can smoothly and successfully change from one leadership style to another without creating a negative or inconsistent image. This viewpoint is shared by Conger (2004) who argues that "we have been losing an appreciation for the fact that leadership approaches do indeed depend upon the situation." He further believes that one of the most intriguing issues in leadership studies is whether managers and executives can develop a leadership capacity with versatility in style or approach. Simply put, can managers vary their leadership styles significantly, given the changing circumstances? Or is leadership style such a product of personality and inbuilt personal behaviors that variations in style are limited to incremental changes (ibid.)?

Notes

1 See https://apps.bea.gov/industry/factsheet/factsheet.cfm
2 See www.bls.gov/iag/tgs/iag23.htm
3 See www.statista.com/statistics/1068213/china-construction-industry-gdp-contribution-share/
4 See www.investindia.gov.in/sector/construction
5 See https://outlook.gihub.org/
6 See www.irena.org/financeinvestment/Investment-Needs
7 See www.housingeurope.eu/blog-1257/the-housing-affordability-gap-is-equivalent-to-1-of-global-gdp
8 See www.transparency.org/news/pressrelease/a_world_built_on_bribes_corruption_in_construction_bankrupts_countries_and
9 See https://fortune.com/2019/05/16/fortune-500-female-ceos/
10 See Labor Force Statistics from the Current Population Survey (2019)< www.bls.gov/cps/cpsaat11.htm

3 Understanding leadership

Introduction

A large and growing volume of literature has been produced on leadership, and there is a distinct body of knowledge on the subject. Even so, there is still confusion and misunderstanding of several of the concepts and aspects. Many of them are still not settled, and there is a fierce debate among researchers and practitioners on each of them. There is a need to explain and illustrate leadership. An attempt is made in this chapter to cover the history of leadership research.

Defining leadership

What is leadership? Many works on leadership start with this question. Bennis (2007) laments that it has almost become a cliché that there is no single definition of leadership. There is no established and commonly agreed definition of leadership by which it can be evaluated, no dominant paradigms for studying it, and little agreement about the best strategies for developing and exercising it (Barker, 1997; Higgs, 2003; Hackman and Wageman, 2007). Several authors note that there are as many definitions of leadership as the number of people who have attempted to define it (Bass, 1990), while some state that the large volume of literature on leadership makes it difficult to present the concept in a single definition (Goethals et al., 2004). A few definitions in the literature are now set out:

> Leadership is the behavior of an individual while he is involved in directing group activities.
>
> (Hemphill, 1949, p. 4)

> Leadership behavior means particular acts in which a leader engages in the course of directing and coordinating the work of his group members.
>
> (Fiedler, 1967, p. 36)

> Leadership is the reciprocal process of mobilizing by persons with certain motives and values, various economic, political and other resources, in context of competition and conflict, in order to realize goals independently or mutually held by both leaders and followers.
>
> (Burns, 1978, p. 425)

[Leadership is] … the capacity to create a compelling vision and translate it into action and sustain it.

(Bennis, 1989, p. 65)

Leadership is the process of persuasion or example by which an individual (or leadership team) induces a group to pursue objectives held by the leader and his or her followers.

(Gardner, 1990, p. 1)

Leadership is a process of social influence in which one person is able to enlist the aid and support of others in the accomplishment of a common task. The main points of this definition are that leadership is a group activity, is based on social influence, and revolves around a common task.

(Chemers, 1997, p. 1)

Leadership is a process—a dynamic process in which the leader(s) and followers interact in such a way as to generate change.

(Kellerman and Webster, 2001, p. 487)

[Leadership is] a process of motivating people to work together collaboratively to accomplish great things.

(Vroom and Jago, 2007, p. 18)

These definitions, offered over different points in time, show that the understanding of leadership has changed from "behaviors" (how a leader behaves) to "actions" (what a leader does) to eventually a "social process" that involves an interaction among the leader, the followers, and situations. Despite the large number of leadership definitions available and the substantial volume of literature on it, Burns (1978) argues that leadership is the most observed but least understood phenomenon. It has both fascinated and perplexed researchers and practitioners, creating a significant amount of research to conceptualize and explain it (see Ayaman, 2000, cited in Goethals et al., 2004).

Scholars from many fields of knowledge such as political science, psychology, education, history, agriculture, public administration, business management, anthropology, medicine, military sciences, philosophy, and sociology have all contributed to the understanding of leadership. However, the authors of *Stogdill's Handbook of Leadership* concluded that the huge accumulation of empirical data has not produced an integrated understanding of leadership (Stogdill, 1974). DuBrin (1995) claims that about 30,000 research articles, magazine articles and books on leadership had been written by the mid-1990s. Goffee and Jones (2000) observe that nearly 2,000 books were published on leadership in the year 1999 alone. Yet new research continues to be undertaken and researchers are now more interested to break out from the confinement of social science laboratory experiments to observe real leaders in action in various areas of endeavor in the increasingly complex world.

Review of leadership research: classification

Leadership is one of the topics in modern research which originated centuries ago. It began when people started to seek an understanding of the importance of the leader's role in various facets of life such as politics, governmental issues, foreign policy, and war.

Philosophers, historians, military leaders, and rulers have paid much attention to this subject to bring improvement to leadership practices during their time. The development of leadership research can be divided into the following stages:

1   Ancient approaches
2   Classical approaches – motivation and trait theories – during the first half of the twentieth century
3   Transactional approaches – behavioural and contingency theories – during the 1950s and 1960s
4   Transformational and charismatic theories – during the 1970s and 1980s
5   Integrative approaches
6   Miscellaneous approaches
7   Developments in recent years.

Ancient approaches

Social and political science scholars have recognized the importance of leadership throughout human history (Chemers, 1997). The ancient leadership approach comprises the writings of early philosophers and thinkers on leaders, leadership, and the importance and necessity of leadership development. The *Encyclopedia of Leadership* (Goethals et al., 2004) mentions Confucius and Sun Tzu, Aristotle, Plato, Niccolò Machiavelli, Pareto, Thomas Hobbes, Mary Parker, Bertrand Russell, and several other philosophers and thinkers who have written on leaders and leadership, and contributed to the development of the theoretical base of leadership. These efforts and other philosophical works constitute a normative approach to understanding leadership which seeks to provide ethical and constructive views of good leadership. Many of the modern theories of leadership are based on ideas from the classical thoughts on the subject which remain relevant. These narratives mostly discuss leadership in general terms, often narrating anecdotes of the lives of leaders in government, the military, and other spheres of public service.

The Republic by Plato appears to be the first attempt to discuss the theory of politics and leadership and was written over 2,000 years ago. *Nichomachean Ethics* and *Politics* are two of Aristotle's books which discuss politics and leadership and are among the earliest writings on the subject. The other famous writings are by Confucius, Sun Tzu (*The Art of War*), Xenophone of Athens (*Kyropaedia*), Niccolò Machiavelli (*The Prince*), and Vilfredo Pareto (*The Treatise on General Sociology*). Many modern scholars of leadership agree that these ancient works offer deep insights into leadership. Several ideas outlined in these works still hold true. The increasing complexity of the business world has been rejuvenating interest in the scholarship of leadership. The following sections present the theories that have been developed after the dawn of the twentieth century.

Classical approaches

Early analyses of leadership, from the 1900s to the 1950s – the classical management period – differentiated between leader and follower characteristics. Frederick Winslow Taylor, who is considered to be the founder of scientific management, wrote *The Principles of Scientific Management*, published in 1911, which opened up the horizons of modern management research. He explained that the best way to increase efficiency was to improve the techniques and methods used by workers. People were seen as instruments or machines to be manipulated by their managers. Also, the organization was seen as a bureaucratic, well-planned and structured big machine. Taylor initiated time and motion studies to analyze work tasks to improve performance in every aspect of the organization.

In the 1920s, Elton Mayo and his colleagues developed the human relations movement which emphasized that it was beneficial for management to look also into human affairs. In the famous Hawthorne studies, they were able to demonstrate the effect of the human factor on efficiency (Mayo, 1933). The scientific management movement emphasized a concern for task (output), whereas the human relations movement stressed a concern for relationships (people). The recognition of these two concerns has characterized the discussion about leadership ever since.

Robert Tannenbaum is famous for proposing the continuum of leader behavior, the extremes of which are authoritarian and democratic leader behavior (Tannenbaum and Schmidt 1958). Kurt Lewin and his colleagues extended this continuum beyond the democratic leader behavior to include the *laissez-faire* leadership style. Rensis Likert (1967) proposed four management styles on a continuum from task-oriented (highly structured authoritarian management style) to a relationships-oriented management style (based on teamwork, mutual trust, and confidence).

Motivation theories

Many classical motivation theories form the foundation of management approaches. Abraham Maslow's (1954) hierarchy of needs and Frederick Herzberg's (1966) motivation-hygiene theory are among the best known. David C. McClelland et al.'s achievement motive is also important when considering the behavior of leaders (McClelland et al., 1953). These classical approaches formed the basis of the scientific study of management and leadership.

Likert's theory is quite close to McGregor's (1960) classic theories (Theory X and Theory Y), which represent pure archetypes of managerial beliefs about the nature of people that, in turn, influence their managerial and leadership behavior (Goethals et al., 2004). According to Theory X, most people are passive, dislike work, avoid responsibility, and need to be closely supervised and told what to do. They prefer to be directed by their leader, want safety above all, and are not interested in assuming responsibility. Theory X argues that people are self-centered, prone to resist change, and not very clever. Theory Y, on the other hand, suggests that people are self-driven, can focus on the objectives they pursue through self-direction and self-control, take full responsibility for tasks given to them, and strive for excellence without close supervision. According to Theory Y, leaders view their employees as valuable assets who are key to the success of the company.

Theory Z is yet another perspective from the classical approaches of leadership and management. Indeed, several researchers have designated their theories as Theory Z. These theorists include Abraham Maslow and William Ouchi. Maslow's hierarchy of needs theory (Maslow, 1954) is considered a classic work in the management sciences. He later posited a transcendent Theory Y leader who epitomized what Maslow called B-values – these values include truth, beauty, wholeness, dichotomy-transcendence, uniqueness, perfection, necessity, completion, justice, order, simplicity, richness, effortlessness, and playfulness (Maslow, 1971). Such leaders are very rare, almost exceptional. They can scan the full potential of people in order to transform them into the ideal, but they are frustrated by the mediocre, the short-sighted, the fearful, and the unimaginative.

William Ouchi's (1981) Theory Z was presented in his book, *Theory Z: How American Companies Can Meet the Japanese Challenge*. Theory Z essentially advocates a combination of all that is best about Theory Y and modern Japanese management, which advocates for a large amount of freedom for, and trust in, workers, and assumes that workers have strong loyalty and interest in team-working and the organization. Theory Z also places more reliance on the attitudes of the workers and their ability to take responsibility, whereas McGregor's Theories X and Y are mainly focused on management and motivation from the managerial and organizational perspective.

Trait theories

The underlying assumption of trait theories, first propounded by Thomas Carlyle in 1840 (Carlyle, 1841) was that leaders have certain characteristics that are utilized across time to enhance organizational performance and leader prestige. The idea was that traits affected behaviors and behaviors affected effectiveness. These traits are the distinguishing personal characteristics of a leader, such as physical features, aspects of personality, and aptitudes. Research on leadership in the early years of the twentieth century examined leaders who had achieved a level of greatness, and later on, this approach became known as the Great Man theory. The underlying idea behind this approach was that some individuals are born with certain extraordinary characteristics and qualities which make them leaders eventually. Bass (1997), for example, argued that during the early twentieth century, leaders were considered to be superior individuals, different from the others around them because of their skills, capabilities, inherited money, and social standing. The aim of trait theories was to prepare a master list of traits which would eventually result in an ideal leader.

Stogdill (1948) studied 124 trait studies conducted in the first half of the twentieth century and observed that the pattern of results was consistent with the conception of a leader as an individual who acquires the leadership role through the demonstration of the ability to facilitate the efforts of the group in attaining its goals. Relevant leadership traits include intelligence, alertness to the needs of others, understanding of the task, initiative and persistence in dealing with problems, self-confidence, and a desire to accept responsibility and occupy a position of dominance and control (Yukl, 2002). However, he noted two problems with this master list approach to leadership: (1) most traits are somewhat abstract and hence not entirely useful in explaining leadership effectiveness; and (2) trait studies usually examine how a single trait is related to leadership effectiveness, but do not explain how different traits interact or are interrelated to influence the leader's behavior (Yukl, 2006). He further argued that no trait was universally required for leadership which varied extensively according to the characteristics, activities ,and goals of the followers.

A meta-study by the Centre for Excellence in Management and Leadership (2002) identified over 1,000 leadership traits in the literature, from which they identified 83 distinct attributes. Of the many lists and categorizations, such as that of Boyatzis (1982), Stogdill's (1974) is considered the clearest. It comprises six primary factors, and the associated leadership traits under each of them:

1   *Capacity* – Intelligence, alertness, verbal facility, originality, judgment
2   *Achievement* – Scholarship, knowledge, athletic accomplishment
3   *Responsibility* – Dependability, initiative, persistence, aggressiveness, self-confidence, desire to excel
4   *Participation* – Activity, sociability, co-operation, adaptability, humor

5   *Status* – Socio-economic position, popularity
6   *Situation* – Status, skills, needs and interests of followers, objectives to be achieved.

Kirkpatrick and Locke (1991) explored several traits which distinguished leaders from non-leaders, including some of those found by Stogdill (1948). Traits such as emotional intelligence, social intelligence, self-awareness, the capacity to be optimistic and hopeful despite obstacles, the ability to empathize with others, and strong social and interpersonal skills appear in the list of some significant leadership attributes. Nevertheless, one of the criticisms of this approach is that it does not tell when selected traits are critical or when some can be omitted without extensive situational analysis (Van Wart, 2005). This is because two leaders can use different sets of traits to attain a goal. Even a similar set of traits may be beneficial in one situation and problematic under some other conditions. Even today, trait theories have still not been able to present that greatly desired "master list" for the ideal leader and this quest still continues (Goethals et al., 2004).

The trait approach to leadership has regained some attention in the recent literature on leadership (Lim and Daft, 2004). Stephen Zaccaro has championed the field by adding more depth and thoroughness to the understanding of traits. In his works (Zaccaro et al., 2001; Zaccaro and Horn, 2003; Zaccaro, 2007), he argues that the previous rejection of trait-based approaches is not justified, considering the empirical evidence in the literature. Traits are significant determinants of leadership effectiveness, and combinations of traits and attributes, integrated in conceptually meaningful ways, are more likely to predict leadership than are independent contributions of multiple traits. Zaccaro (2007) also observes that individuals have different patterns of traits that reflect a person's stable tendency to lead in different ways across different organizational domains. He also argues that the traits of some leaders have more distal influences on leadership processes and performance while some others have more immediate effects. This is essentially mediated by situational parameters (ibid.). According to the modern understanding of traits, a combination of distal attributes (personality, cognitive abilities, motives, and values) and proximal attributes (social appraisal skills, problem-solving skills, and expertise or tacit knowledge) combine to engender leader emergence, effectiveness, advancement, and promotion (see Zaccaro et al., 2004).

The Great Man or trait theory has been criticized by many authors (such as Stogdill, 1948; Levine, 2008), but it is still cited and studied, and its roots are evident in the assumptions about leaders, and many of the theories of leadership today. Luthans et al. (2007) further the discussion on traits and argue that there are two types of traits: trait-like-traits and state-like-traits. Trait-like-traits are more rigid and difficult to learn. Examples of these are: intelligence, coping, and interpersonal needs. State-like-traits are malleable and therefore can be developed and hence learnt through interventions. Examples are: hope, optimism, resilience, and self-efficacy. These developments in leadership research show that traits are still important as far as leadership is concerned. Trait theory remains vibrant. Researchers today, studying modern models of leadership, are finding correlations between certain personality characteristics (charisma, extroversion, conscientiousness, integrity, and achievement motivation) and leadership. With advanced research methods, including the medical sciences, there is a renewed sense that defined traits can be used to identify potential leaders, explain leadership, and play a part in its development. Like Luthans et al. (2007), there is agreement that there is a need to focus on identifying traits that can be learned through interventions.

Transactional approaches

After classical research on leadership, a new era of explorations started in the second half of the twentieth century. Most of the popular theories, concepts, and frameworks of leadership literature were presented in this period which deserves the name of the "golden period" or "modern period" of leadership research. Trait studies, behavioral studies, and contingency studies came up during this "golden period." Wart (2005) notes that basic research at Ohio State University, the University of Michigan, and other settings during the 1950s started challenging some of the implicit leadership assumptions of the early management and trait theories. The leadership theories developed during the 1960s were later referred to as the transactional approaches, which can be further divided into behavioral and situational approaches.

Behavioral theories

According to the behavioral approach to leadership, anyone who adopts an appropriate behavior that appeals to followers can become a good leader. Behaviors can be learned more readily than traits; this underlines the importance of behavioral approaches in leadership studies. As the notion of inherited or inherent leadership was dispelled, behavioral scientists turned their attention to the measurable behaviors of leaders. The idea was to see what leaders actually do rather than what they have in the form of traits and attributes. Behavioral approaches were criticized for putting emphasis on behavior and not the situation. That is why some individuals may exhibit leadership in one situation but not in another.

The Ohio State Leadership Studies (see, for example, Halpin and Winer, 1957) laid the foundation for understanding the difference between successful and unsuccessful leaders and led to the modern conception of leadership styles. The research identified two dimensions of leadership behavior: consideration and initiating structure. *Consideration* means that a leader acts in a friendly and supportive manner, shows concern and consideration for subordinates, and looks after their welfare. The leader creates an environment of emotional support, warmth, friendliness, and trust for the followers. On the other hand, *initiating structure* means that the leader defines and structures his/her own role and the role of the subordinates for undertaking formal tasks and the attainment of goals. Leaders scoring high on this dimension define the relationship between themselves and their employees. They criticize the poor performance of their subordinates, are strict about the keeping of deadlines, maintain the set standards of performance, and offer new approaches to problem-solving.

Researchers at the University of Michigan (see, for example, Katz and Kahn, 1952) also conducted a major program of leadership research almost at the same time as the Ohio State leadership studies. Their classification of leaders consisted of *production-centered leaders*, *employee-centered leaders*, and *participative leaders*. This classification is similar to that of the Ohio State studies. Production-centered leaders are those who put more emphasis on planning, scheduling, and coordinating the activities of their subordinates, and on providing the necessary resources and support for the attainment of the set goals. Employee-centered leaders are more inclined to focus on their relationships with subordinates. Participative leaders use group supervision instead of micromanaging each subordinate. They encourage the participation of employees in decision-making, problem-solving, and other activities in the workplace.

One of the first theories that tried to make sense of the then new behavioral orientation of studies on leadership was proposed by Blake and Mouton (1964). Their managerial grid theory places five leadership styles on a grid constructed with two behavioral axes. It is a framework for simultaneously specifying the concern for production and people dimensions of leadership. The five leadership styles on the grid are: Authority-Compliance, Impoverished Management, Country Club Management, Team Management, and Middle of the Road Management. The grid theory was the first highly popular theory of leadership that used the task-people duality for effective leadership. Many researchers criticize the managerial grid theory for dictating one best style, although it is pertinent to note that the team management style includes the leader adapting to the situation.

Contingency theories

Contingency theories were championed by researchers who started thinking about leadership in relation to situation. In an empirical sense, contingency theories guided research into the kinds of persons and behavior who are effective in different situations. Fred Fiedler (1963) was the first to introduce the term "contingency" in leadership studies through his contingency model. Later, many others contributed to the field and discussions on the situational aspect of leadership are omnipresent in the leadership discourse. For example, Vroom and Jago (2007), in their discussion of the role of situation in leadership, note that "viewing leadership in purely dispositional or purely situational terms is to miss a major portion of the phenomenon. The task confronting contingency theorists is to understand the key behaviors and contextual variables involved in this process" (p. 23). This shows that interest in the contingency or situational approach remains alive, although the modern literature has embraced the broader term of "context" (Avolio, 2007). It is plausible to believe that contingency, situation, or context will always be a relevant consideration in any discussion, framework, or theory of leadership. The discussion below highlights some frameworks presented under the broad label of the "contingency or situation approach."

Fiedler (1963) was the first researcher to respond to Stogdill's (1948) call to formulate trait contingency models (see Goethals et al., 2004). He developed the most widely researched and quoted contingency model which advocates that the best style of leadership is determined by the situation. His model was the first in leadership research to integrate leader, follower, and situational characteristics. More specifically, Fiedler's model predicts that relationship-oriented leaders are more effective in medium-control situations, and that task-oriented leaders are more effective in high- and low-control situations. The leader's orientation determines if he or she is in synchronicity with the situation or out of synchronicity with the situation. If the leader's orientation matches the situation, he or she is predicted to perform more effectively and vice versa. Some criticize this theory on the basis of its conceptual weaknesses and methodological controversy (see, for example, Yukl, 1970; Schriesheim and Kerr, 1977). However, despite the controversies and criticism of this theory, Fiedler was a pioneer in taking leadership research beyond the purely trait or purely situational perspectives that preceded his contribution (Vroom and Jago, 2007).

Hersey and Blanchard (1982) belong to the group of researchers who advocated a contingency approach in which different leadership styles hinged upon different factors, mostly situational. Their model explains how to match the leadership style with the readiness of the group members. Readiness of situational leadership is defined as the extent to which a group member has the ability and willingness or confidence to accomplish a particular task or activity. Based on follower capacity, ability, and motivation, the situational leadership model prescribes four different leadership styles: (1) directing; (2) coaching; (3) supporting; and (4) delegating. Some authors have used other words such as: telling, selling, participating, and delegating. Here, the key point is that as group member readiness increases, a leader should rely more on relationship behavior and less on task behavior. For example, followers who are low in competence but high in commitment, such as new employees, are eager for instructions and structure but do not need much supportive behavior. A directive style would be more useful in such a situation. Moderate competence and low commitment call for a directive and supportive style. However, competent subordinates need less specific direction than do less competent ones. These illustrations show that this model is useful for the training and development of leaders.

The path-goal theory emerged and was further developed in the 1970s (see, for example, House, 1971; House and Dessler, 1974; House and Mitchell, 1974). The theory illustrates how the behavior of a leader influences the satisfaction and performance of subordinates. It describes what a leader must do to achieve high productivity and morale in a given situation. In general, a leader attempts to clarify the path to a goal for a group member so that he or she receives personal payoffs. The theory went through a refining process by a number of researchers in the subsequent years. The major proposition of the theory is that the leader should choose a leadership style that takes into account the characteristics of the team members and requirements of the task.

The path-goal theory has been expanded to develop the "leadership substitute theory." Kerr and Jermier (1978) first introduced the concept of leadership "substitutes" and "neutralizers." Their theory identifies the aspects of certain situations when there is almost no need for leadership, or where leadership is of little importance. Substituting followers' skills for leadership is beneficial because formal leaders have limited time. Less need to perform the leadership function allows followers to concentrate on other possibly more critical issues and thus to enhance their effectiveness in certain crucial areas. At the same time, reduction in direct leadership supervision or oversight allows the subordinates to become self-reliant, more responsible, and innovative in their tasks. According to this theory, "neutralizers" are the characteristics, which make it effectively impossible for leadership to make a difference. Kerr and Jermier (1978) proposed 13 dimensions, which, they hypothesized, would neutralize the effectiveness of leaders. There are six organizational characteristics (the degree of organizational formalization; inflexibility of rules; cohesiveness of work groups; amount of staff and advisory support; organizational rewards not controlled by the leader; and the spatial distance between supervisors and their subordinates); three types of task characteristics (routine, methodologically invariant tasks; intrinsically satisfying tasks; and task feedback); and four subordinate characteristics (abilities, experience, training, and knowledge; need for independence; professional orientation, and indifference to rewards) (Podsakoff and Mackenzie, 1997). This theory has generated a considerable amount of interest (see Howell, 1997; Chemers, 2000) because it offers an intuitively appealing explanation why leader behavior impacts subordinates in some situations but not in others. However, some of its theoretical propositions have not been adequately tested. The theory continues to generate empirical research.

Another popular theory in leadership research is the leader-member exchange theory, mostly known as the LMX theory (Van Wart, 2005). It was first presented in the early 1970s in reaction to the dominant behavioral and contingency models of leadership. It was originally called the vertical dyad linkage (VDL) theory, and it suggested that leaders adopt different leadership styles with different subordinates. Leaders also develop different dyadic exchange relationships with different specific subordinates. Such relationships can be ones that treat the subordinate as being in a close relationship with the leader or ones that treat the subordinate as a more distant individual. The LMX theory focuses on the ongoing relationship that leaders and members of their group experience as they exchange mutual perceptions, influence, types and amount of work, loyalty and prerequisites, and so on (ibid.). An advanced version of LMX advocated the notion that good leaders create as many high-exchange relationships as possible. Good leaders need loyal, committed, hard-working, productive, flexible, and competent subordinates to advance the group goals and achieve higher levels of accomplishment and innovation, and they need to cultivate these in their subordinates. The LMX theory explains this dynamic well.

Although it was originally developed earlier (Vroom and Yetton, 1973), the theory of decision-making later went through a refining process (see Vroom and Jago, 1988). This is why the model is known as the Vroom-Yetton-Jago model in management studies today. The theory is much narrower in its focus and it deals with the form in, and degree to, which the leader involves his or her subordinates in the decision-making process (Vroom and Jago, 2007). Leaders undertake decision-making on a regular basis and they have to set certain parameters for this in the organizational set-up. Therefore, leaders must choose a style that elicits the correct degree of group participation while making decisions. The Vroom-Yetton-Jago model perceives leadership as a decision-making process.

The effect of decision procedures on the quality and acceptance of the decision depends on various aspects of the situation. It is natural that a procedure which is very effective and efficient in one set of conditions and circumstances may prove to be a failure in another set of conditions. The Vroom-Yetton-Jago theory has a number of strong points. It delimits the aspects of leadership it endeavors to elaborate upon. It does not over-simplify the conditions for a phenomenon as complex as decision-making. Vroom and Jago (2007) admit that their theory does not encompass all or even most of what a leader does. However, they believe that the sharpness of their focus in the framework allows a high degree of specificity in the predictions that are made.

The multiple linkage model was presented by Yukl in 1981 and further refined by him in 1989. It is called an "ambitious integrative theory" by Chemers (1997). The model includes four types of variables: (1) managerial behaviors; (2) intervening variables; (3) criterion variables; and (4) situational variables. The model suggests how different variables combine and affect each other to determine an organization's performance. The emphasis in the multiple linkage model is on the intervening variables and the leader behaviors that affect them. The weakness of the model is that these linkages are not very comprehensive; its strength is that it considers the intervening process as the link between leader behaviors and group outcomes. This model brings the leader, situation, process, and outcome together.

Fiedler's scientific curiosity was once again aroused when he came across some empirical findings that did not agree with either common sense or with accepted scientific wisdom. Therefore, as an extension of his contingency model, Fiedler presented the cognitive resources theory in 1986 and refined it in 1987 with his colleagues. This theory examines the conditions under which cognitive resources, such as intelligence and experience, are related to job performance. The theory argues that group performance is determined by a complex interaction among two leader's traits (intelligence and experience), one type of leadership behavior (directive leadership), and two aspects of the leadership situation (interpersonal stress and the nature of the group's task). Fiedler and Garcia (1987) presented a causal chain in which a leader's cognitive resources have a profound impact on group performance when the leader actively directs follower activity. This impact is positive for intelligence under low stress conditions and experience under high stress conditions.

Transformational and charismatic approaches

Transformational approaches

One of the most influential developments of leadership research is the concept developed by James MacGregor Burns, first presented in 1978 under the title "transformational leadership." Writing in the political science tradition, Burns discusses various types of leadership, especially contrasting transactional leadership, which largely appeals to the self-interested motivations of followers, with transformational leadership, which attempts to raise the followers' consciousness to reform and improve the organizations (Van Wart, 2005). Burns (1978) makes a central distinction between what he calls "transactional" and "transforming" leadership.

According to Burns, transactional leadership takes place when "one person takes the initiative in making contact with others for the purpose of an exchange of valued things" (ibid., p. 19). This type of leadership is best described as the politics of exchange, in which, for example, a public official bargains jobs for votes. Transformational leadership, in contrast, has a moral dimension. Burns believes that transformational leadership occurs when "one or more persons engage with each other in such a way that leaders and followers raise one another to higher levels of motivation and morality" (ibid., p. 20). Burns defines transformational leadership as a dynamic, two-way relationship between leaders and followers. Leaders must connect with the needs and wants of the followers and establish motivation to accomplish collective goals that satisfy the needs of both the leader and the followers. Mutual need and empathy are the key characteristics of transformational leadership. He also believes that every person is engaged in the leadership process in one way or another at different times and in different situations (ibid.).

In 2003, James MacGregor Burns published a follow-up book to his earlier works, entitled *Transforming Leadership*, to explore and expand his theory nearly 30 years after his book *Leadership* in 1978. He believes that all leaders have a social responsibility to empower people to pursue their own happiness by effecting social change. He states: "Leaders working as partners with the dispossessed people of the world to secure life, liberty, and the pursuit of happiness – happiness empowered with transforming purpose – could become the greatest act of united leadership the world has ever known" (Burns, 2003, p. 4).

In his (2003) book, Burns is of the opinion that a transformational leader not only speaks to immediate wants but elevates people by vesting in them a sense of possibility, a belief that change can be made, and that they can make the change. Motivation, according to Burns, is what powers leadership. Creativity is another key element of transformational leadership. Transforming leaders have the ability to see possibility and innovation and to share that vision with others. He believes that leaders seize opportunities, overcome obstacles, and change how the rest of the world acts, thinks, and lives. Burns believes that a crisis can often be a source of creativity. He cites examples of skillful leaders, including presidents, military commanders, and leaders of corporate organizations who have applied creativity in times of crisis to effect change.

Burns (1978) believed that leaders were either transactional or transformational. However, seven years later Bernard Bass (1985) proposed that both types of leadership are necessary and that transformational leadership actually enhances transactional behaviors. Bass perceives leadership as a single continuum. It progresses from non-leadership to transactional leadership to transformational leadership. Non-leadership leads to haphazard outcomes; transactional leadership gives improved and better results which are mostly conventional; but transformational leadership provides the best outcomes.

Bass is of the view that transformational leadership is a widespread phenomenon across all levels of management, types of organizations, and nations. He characterizes transformational leaders as having four significant attributes:

1 *charisma or idealized influence* – they have conviction and values and they emphasize the importance of purpose, commitment, and ethical components of decisions;
2 *inspirational motivation* – they articulate an appealing vision of the future, challenge followers with high standards, speak with optimism and with enthusiasm, and provide encouragement and meaning for what needs to be done;
3 *intellectual stimulation* – they push followers to consider new points of view, to question old assumptions, and to articulate their own views;
4 *individualized consideration* – they take into account the needs, capacities, and aspirations of each individual follower in the effort to treat followers equitably.

Of all the transformational leadership theories, Bass's is the most highly researched, and it has a good deal of positive support. His approach is appealing as well as relatively elegant, considering the large number of styles that it incorporates. Nevertheless, fuzziness and overlap of the transformational concepts are problems with it. Bass and Avolio (1995) developed the Multifactor Leadership Questionnaire (MLQ) to measure transformational leadership. It has become the most popular tool for leadership assessment.

Presented by Tichy and Devanna (1986), and refined later (Devanna and Tichy, 1990), transformational leadership is about change, innovation, and entrepreneurship. According to this view, managers can be found more commonly but transformational leaders are rare and they engage in a process which includes a sequence of phases: recognizing the need for change, creating a new vision, and then institutionalizing the change. Leaders who are of the transformational type are individuals who create new approaches and imagine new areas to explore; they relate to people in more intuitive and empathetic ways, seek risk where opportunities and rewards are high, and project ideas into images to excite people. These leaders must bring a change in organizations in three stages. First is recognizing the need for revitalization; followed by the second stage, where the leader should create a new vision. In the third and final stage, institutionalizing the change is imperative; as the new vision is understood and accepted, new structures, mechanisms, and incentives must be in place.

Kouzes and Posner (1987; 1988) adopted a novel approach to formulate their ideas about transformational leadership. They asked leaders "What leads to excellent leadership based on your personal experiences?" They collected responses from over 1,000 leaders using a critical incident methodology and focusing on the "personal best" experiences of respondents. They found five major practices of transformational leaders. These leaders: (1) challenge the process; (2) inspire a shared vision; (3) enable others to act; (4) model the way; and (5) encourage the heart. Kouzes and Posner also developed the Leadership Practices Inventory (LPI), an instrument for measuring the practices of transformational leaders.

Despite the recognition that the transformational leadership theories have gained, there have been criticisms as well. Yukl (1999) presents a strong case in this regard and highlights several conceptual weaknesses in the concepts of transformational and transactional leadership. He believes that the underlying influence processes for transformational and transactional leadership are still unclear. He further argues that each transformational behavior includes diverse components, which makes the definition ambiguous. The partially overlapping content and the high inter-correlation found among the transformational behaviors raise doubts about their construct validity. Moreover, some important transformational behaviors (such as inspiring, developing, and empowering) are missing in the Bass (1996) version of the theory and in the MLQ, which was designed to test the theory (Bass and Avolio, 1990). The transformational leadership concept also fails to identify any situation where it can prove to be detrimental. Extending his critique, Yukl (1999) argues that transactional leadership includes a diverse collection of (mostly ineffective) leader behaviors that lack any clear common denominator.

It can be concluded that transformational leadership differs from transactional leadership in many ways. For example, to Price (2003), first, it focuses more on ends than on means; hence, the focus is more on the vision or purpose than the mechanisms for achieving them. Second, it emphasizes the role of the leader's charisma, and therefore the ability to inspire and stimulate is more important than controlling followers. Finally, it assumes that leaders have a moral responsibility; thus, they are committed to altruistic values, and focus on the interests of followers.

Charismatic approaches

The credit for introducing the word "charisma" goes to the German sociologist Max Weber (see Weber, 1968). Charisma is a Greek word meaning "divinely inspired gift" which imparts an extraordinary quality to charismatic individuals by which they can influence the hearts and souls of others, perform miracles, or predict the future. According to Weber, charisma is a certain quality of an individual personality, by virtue of which the person is set apart from ordinary humans and treated as endowed with supernatural, superhuman, or at least specifically exceptional powers or qualities. Max Weber believed that charismatic leaders were more likely to emerge during times of crisis and social upheaval. In organizations, crisis may be in the form of some major reform, financial problems, or even poor performance causing damage to organizational reputation and prestige in the market.

Another definition of charismatic leadership is in terms of the degree to which the leader engages in the following behaviors: articulating a captivating vision or mission in ideological terms; showing a high degree of confidence in themselves and their beliefs; setting a personal example of involvement in, and commitment to, the mission for followers to emulate; behaving in a manner that reinforces the vision or mission; and communicating high expectations to followers and having confidence in their ability to meet such expectations (see Conger and Kanungo, 1987; Shamir et al., 1993).

Several premises of charismatic leadership have been proposed. Robert House was the first to present a fully developed theory of charismatic leadership, in 1977. House believed that charismatic leaders had a strong influence on their followers. He posited that charismatic leaders have their major effects on the emotions and self-esteem of followers – the affective motivational variables rather than the cognitive variables (House et al., 1988). Once followers are convinced of the ideology of the leader, they follow him or her willingly, become fully involved in the task, obey the commands of the leader in their entirety, feel an emotional attraction toward the leader, consider the leader's goals as their own, and believe that they are part of a mission which must be accomplished under the guidance of their beloved leader. The limitation of House's initial theory was the ambiguity surrounding the influence process.

House et al. (1991) refined House's (1977) original theory of charismatic leadership and presented a more complete conceptualization of the theory. They defined charismatic leadership in terms of three constituents: (1) effects on followers; (2) the leader's personality and behavior; and (3) attributions of charisma to leaders by followers and observers.

Charismatic leadership is described as an interactive process between followers and their leader in the first constituent. This interaction results in the attraction of followers to the leader and strong internalization of the leader's values and goals by the followers. Over time, the followers develop unquestioning acceptance of, and commitment to, the leader. The followers trust fully in the correctness of the leader's beliefs and are willing to obey the leader. The next constituent involves specific leadership traits and behaviors that give rise to charismatic leadership. The traits that distinguish charismatic leaders from non-charismatic leaders are self-confidence, the need to influence, dominance, and strong conviction in the moral rightness of their beliefs. More recently, from the GLOBE study (House et al., 2004), charismatic or value-based leadership reflects the ability to inspire, to motivate, and to expect high performance outcomes from others based on firmly held core values. It includes these leadership subscales which are similar to leadership traits or styles: visionary; inspirational; self-sacrificing; showing integrity; decisive; and performance-oriented.

The theory of charismatic leadership presented by Conger and Kanungo (1987) focuses on how charisma is attributed to leaders. Its underlying assumption is that charisma is an attributional phenomenon. In later versions, the authors of the original theory refined it and proposed that the followers' attribution of charismatic qualities to a leader is jointly determined by the leader's behavior, skill, and aspects of the situation. Conger and Kanungo were of the view that the context for the emergence of charismatic leadership has to be problematic in some way, which means a situation of crisis, social unrest, or unusual circumstances. Therefore, situation demand is a moderating factor for the emergence of charismatic leaders, though not necessarily always. Because of their emphasis on deficiencies in the system and their high levels of intolerance of them, charismatic leaders are perceived as organizational reformers, change agents, visionaries, and entrepreneurs.

Charismatic leaders enjoy enormous magnitude of position and expert and personal power; and are highly vulnerable to narcissism which may lead to self-serving and self-centered behavior. Narcissism makes such leaders forgetful of the organizational needs and they start pursuing their personal goals (Toor et al., 2007). An overwhelming desire for power and importance leads them to ignore their followers, the viewpoints of others, realities and facts, open feedback, the followers' need of leadership development, and the requirement of coaching and mentoring of the subordinates for the future (Conger, 1989; Van Wart, 2005). Toor et al. (2007) note that there has been widespread criticism of charismatic leadership; it is suggested that there is confusion about the meaning of charismatic leadership, partly due to how different theorists define it (Bryman, 1993). Therefore, a consistent definition of charismatic leadership is required. Yukl (1999) notes that the creators of charismatic theories also disagree about the relative importance of the underlying influence processes and behaviors in charismatic leadership. He further points to a related issue that the charismatic leadership concept seems to advocate socially acceptable behaviors that increase the followers' perception of leader expertise and dependence on the leader.

Leaders can use manipulative behaviors to exploit their position and authority. These behaviors are known as the negative side of charisma, as suggested by Conger and Kanungo (1987) who believe that there are two sides of charisma: the positive and the negative. Charismatic leadership depends on the leader's passion, confidence, and exceptional ability to persuade and convince people.

These perspectives of charismatic leadership describe both its positive and negative sides and its potential outcomes. They provide details of negative charisma which some leaders use for their own personal benefit and which may be damaging under certain situations. Nevertheless, charismatic leadership theories do not consider non-charismatics and leadership situations which do not encourage change. Moreover, they describe the leaders as heroic personalities and therefore do not portray the full picture of leadership.

Integrative theories of leadership

In 1997, Martin Chemers presented his idea of the integrative theory of leadership. He first argues that the literature on leadership and on other organizational theories is fragmented and contradictory (Chemers, 1997). He points out that there are obvious controversies between popular and scientific approaches which make an integrative framework that transcends the apparent contradictions necessary. One integration adopts a functional perspective that hypothesizes that apparent divergences between theoretical approaches result from attempts to use one theoretical orientation to explain separate and distinct leadership functions with various processes and effectiveness criteria. A second integration ties together the central processes that underlie team leadership in an attempt to provide a coherent understanding of the dynamic qualities of effective leadership (Chemers, 1997). He asserts that the first step in sorting out the commonalities and contradictions among leadership theories is to recognize that leadership is a multifaceted process. Leaders have to deal with a wide range and variety of tasks; they should be able to analyze information, solve problems, motivate subordinates, direct group activities, inspire confidence, and so on. In his integrative model of leadership, Chemers maintains that there are three functional aspects of leadership: (1) image management; (2) relationship development; and (3) resource utilization, even though all functions of leadership may overlap in multidimensional, reverberating, and dynamic ways.

These three functional facets of effective leadership help to provide some common principles that can be integrated into a comprehensive theory of leadership, while an additional perspective can be provided by analyzing the process through which leadership is integrated at intrapersonal, interpersonal and situational levels (ibid.). Chemers limits the styles to three major types: structuring, consideration, and prominence. Moreover, the overall integrative process model attempts to address individual, dyadic, group, and organizational interactions. The integrative process is divided into three zones, each one of which examines a predominant interface of persons and environment. These zones are:

- the zone of self-deployment (here, individuals assess personal characteristics and situational demands);
- the zone of transactional relationship (here, a socially constructed reality between leader and followers is important);
- the zone of team development (here, reality-based outcomes are emphasized).

According to Chemers, integration across perspectives may allow researchers to construct a base that can provide a stepping stone for moving to the next level of study. He concludes that it is a challenging task; nevertheless, accepting this challenge is worth it, considering the importance of leadership to organizational and societal success. Other authors have discussed team-oriented leadership which, to Hollander (2008), also draws on the notion of the leader as a servant.

Another attempt at integrating various perspectives on leadership was made by Goethals and Sorenson (2007) in a book entitled: *The Quest for a General Theory of Leadership*. The book aimed to draw the contours of a grand theory of leadership by bringing together cross-disciplinary perspectives on leadership. It discusses a variety of topics, including power, leader-follower relations, ethics, constructivist perspectives on leadership, causality, and social change; and the historical and cultural contexts that impact leadership. The book does not really offer a single unified, general or grand theory of leadership but presents a useful framework for understanding leadership from a cross-disciplinary perspective.

Miscellaneous approaches

Self-leadership theory

The concept of self-leadership was presented by Manz (1986). Manz (1992) defines self-leadership as "the process of influencing oneself." Various researchers such as Bass (1990); Manz (1992), and Sims and Lorenzi (1992) have made contributions to the research on this concept of leadership. The meaning and importance of self-leadership have become evident lately. Self-leadership is a universal style and is composed of self-direction, self-support, self-achievement, and self-inspiration. This concept does not consider any intervening variables. Self-leadership puts emphasis on self-reliance, understanding the self in depth, and conceiving of self-potential, strengths, and motives. According to Bryant and Kazen (2012), self-leadership is about intentionally influencing one's thinking, feelings, and actions to attain one's objectives. The concept of self-leadership focuses on the specific strategies that produce an effective form of self-leadership in individuals. These strategies are: behavior-focused strategies, natural-reward strategies, and constructive thought-pattern strategies (Van Wart, 2005). A combination of these strategies would help individuals to become more effective leaders. The performance variables of self-leadership include enhanced self-efficacy, higher personal standards, greater determination and focus, and improved self-satisfaction and fulfillment.

One notion of self-leadership is that the best organizations subscribe to and promote the concept of leadership at all levels within the organization. Each individual provides leadership for those responsibilities that have been assigned to them. For the highest-performing organizations, even the lowest-ranked personnel must assume leadership and attention to detail for their responsibilities in a manner similar to the most senior and powerful. Complying with this intention, followers become self-leaders and the organization benefits from individuals at all levels. Self-discipline, self-analysis, self-goal setting, self-improvement, self-management, self-monitoring, and self-fulfillment are crucial for success and the long-term achievement of leaders. The model of self-leadership is generally useful for the development and training of effective leaders. It not only lists important leadership traits but also suggests methodologies to help leaders develop those traits so that it is possible for one trait to incrementally influence in order to become more effective and efficient.

Servant leadership theory

Servant leadership is a philosophy which supports people who choose to serve first, and then lead as a way of expanding their service to individuals and institutions. Servant-leaders may or may not hold formal leadership positions. Servant leadership encourages collaboration, trust, foresight, listening, and the ethical use of power and empowerment. First presented by Robert Greenleaf, who coined the term "servant leadership," in 1977, this approach suggests that leaders must place the needs of subordinates, customers, and the community ahead of their own interests in order to be effective. It puts emphasis on upside-down leadership because leaders transcend their self-interest to serve others and their organizations. To Greenleaf (1977), the servant leader must pass the following test:

> Do those served grow as persons? Do they while being served become healthier, wiser, freer, more autonomous, more likely themselves to become servants? And what is the effect on the least privileged in society; will they benefit, or, at least, will they not be further deprived?
>
> (p. 9)

The characteristics of servant leaders include empathy, stewardship, healing, and commitment to the personal, professional, and spiritual growth of their subordinates. They operate on realizing achievements on two levels: (1) the fulfillment of their followers' goals and needs; and (2) the attainment of the larger mission of the organizations they work for. They value their followers, respect their ideas and opinions, encourage their participation, share power with them, and provide information to them; they build communities (Spears, 1995). They hunt for talent and appreciate it, provide coaching and mentoring, and create the sense of mutual respect among themselves and their followers. New perspectives and refinements of servant leadership continue to emerge. For example, Nathan et al. (2019), after a review of articles and studies published on servant leadership during the last two decades, define servant leadership as an "other-oriented approach to leadership manifested through one-on-one prioritizing of follower individual needs and interests, and outward reorienting of their concern for self towards concern for others within the organization and the larger community." This definition is built around three features – motive, mode, and mindset – that are the *sine qua non* of the proper understanding of servant leadership.

Collins's model of Level-5 leadership

The Level-5 leadership model was presented by Jim Collins (2001) and his team who conducted a five-year study of 11 companies selected from more than 1,400 that had been listed in the Fortune 500 from 1965 to 1995. Each of the selected companies had ordinary results for 15 years and then they went through a transition. From that point, they outperformed the market by at least 3 to 1 and sustained that performance for at least 15 years. Each of them was compared with companies in the same industry and about the same size. The researchers identified the critical importance of what Collins calls "Level-5 leadership" in transforming companies from merely good to truly great organizations. Collins describes his findings in his famous book, *Good to Great*.

Level-5 leaders are those who build enduring greatness through a blend of personal humility and professional will. Such leaders are timid and ferocious, shy and fearless, and modest with a fierce, unwavering commitment to high standards. They are resolute and humble and they do their work conscientiously, responsibly, and successfully. They care about the people they work with, they care about the organization, and about the community; they shun the limelight and any publicity, and live normal and quiet lives. They rely on instilling inspired standards and not inspiring charisma to motivate. They build a culture of discipline in the organization. This culture is not a tyrannical disciplinarian one but one that enables freedom and responsibility among the subordinates. Level-5 leaders channel their ego needs away from themselves and toward building a great organization. They often will sacrifice their own gain for the gain of the organization. When things do not go well, Level-5 leaders take responsibility for the failures and never blame other people, external factors, or bad luck. On the other hand, when things do go well, they attribute the success of their organizations to external factors, their team, or luck.

Shared leadership theory

Shared leadership is a loose model showing the relationship of various distributed and vertical styles rather than a fully developed and understood theory (Van Wart, 2005). A shared approach to leadership emphasizes lateral, peer influence rather than the downward influence of an appointed leader upon subordinates (Pearce and Conger, 2003). Proponents of shared leadership define it as

> a dynamic, interactive influence process among individuals in groups for which the objective is to lead one another to the achievement of group or organizational goals or both. This influence process often involves peer, or lateral, influence and at other times involves upward or downward hierarchical influence.
>
> (ibid., p. 1)

Models of distributed leadership (Gronn, 2000; Spillane et al., 2004) such as shared leadership, challenge theories which are focused on the individual leader (such as the trait, situation, style, and transformational theories) by arguing for a more systemic concept of leadership which is collectively embedded. They contradict the traditional hierarchical view of power and influence, and suggest that leaders are distributed throughout organizations, at many levels and in varied contexts. Pearce et al. (2008) note that the shared leadership theory is an explicit attempt at integrating two important perspectives: integrating the view of leadership as a role performed by an individual with the view of leadership as a social process.

The style offered by shared leadership is a collective one, based on both vertical and distributed forms of leadership occurring alongside each other. Since various members of an organization have varied roles in it, shared leadership presents a multilevel model. A vital component of shared leadership is the empowered team which works on the ideology of self-organization and distributed leadership, such as accountability and role assignments. In shared leadership, members of the team exhibit leadership behavior to influence the team and to maximize their collective effectiveness, leading to less conflict, more consensus, more trust, and more cohesion (Northouse, 2016). The factors which moderate the success of shared leadership are: the capability of subordinates, the capability of leaders to develop and to delegate, and the general willingness of the organization to embrace the idea of shared leadership. Researchers have noted a number of benefits that shared leadership offers. For example, Pearce and Sims (2002) suggest that shared leadership among peers accounts for more variance in team self-ratings, manager ratings, and customer ratings of change management team effectiveness than the leadership of formally designated team leaders.

According to the GLOBE study (House et al., 2004), team-oriented leadership emphasizes effective team-building and implementation of a common purpose or goal among team members. It includes five leadership subscales: collaborative team orientation; team integrator; diplomatic; malevolent (reverse-scored); and administratively competent.

Cross-cultural perspectives on leadership

After the publication of Geert Hofstede's seminal work on culture (1980; 1991), a considerable amount of research has been undertaken in the field of cross-cultural leadership (see House et al., 1997; Den Hartog et al., 1999; House et al., 2003). These studies provide substantial support for the culture-specific nature of leadership (Dickson et al., 2003;

Gelade et al., 2006). Many argue that the prototypical traits of a leader may be different across cultures. Gerstner and Day (1994) assert that the differences in perceptions of the leadership prototype may sometimes reflect conventional wisdom, but this may not necessarily hold true. In their study of two different cultures, Gerstner and Day found differences among the perceptions of leadership attributes in different countries. Similar assertions are noted by Chong and Thomas (1997), who ascertained that leader and follower ethnicity interacted to affect follower satisfaction. Their study showed that leadership prototypes were also different in groups from different cultures. Therefore, for example, to be seen as a leader in Chinese culture may be entirely different from what would be expected in the West, African, or even other Asian societies. Den Hartog et al. (1999) suggest that in a culture where the authoritarian style is prototypical for a leader, people may perceive the leader's sensitivity as a sign of weakness. On the other hand, the same sensitivity may be seen as an important attribute in a culture that endorses the nurturing style.

Major cross-cultural projects on leadership have focused on implicit leadership theories across various cultures (see Den Hartog et al., 1999); the relationship of autocratic and democratic leadership with the economy, geoclimate, and bioclimate (Van de Vliert, 2006); culture against psychological factors and organizational outcomes (Gelade et al., 2006); the attitude of self-managing work teams in relation to their cultural values and team effectiveness (Kirkman and Shapiro, 2001); and the correlation of prevailing values and sources of guidance in organizations (Smith et al., 2002). These examples show that cultural differences do matter in cross-cultural leadership. Interest in cross-cultural leadership is on the increase. Most leadership theories now put emphasis on the importance of culture while describing effective leadership dispositions or strategies. However, in the globalizing world, researchers now underscore the importance of cultural capital (London and Chen, 2007) and cultural intelligence (Earley and Ang, 2003; Early, 2006). Therefore, it is pertinent to consider these variables in any attempt in future to build a theory of leadership.

Despite the stress on the need to consider cultural differences in the approach to leadership in different contexts, it has been found that some features of good leadership are universal. University of Cambridge Institute for Sustainability Leadership (2017) notes that Gallup's Executive Leadership Research Programme has, over the last four decades, studied more than 50,000 prospective leaders and senior executives around the world across government and business; it identified 12 universal traits which help to distinguish leadership styles, and set apart successful from unsuccessful leaders, and that is, whether they are: intense, competitive, inspiring, courageous, prepared, consistent, enthusiastic, caring about individuals, success-oriented, analytical, focused, and visionary (Newport and Harter, 2016). The University of Cambridge Institute for Sustainability Leadership (2017) presents summaries of other studies. The WEF's Survey on the Global Agenda found that, from the USA to Europe and Asia, there is agreement that having a "global perspective" is the number one skill which makes for strong leadership in 2015 (World Economic Forum, 2014). "Collaboration" and "communication" were also highly favored across the various geographical groups. Similarly, research by the Thunderbird School among over 200 global executives and over 6,000 managers found that the qualities that are critical for the leaders of tomorrow could be categorized under a "global mindset" (Javidan, 2010). To the research team, having a global mindset requires:

- intellectual capital – global business savvy, cognitive complexity, cosmopolitan outlook;
- psychological capital – passion for diversity, quest for adventure, self-assurance;
- social capital – intercultural empathy, interpersonal impact, diplomacy.

Other developments

Ethical leadership

Scholars of leadership almost always mention the attributes and characteristics that pertain to "good leadership" or the ethical conduct of leaders. Good character, honesty, integrity, altruism, trustworthiness, collective motivation, encouragement, and justice have often been mentioned as being universally endorsed positive attributes of leadership (see Den Hartog et al., 1999; Palanski and Yammarino, 2007). These attributes are important for a leader to be perceived as an ethical leader. However, research has shown that there is more to ethical leadership than just attributes such as these (Treviño et al., 2000). Ethical leadership, in a true sense, promotes ethical conduct by practicing as well as consciously managing the ethics, and holding everyone within the organization accountable for it (Treviño and Brown, 2004). In interviews with executives and ethics managers, Treviño et al. (2000) found that a reputation of being ethical is also salient for an ethical leader. Their research found that ethical leadership has two dimensions: moral persons and moral managers. Brown et al. (2005) explain that "moral persons" are those who model "normatively appropriate" conduct so that they appear honest, trustworthy, and credible to others. Moral persons are perceived as fair and just decision-makers, ethically principled, and caring and altruistic (Treviño et al., 2000). Consideration as a "moral person" has to do with how others perceive the leader's character, traits, attributes, and personal characteristics (Brown and Treviño, 2006).

On the other hand, the "moral manager" dimension of ethical leadership means that the leader openly and explicitly talks about ethics and also empowers the employees to be just and seek justice (see Brown et al., 2005). The "moral manager" aspect characterizes the proactive efforts by which he or she influences the followers' actions and beliefs about ethics. Moral managers frequently mention ethics in their messages to their followers and use a reinforcement mechanism (reward and discipline) and make them accountable for their actions and decisions. Moral managers demonstrate to their followers the importance of ethics in their daily work and make this message clear so that everyone practices ethical conduct (see Treviño et al., 2000; 2003; Treviño and Brown, 2004). Based on these conceptualizations, Brown et al. (2005) took up the task to define and empirically test the construct of ethical leadership. They define ethical leadership as "The demonstration of normatively appropriate conduct through personal actions and interpersonal relationships, and the promotion of such conduct to followers through two-way communication, reinforcement, and decision-making" (ibid., p. 120). The first part of this definition ("demonstration of normatively appropriate conduct") refers to the moral person aspect, whereas the second part ("the promotion of such conduct to followers") refers to the moral manager aspect of ethical leadership.

In their conceptualization of ethical leadership, Treviño and colleagues present a matrix comprising unethical leadership (weak moral person, weak moral manager), hypocritical leadership (weak moral person, strong moral manager), ethical leadership (strong moral person, strong moral manager), and ethically "silent" or "neutral" leadership (weak/strong moral person, weak moral manager). Ethically silent/neutral leaders, Treviño and Brown (2004) note, may be ethical persons but they fail to provide leadership in crucial areas where ethics are vital. Such leaders typically focus on the bottom-line gains without explicitly setting ethical standards for followers. Ethically silent/neutral leaders are self-centered, focused on short-term gains, and are either unaware or less concerned about improving the state of affairs on ethics (see Treviño et al., 2000).

Interest in ethical leadership has increased in the last decade. A recent study contends that ethical leadership should be seen as a process rather than a style and that doing so helps readers to understand the role ethics plays in leadership (Shakeel et al., 2019). Van Wart (2014) presents an intriguing perspective which brings together three pillars that must function for good or ethics-based leadership: (1) proper intent (or character); (2) proper means (or duty); and (3) proper ends (greatest good). Van Wart categorized values-based leadership theories (or styles) under several pillars as follows:

- proper intent (or character): virtuous leadership; authentic and positive leadership;
- proper means (or duty): moral management leadership; professionally-grounded leadership;
- proper ends (or greatest good): social responsibility leadership; servant and spiritual leadership; transformational leadership.

In other words, ethical leadership encompasses many other forms of leadership, and for it to function well, several leadership styles interact in some ways at the same time, and complement each other.

Spiritual leadership

The concept of spirituality in modern leadership theories is fairly recent. Most leadership theories had not addressed the role of religion or spirituality in political or workplace institutions (Hicks, 2002). Fry (2003) proposed a causal theory of spiritual leadership that incorporates vision, hope and faith, and altruistic love, theories of workplace spirituality, and spiritual survival. He further argues that spiritual leadership theory is not only inclusive of other major forms of motivation-based theories of leadership, but that it is also more conceptually distinct, parsimonious, and less conceptually confounded. Moreover, "by incorporating calling and membership as two key follower needs for spiritual survival, spiritual leadership theory is inclusive of the religious- and ethics and values-based approaches to leadership" (ibid.).

Fry's work on spirituality was followed by various studies which emphasized workplace spirituality and its positive outcomes in the form of organizational performance and employee well-being (see, for example, Dent et al., 2005; Fry et al., 2005; Kriger and Seng, 2005; Parameshwar, 2005; Reave, 2005). Some other works on spirituality in the workplace include Mitroff and Denton (1999) and Ashmos and Duchon (2000). These works primarily emphasize the significance of workplace spirituality and its impact on work outcomes. The central idea that each of these works advocates is that spirituality in the workplace is a positive phenomenon that leads to positive organizational culture and transcendent values in the organization's members.

In the opinion of Hicks (2002), work on spirituality in leadership studies is a reflection of the attempt by the social sciences to model public life as free from the duties and trappings of religion. Most of these works mention a common set of spiritual characteristics which include vision (a broad appeal to key stakeholders which defines the destination and journey, reflects high ideals, encourages hope and faith, and establishes a standard of excellence), altruistic love (forgiveness, kindness, integrity, empathy/compassion, honesty, patience, courage, trust and loyalty, and humility), and hope/faith (endurance, perseverance, do what it takes, stretch goals, and expectation of victory and reward). Although many of these dimensions have been derived from various theologies, religion has been kept separate from workplace studies.

There remains a debate on whether spirituality and religiosity represent one construct or if they are two different constructs that cannot be combined (Toor, 2008).

The challenge for the spiritual leadership theory is how to measure spirituality. Researchers have noted that spirituality is a complex construct which is hard to measure (see Dent et al., 2005; Fry, 2003; Fry et al., 2005). Although some methods of measurement have been proposed and tested by some researchers (see Ashmos and Duchon, 2000; MacDonald, 2000), there are doubts about their validity. Spiritual leadership theory has continued to attract attention and is expected to grow further in future.

Aesthetic leadership

Another notable development in leadership research is emphasis on the "aesthetics" of leadership (see Hansen et al., 2007; Ladkin, 2008). In their conceptualization of aesthetic leadership, Hansen et al. (2007) refer to leadership as "sensory knowledge and felt meaning of objects and experiences." They note that "aesthetics involves meanings we construct based on feelings about what we experience via our senses, as opposed to the meanings we can deduce in the absence of experience, such as mathematics or other realist ways of knowing" (ibid., p. 545). Hansen et al. also believe that aesthetic leadership is at the conjunction of both "management of meaning" and "follower-centric models" of leadership. It is concerned with the experiential, but considers reality to be a subjective experience, a human-made interpretation based on people's awareness, perception, and the subjective materials people use to make those interpretations.

Aesthetic leadership also takes into account a rather more holistic perspective by including the skills and competencies of people interacting in complex contexts instead of just relying on the cognitive faculties of leaders. It puts emphasis on direct experience and is constructed – made, shared, transformed, and transferred – in relationships between people through interactions. In this sense, it can also be argued that aesthetic leaders, being managers of meanings and developers of followers, reconcile these with the social realities around them in order to achieve social order and continuation of leadership.

Political leadership

Proposed by Ammeter et al. (2002), political leadership theory reminds one of a forgotten domain in which leadership research should have been directed in the past. Ammeter et al. describe politics in organizational leadership as constructive management of shared meanings. Recognizing the negative connotation attached to organizational politics, Ammeter et al. perceive political leadership as an essential phenomenon in a positive sense. They argue that leaders must adopt political behavior that aligns with the established political norms, because their behavior must match the situational assessments of their followers. They also explain two kinds of political behavior: proactive and reactive. Proactive behaviors are those which leaders exhibit in response to a perceived opportunity for the organization, whereas reactive behaviors are exhibited in response to a threat to the organization. By exhibiting such behaviors, leaders are able to reconcile between the competing demands of stakeholders and develop coalitions and alliances which are rather supportive of the leadership purpose.

Butcher and Clarke (2006) support the notion of political leadership in democratic systems. They note that democracy intends to realize the desires of individuals for meaningful control of their lives by treating them equally. It is natural for this to result in the competing interests of people and diversity of opinions. However, political leadership is instrumental in attempts to reconcile the drive for cohesion and the productive exploitation of differences. Political leaders also reconcile the exercise of bureaucratic politics and the civic virtue in which they balance their personal interests with those of wider concern. They coalesce and distribute power in a micro-political process in organizations. Therefore, they embrace political behaviors, such as debating, lobbying, and coalition-building, and information management. Such micro-political behavior in organizations is far from being dysfunctional. On the other hand, it is beneficial for the achievement of managerial goals in the organizations (ibid.).

Political leadership theory has its historical roots in the discourse on leadership. Plato, Aristotle, Confucius, Machiavelli, and many others had discussed leadership from the political perspective, directly or indirectly. More work on the theory and its integration with other theories can provide a comprehensive understanding of how leaders use political tactics to be more effective.

Recent developments in leadership theory

There have been many other recent developments in research on leaders and leadership which are now discussed. For example, Blanchard et al. (2015) identified two types of leadership: strategic (which entails setting the vision and direction), which is usually what is associated with leadership; and operational (which entails how the vision is implemented), which is thought of as management, but state that the two are strongly connected. Ireland and Hitt (2005) discuss strategic leadership which focuses on senior executive leaders at the top of organizations and business units. It is defined as: "a person's [or executive team's] ability to anticipate, envision, maintain flexibility, think strategically, and work with others to initiate changes that will create a viable future for the organization."

Authentic leadership

Authentic leadership is an emerging research area. "Authenticity" has been discussed in philosophical and psychological terms (see Erickson, 1995; Harter, 2002). In relation to leadership, "authenticity" was popularized by Bill George, a professor at Harvard Business School and former Chairman and Chief Executive Officer of Medtronic – a leading organization in medical technology. George's best-selling books, *Authentic Leadership: Rediscovering the Secrets to Creating Lasting Value* (2003) and *True North: Discover Your Authentic Leadership* (George and Sims, 2007) focus on the role of authenticity in leadership and leadership development. Inspired by the ideas of George, researchers on leadership have realized that leadership is not merely a style, charisma, motivation, inspiration, or strategy. It should be looked at as character, positive behavior, and authenticity.

Authentic leaders are described as possessing the highest level of integrity, a deep sense of purpose, the courage to move forward, a genuine passion and skillfulness for leadership (George, 2003; George et al., 2007). They are ready to embrace the challenges they face by growing themselves as authentic leaders and their followers as authentic followers. Extracted from positive psychology, ethical leadership and positive organizational behavior, the authentic leadership construct emphasizes authenticity of character, awareness of the self, regulation of the self, the faithfulness of individuality, a genuineness in beliefs, truthfulness of convictions, the practicality of ideas, veracity of vision, sincerity in actions, and openness to feedback (see Avolio and Luthans, 2006; George and Sims, 2007; Toor and Ofori,

2008a; Walumbwa et al., 2008). These characteristics may be similar to some of the features of transformational, charismatic, servant, spiritual, and ethical leaderships. However, the proponents of authentic leadership contend that it is distinct from other forms of leadership in many respects (see Avolio and Gardner, 2005) and it lies at the roots of all forms of positive leadership.

Authentic leadership has received considerable interest from scholars around the world. The newness of the construct has led to different developmental perspectives which have been proposed for authentic leadership development (see, for example, Luthans and Avolio, 2003; May et al., 2003; Avolio et al., 2004; Gardner et al., 2005; Ilies et al., 2005; Michie and Gooty, 2005; Shamir and Eilam, 2005). In addition, Cooper et al. (2005) underline several challenges which the area may face in subsequent stages of research on it and emphasize the need to take further steps in defining, measuring, and rigorously researching the authentic leadership construct before embarking on designing the interventions to develop authentic leaders. Measuring authentic leadership has been a challenge but researchers have developed and tested instruments such as the Authentic Leadership Questionnaire (Walumbwa et al., 2008) and the Authentic Leadership Inventory (Neider and Schriesheim, 2011). Authentic leadership has also been subject to criticism, as discussed in Chapter 4.

Toxic leadership

The recent literature has also made attempts to explore the negative personal attributes of the leader that contribute to leadership ineffectiveness. At the lowest level, ineffective leadership can be regarded as passive or "laissez-faire leadership" where the leader takes a passive approach toward leading and does not show any interest in fulfilling his or her responsibilities and duties (Lewin et al., 1939; Bass, 1990; Avolio and Bass, 1995). "Laissez-faire leadership", in the opinion of Einarsen et al. (2007), is in clear violation of organizational interests as it results in poor efficiency and the possible undermining of the motivation, well-being, and job satisfaction of the subordinates. At a further level, researchers argue that the leader may become obsessed by the power and personal authority and therefore may resort to narcissism, self-serving and self-centered behavior, the wrong use of power, manipulation, intimidation, coercion, and one-way communication (see Conger and Kanungo, 1987; Conger, 1989; Howell and Avolio, 1992; O'Connor et al., 1995; Yukl, 1999; Toor and Ofori, 2006a). This dimension of the leadership can truly be regarded as the "dark side of the charisma" (Hogan et al., 1990; Howell and Avolio, 1992; Padilla et al., 2007) or derailed leadership (Bentz, 1987), which is mostly a result of the absence of positive characteristics and the presence of negative characteristics (Lombardo et al., 1988).

At an advanced level, ineffective leadership may also be due to many negative personal attributes of leaders actually changing the perception of leadership. Researchers refer to these as negative attributes or impediments to effective leadership (see Den Hartog et al., 1999); and the resulting leadership is known as toxic leadership (see Frost, 2004; Padilla et al., 2007), abusive leadership (Lipman-Blumen, 2005; Tepper, 2000; Harvey et al., 2007) or destructive leadership (Kellerman, 2004; Mumford et al., 2007; Schaubroeck et al., 2007). The consequences of laissez-faire leadership, negative leadership, toxic or destructive leadership can affect the followers, organizations, external stakeholders, and even the leaders themselves (House and Howell, 1992; O'Connor et al., 1995; Kellerman, 2004; Zaccaro et al., 2004).

In a meta-analysis of 57 studies on destructive leadership and its outcomes, Schyns and Schilling (2013) found a strong correlation between destructive leadership and attitudes toward the leader as well as counterproductive work behavior. As a consequence of negative leadership in the workplace, followers may suffer from poor psychological health, lack of interest, low job performance, poor organizational citizenship, and low self-confidence; organizations may suffer from high turnover rates and low productivity; and leaders may suffer from personal problems, such as lack of personal influence, derailment, demotion, and personal psychological suffering (see Bentz, 1987; Lombardo et al., 1988; Ashforth, 1994; 1997; Padilla et al., 2007). Judge et al. (2002) also showed that individuals with the trait of negative affectivity had a lower chance of emerging as leaders. Even if they were able to reach leadership positions, they were rated as less effective leaders. Zaccaro et al. (2004) also noted that destructive personal attributes contribute to harmful and negative leadership influences.

In relation to the negative side of leadership, conceptualizations such as the "toxic triangle" (Padilla et al., 2007) offer useful insights into how negative leaders, susceptible followers, and favorable environmental factors can underpin destructive leadership. Mumford et al. (1993) suggest that the interaction of the leader's characteristics and situational factors promotes discretionary actions on the part of the leader that harm the well-being of organizational members and long-term organizational performance. To avoid the harms of negative leadership, organizations must take appropriate actions to control such behavior by the leader.

Science and leadership

Arguably the most exciting new trend is the emergence of science, and in particular, biological science as a field addressing the subjects of leaders and leadership. Antonakis and Day (2018) discuss this new perspective of leadership research, which is related to the trait approach with respect to measuring individual differences. It is a hard-science approach with regard to measuring observable individual differences, such as biological variables or processes, and also considering why certain variables might provide an evolutionary advantage to an organism. Antonakis and Day (2018) note that this research stream ranges from considering the behavioral genetics of leadership emergence (Ilies et al., 2004) to leadership role occupancy (Arvey et al., 2007). They also suggest that there are studies on the effect of hormones on correlates of leadership, such as dominance (Gray and Campbell, 2009; Zyphur et al., 2009); identifying specific genes associated with leader emergence (De Neve et al., 2013); neuroscientific perspectives of leadership and leader outcomes (Antonakis et al., 2009; Diebig et al., 2016), or evolutionary points of view (Kramer, Arend, and Ward, 2010; Von Rueden and Van Vugt, 2015). Another trendy topic concerns the effects of physical appearance on leader outcomes (Spisak et al., 2014; Antonakis and Eubanks, 2017). Rock and Ringleb (2013) first came up with the term "neuroleadership." They drew from research related to the brain to improve the quality of leadership and leadership development. The subject covers the neuroscience of four leadership activities: (1) how leaders make decisions and solve problems; (2) how leaders regulate their emotions; (3) how leaders collaborate with others; and (4) how leaders facilitate change.

Antonakis and Day (2018) suggest that this new area of leadership research might contribute to understanding the socio-biology of leadership. Interest in the new perspective is growing, as indicated by the recent special issue of *The Leadership Quarterly* entitled "The Evolution and Biology of Leadership," published in April 2020. In the editorial, Van Vugt and Von Rueden (2020) discuss the insights the articles provide in understanding leadership under four themes: (1) the evolved functions of leader-follower relationships; (2) the importance of context; (3) the interaction of biological and cultural evolution in shaping leadership; and (4) the evolved and cultural roots of gender differences in leadership.

Leadership style: the eternal dilemma?

The discussion so far has shown that leadership style is a dilemma which business leaders face daily. With several leadership styles discussed in the literature, during training workshops, and in the popular press, business leaders are at a loss as to which style suits them the best and how it can bring success in their positions. George et al. (2007) observe that many studies have attempted to determine a prototype of leadership – the definitive styles, characteristics, or personality traits of great leaders. All such attempts have been unable to produce a clear profile of the ideal leader or "a cookie-cutter leadership style" (ibid., p. 129).

It is suggested that leadership training programs often focus on developing a particular leadership style in the trainees without reference to the sort of organizations they work for and their personal motives, convictions, aspirations, self-knowledge, and self-awareness. This leads to confusion as the trainees cannot find sufficient congruence between their inner self and the leadership style which is being advocated for successful performance. As a result, they can neither discover their own leadership style nor do they benefit from any leadership training program. That is why Gardner et al. (2005) ask leadership training consultants if they have any evidence that they have developed even a single leader. Another issue regarding the contemporary taxonomies of leadership styles is that most of them are self-centered, task-centered, relationship-centered, or change-centered (Toor and Ofori, 2006a). These styles do not tell if the effort behind the leadership is genuine, reliable, and earnest. Leaders can pretend, and put such styles on show for certain purposes; for example, they can pretend to be charismatic and transformational while being different in reality (Yukl, 2002). However, recent developments indicate that adaptive capacity (Bennis, 2007), psychological capital (Luthans et al., 2007), cultural intelligence (Early, 2006), social capital (Adler and Kwon, 2002), and authenticity (Avolio and Luthans, 2006; George and Sims, 2007) are key to effective and successful leadership.

Conger (2004) argues that the "chameleon leadership capability" is important for leaders to adapt to new situations and handle the dynamism of their working environment. However, this adaptive capacity should be well regulated and must not send out a message that the leader is being tricky, misleading, or inauthentic. A rapid change in leadership style can result in lack of trust in the followers. The answers to these questions lie somewhere in a natural, humanistic, authentic, and genuine leadership which is based on personal values, convictions, aspirations, and motivations. Authentic leaders do not have to change their inner values and principles in varying situations. They do not have to copy others to pretend to have a certain form of disposition. They remain authentic and genuine, but they can still change their approach to suit the circumstances. Goffee and Jones (2005) describe the adaptive capacity of leaders as effective management of personal authenticity. They argue that:

> Authenticity is not the product of pure manipulation. It accurately reflects aspects of the leader's inner self, so it can't be an act. But great leaders seem to know which personality traits they should reveal to whom and when … Authentic leaders remain focused on where they are going but never lose sight of where they came from … They retain their distinctiveness as individuals, yet they know how to win acceptance in

strong corporate and social cultures and how to use elements of those cultures as a basis for radical change

(ibid., p. 88)

This leads to an important premise that leadership is not a style; it is, in fact, authenticity of character (George, 2003; Luthans and Avolio, 2003; Avolio and Gardner, 2005). Authentic leadership is posited as the root construct of all forms of positive leadership. Thus, it provides a broader base for understanding leaders, leadership, and leadership development (Luthans and Avolio, 2003). George (2003) and Luthans and Avolio (2003) present the construct of "authentic leadership" as a solution to contemporary leadership challenges and future leadership demands.

While advocating the need for a new form of leadership, George (2003) argues that every individual is unique, and has a distinctive set of personal values, life history, professional and personal experiences, and current and future motivations. Based on this, one needs to have a unique and authentic leadership that is coherent with one's personality and is consistent with one's personal values and motivations. Adhering to a normative style due to organizational demands cannot help in achieving the goal of authentic leadership development. Authentic leaders hone their style to be effective in different environments to lead different types of people. They are able to develop the capability of working with different people under varying situations.

The important feature that the leader needs to develop in this regard as an intrinsic personality trait is "authenticity" which means "to thine own self be true" in Greek philosophy (Avolio and Gardner, 2005). Terry (1993) believes that to be authentic is to act, to embody, to engage, and to participate in life. Authenticity of character helps to develop the individuals as authentic leaders who are defined as

deeply aware of how they think and behave and are perceived by others as being aware of their own and others' values/moral perspectives, knowledge, and strengths; aware of the context in which they operate; and who are confident, hopeful, optimistic, resilient, and of high moral character.

(Avolio et al., 2004, p. 4)

Avolio and Gardner (2005) argue that authentic leadership can make a fundamental difference in organizations by helping people find meaning and connection at work through greater self-awareness and self-regulation. Authentic leadership produces extraordinary results by inculcating positive psychological capacities; by promoting transparent relationships and decision-making that build trust and commitment among followers; and by fostering inclusive structures and positive ethical climates. This introduction of authentic leadership explains that leaders do not have to worry about the "style dilemma" any longer. They can be authentic leaders by following their own humanity, purpose, and convictions that are grounded in their life stories.

No one can be authentic by trying to imitate someone else. You can learn from others' experiences, but there is no way you can be successful when you are trying to be like them. People trust you when you are genuine and authentic, not a replica of someone else.

(George et al., 2007, p. 129)

Chapter 4 is dedicated to a detailed discussion of the concept of authenticity and authentic leadership.

Discussion and identification of the knowledge gap

Yukl (1994) claims that, after thousands of studies on the subject, a general theory of leadership that can explain all aspects of the process adequately has not been developed. Most leadership studies have also focused on a single level of analysis, ignoring the influence of intra-psychic, group, or organizational factors. There seems an excessive focus on "the leader" rather than leadership (Parry, 1998), especially in North American research that remains largely objectivist in nature (Bass and Avolio, 1990; Parry, 1998; Bryman, 2004). Ciulla (1995) proposes that "What is leadership?" is not the ultimate question that researchers and scholars should be asking. Instead, he argues that "What is good leadership?" or "What is ethical leadership?" should be the questions that are addressed in theories and debates on leaders (see Ciulla, 2004). Ciulla (1995) refers to both ethics and competence as part of "good leadership." Kodish (2006), while discussing Aristotle's philosophy on leadership, also argues that

> leadership is more than a skill, more than the knowledge of theories, and more than analytical faculties. It is the ability to act purposively and ethically as the situation requires on the basis of the knowledge of universals, experience, perception, and intuition. It is about understanding the world in a richer and broader sense, neither with cold objectivity nor solipsistic subjectivity.
>
> (p. 464)

Bass and Steidlmeier (1999) take this discussion further and argue that leaders can be "authentic" and "pseudo" transformational. In their view, pseudo transformational leaders are self-centered, unreliable, power-hungry, and manipulative. On the other hand, authentic transformational leaders have a moral character, a strong concern for self and others, and ethical values – that are deeply embedded in the vision. Ladkin (2008) believes that "leading beautifully" has three major dimensions: (1) mastery – in understanding the self and the context; (2) coherence – congruence between various forms of self and with one's purpose and message; and (3) purpose – attending to one's goal. Ladkin argues that leading beautifully "brings into play the ethical dimension of a leader's endeavor" and questions whether one's purpose "serves the best interests of the human condition" (ibid., p. 33). Similarly, Kanungo (2001) notes that ethical leaders engage in acts and behaviors that benefit others and at the same time, they refrain from behaviors that can cause any harm to others.

The majority of recent theories on leadership originate from the USA. Evidence to support this assertion is that the vast majority of empirical research on the subject is North American in character (House and Aditya, 1997). This means that it either originates from North America (Avolio et al., 2003) or has been substantially contributed to by authors residing in the United States (Lowe and Gardner, 2000). Bryman (2004) also observes that over 60 percent of qualitative studies on leadership have been carried out in the United States. Although there are several cross-cultural studies on leadership (see House et al., 1997; Den Hartog et al., 1999; House et al., 2003), and much more work is being done on leadership outside the United States, there is still a paucity of qualitative grounded models that come from other parts of the world.

Also, although there are many powerful and widely researched theories of leadership, there is no completely agreed-upon theory (Yukl, 1994). Yet more theories and conceptualizations are proposed. However, corporate scandals in the early 2000s stirred discussion on the ethics of leadership, and there have also been debates about positive organizational scholarship (Cameron et al., 2003), positive organizational behavior (Luthans, 2002; 2003), positive psychological capital (Luthans et al., 2007), authentic leadership (Luthans and Avolio, 2003; Gardner et al., 2005), spiritual leadership (Fry, 2003; Fry et al., 2005), and servant leadership (Liden et al., 2008). All of these organizational concepts stress ethics, positive attributes, capacities, and virtues for improving human functioning in organizational settings. Many of these concepts are inspired by developments in positive psychology (Seligman, 1999; Seligman and Csikszentmihalyi, 2000), which primarily focuses on building positive qualities in humans. Research in these fields is still emerging.

Theories such as those on ethical leadership, authentic leadership, servant leadership, transformational leadership, and spiritual leadership, are all positive forms of leadership and underscore positive attributes in leaders as well as in followers, to help build successful and sustainable organizations. However, a pertinent point is to explain how these forms of leadership are different from, or similar to, each other. Scholars have attempted to draw parallels as well as lines of distinction among various positive forms of leadership (see Luthans and Avolio, 2003; Avolio and Gardner, 2005; Brown and Treviño, 2006; Walumbwa et al., 2008).

To measure forms of positive leadership, researchers have established scales of measurement such as the MLQ (Bass and Avolio, 2004), a widely accepted and applied measure for transformational leadership. Brown et al. (2005) have also developed and validated their Ethical Leadership Scale (ELS). De Hoogh and Den Hartog (2008) have used already existing measures (sub-scales on morality and fairness, role clarification, and power sharing) in the Multicultural Leadership Questionnaire (MCLQ) to examine ethical leadership. Liden et al. (2008) has developed a measure for servant leadership and Walumbwa et al. (2008) developed a measure for authentic leadership. According to Gardner et al. (2010) and Dinh et al. (2014), among others, the full-range leadership model (Bass and Avolio, 1990) and Leader-Member Exchange (LMX) are the dominant perspectives from which to understand leadership effects. However, as noted above, the empirical research projects have been undertaken mainly by researchers in the United States.

As noted above, what is perceived as positive in the Western culture may not be so in Arab, African, Chinese, or Indian cultures. Therefore, leadership needs to be explored qualitatively in various countries and cultures and grounded frameworks should be developed which can guide the development of culture-specific measurement tools for leadership. Such efforts can then lead to a meta-measurement of leadership.

There is increasing acceptance in the research community that leadership is a social phenomenon (Parry and Meindl, 2002; Bryman, 2004; Hackman and Wageman, 2007) that is socially constructed (Osborn et al., 2002; Chan, 2005) and that achievements, failures, and crises change and reshape the experiences of the leader and the led (Conger, 1998). Osborn et al. (2002) suggest that contextual macro views need greater recognition than they currently receive. They suggest four contexts – stability, crisis, dynamic equilibrium, and edge of chaos – to explain the context of leadership in organizations. They consider leadership as a series of attempts, over time, to alter human actions and organizational systems. This is a departure from the conception of leadership in terms of qualities, behaviors, and attributes of the leader, as advocated in many classical theories. Recent works also argue that leadership is a social process comprising the leader, followers, and situations. The inclusion of historical, distal, and proximal contexts (Avolio, 2007) in understanding leadership development and influence is a significant move toward developing more complete and contextually robust theories (Osborn et al., 2002) of leadership.

However, empirical work in this area is also limited although more studies are considering context as a related dimension in their research designs.

Another issue that has attracted less attention in the literature on leadership is leadership development. The long-standing question of whether leaders are "born or made" continues to appear in the literature on leadership. There have been several assertions that leadership is a choice (Mirvis and Ayas, 2003) and much of leadership is a social construction in nature and is deeply influenced by the people and events in the lives of leaders (Bryman, 2004; McCall, 2004; Parry, 2004 ;Chan, 2005). Avolio (2007) argues that leaders cannot be thought of as standing apart from the historic context in which they arise, the setting in which they function, and the systems they lead

Leadership can be better understood by researching when, where, and how it is activated and how it makes a difference in team performance and process and organizational effectiveness (Graen et al., 2006). These views strengthen the argument that leaders are not born; they are made in the social settings through complex experiential and social learning (Bandura, 1977) within and outside the organizations (see McCall et al., 1988; Day, 2000; Conger, 2004). Yet, empirical evidence on leadership development is scarce. Existing studies mostly focus on student samples and emergent leaders (see Zacharatos et al., 2000; Brown and Gardner, 2006; Toor and Ofori, 2008a) or on transformational and charismatic leadership (see Avolio and Gibbons, 1988; Bass et al., 1996; Bono and Judge, 2004). Given the rising interest in positive forms of leadership, and particularly in authentic leadership, it is pertinent to examine how authentic leaders develop and what antecedents play a significant role in the making of such leaders.

Summary

In this chapter, a brief introduction to various well-known theories in leadership research is presented. These theories were developed from the perspectives of philosophy, politics, psychology, and sociology. A number of theories focus on the traits of leaders, and some on their behaviors; some theories consider the situation in which leaders and followers function, and others concentrate on the relationships between leaders and followers. There are theories which discuss decision-making by leaders, implicit mental models of leadership, cognitive dimensions, and the effect of culture on different types of leadership. In leadership research, very few theories consider how leadership really develops and what encourages an individual to take on a leadership role or exhibit leadership in a given context.

Although some scholars focus on leadership development (see Avolio and Gibbons, 1988; Bennis, 1989; Mumford et al., 1993; Shamir et al., 2005), they do not necessarily integrate leadership development with leadership influence. Moreover, such works also do not explain in much depth how leaders develop and sustain their leadership over time, how effective leaders achieve sustained success, and also how they are able to reconcile the multiplicity of their roles in different situations and conditions and dynamic circumstances over time.

4 Authentic leadership

Introduction

There has been a recent surge of interest in authentic leadership. Researchers have been calling for leaders who are true to their values and principles, and whose behavior is in congruence with their inner selves, making them authentic to themselves as well as to those they lead. The increasing complexity and competition in business, uncertainty in market conditions, and changing operating environment of organizations and teams call for a renewed focus on restoring confidence and optimism; building corporate resilience; helping employees in their search for meaning and connection by fostering a new self-awareness; and relating appropriately to all stakeholders. There is growing recognition among scholars (Luthans and Avolio, 2003; Seligman, 2002) and practitioners (George, 2003; George and Sims, 2007; George et al., 2007) alike that authentic leadership and its development strategies are relevant and needed if organizations are to attain desirable outcomes.

This chapter presents an overview of the authentic leadership construct and a review of research on the construct. The discussion makes clear how authentic leadership differs from other forms of leadership in the literature. It presents a model for authentic leadership development and offers recommendations for future research on the construct.

Authenticity

The concept of authenticity has its roots in Greek philosophy ("to thine own self be true"). The roots of the word "authenticity" can be traced to the Greek words of *authentikos* which means original, genuine, principal, or from *authentes* which means "one acting on one's own authority."[1] Researchers such as Erickson (1995) and Harter (2002) provide reviews of the origins and history of authenticity in the fields of philosophy and psychology. The term authenticity also refers to

> owning one's personal experiences, be they thoughts, emotions, needs, wants, pre-ferences, or beliefs, processes captured by the injunction to know oneself and further implies that one acts in accord with the true self, expressing oneself in ways that are consistent with inner thoughts and feelings.
>
> (Harter, 2002, p. 382)

Martin (1986) perceives authenticity as avoidance of self-deception. According to Martin,

Existentialists, who represent the main current in the Authenticity Tradition, are pre-occupied with the process of decision making. Their concern is not so much with what choices are made as with how they are made. Decisions, they insist, must be made in a fully honest way, based on a courageous willingness to acknowledge the significant features of the human condition, of one's immediate situation, and of one's personal responses.

(ibid., p. 53)

In the opinion of Ilies et al. (2005), "authenticity is a broad psychological construct reflecting one's general tendencies to view oneself within one's social environment and to conduct one's life according to one's deeply held values" (p. 376). At a more specific level, authenticity is manifested in concrete aspects of one's behavior and action in different capacities. Other researchers argue that authenticity cannot be meaningful if the self is empty of character, and it cannot be real if it ignores the dynamics of lived experience. It is the narrative self that unites character and self-constancy: "Narrative identity makes the two ends of the chain link up with one another: the permanence in time of character and that of self-constancy" (Ricoeur, 1992). Goffee and Jones (2005) assert that authenticity is not an innate quality—that a person is either authentic or not. They argue that authenticity is a quality that others must attribute to the person. They contend that:

Authenticity is largely defined by what other people see in you and, as such, can to a great extent be controlled by you. If authenticity were purely an innate quality, there would be little you could do to manage it.

(ibid., p. 88)

The work of Kernis and colleagues (Goldman and Kernis, 2002; Kernis, 2003; Kernis and Goldman, 2005a, 2005b, 2005c; Kernis and Goldman, 2006; Lakey et al., 2008) has significantly influenced the conceptualization of authenticity. Kernis (2003) notes that authenticity is a psychological construct which reflects "the unobstructed operation of one's true, or core, self in one's daily enterprise" (p. 1). Referring to Rogers (1959) and Maslow (1968), Goldman (2006) observes that "psychological authenticity can be conceptualized as a dynamic set of processes whereby one's full inherent nature is discovered, accepted, imbued with meaning, and actualized" (p. 134). Kernis and Goldman (2005b) divide authenticity into four components: awareness, unbiased processing, behavior, and relational orientation. *Awareness* refers to understanding and knowledge of, and trust in, one's motives, feelings, desires, and self-relevant cognitions, strengths and weaknesses, and emotions. *Unbiased processing* involves impartial self-evaluation without denying, distorting, exaggerating, or ignoring, private knowledge, internal experiences, and externally based self-evaluative information. *Behavior* refers to acting in line with one's values, preferences and needs and not conforming to social demands. Finally, *relational orientation* means genuineness, truthfulness, and openness in relationships.

In summary, authenticity is clarity of one's own ideas, beliefs, convictions, motives, and self-knowledge and congruence of these with one's actions, decisions, dispositions, expressions and behaviors. Authenticity is guided by genuineness, trustworthiness, sincerity, integrity, purity of purpose, and clarity of intentions. An individual is authentic if he or she manifests his or her behaviors without adding any artificial element, social bias, or personal prejudice. The dispositions of authentic individuals truly reflect their inner selves, values, beliefs and principles. Authentic individuals are, in other words, the "real them" who do not show a fake persona.

Authentic leaders

Early research on authenticity and authentic leadership emerged within the fields of sociology and education. After Seeman's (1960) focus on inauthenticity, Hoy and Henderson (1983) revived the construct within the field of educational leadership, and revised Seeman's scale by adding new items. They defined a leader as being inauthentic when the person is overly compliant with stereotypes and demands related to the leader role. However, current research on, and conceptions of, authentic leadership are mainly drawn from positive psychology and positive organizational scholarship (Luthans and Avolio, 2003). Authentic leaders are ready to grow as leaders and grow their followers as authentic followers (Toor and Ofori, 2008a). Avolio et al. (2004) define authentic leaders as:

> "those who are deeply aware of how they think and behave and are perceived by others as being aware of their own and others' values/moral perspectives, knowledge, and strengths; aware of the context in which they operate; and who are confident, hopeful, optimistic, resilient, and of high moral character."
>
> (ibid, p. 4)

Shamir and Eilam (2005) believe that authentic leaders have the following characteristics: they are true to themselves rather than conforming to the expectations of others or faking their leadership; they are motivated by personal convictions, rather than attaining status, honors, or other personal benefits; they are originals, which means that they lead from their own personal point of view; and their actions are based on their personal values and convictions.

Instead of trying to imitate leadership role models, authentic leaders make the most of their own leadership assets (Dearlove and Coomber, 2005) and capitalize on what they have learnt through deeper self-awareness and self-regulation. Goffee and Jones (2005) have coined the phrase "authentic chameleon" to describe the capability of authentic leaders to play different roles while remaining true to their own identity. They believe that these qualities must be real to be beneficial. They further maintain that leadership is not about adopting the styles or traits of successful leaders and imitating what others have done. Authentic leadership demands the internalization of beliefs and convictions which comes through introspection and self-awareness. Those who skip the necessary stages of self-development could adopt "false" personas that are not true to their own values or beliefs. The presence of such leaders can be detrimental to organizations, particularly if they are compensating for perceived personal shortcomings through their leadership roles (Dearlove and Coomber, 2005).

Authentic leaders exhibit a higher moral capacity to judge dilemmas from different angles and are able to take into consideration the needs of different stakeholders (May et al., 2003). Such leaders are not necessarily transformational, visionary, or charismatic but they are the leaders who take a stand that changes the course of history for others, be they whole organizations, departments within them, or just other individuals (ibid.). Other researchers posit that authentic leaders, "by expressing their true self in daily life, live a good life (in an Aristotelian way), and this process results in self-realization (eudaemonic well-being) on the part of the leaders, and in positive effects on followers' eudaemonic well-being" (Ilies et al., 2005, p. 376).

Authentic leaders gain the loyalty and commitment of others through trust-based relationships rather than by virtue of their position or through manipulation (Khan, 2010). Cooper et al. (2005) argue that the initial conceptualization of authentic leaders is multidimensional and multi-level. It contains elements from diverse leadership domains – traits, states, behaviors, contexts, and attributions. Moreover, levels of analysis at which authentic leadership may be undertaken include the individual, team, and organizational levels (Avolio et al., 2004). Considering this, Cooper et al. (2005) suggest that such a broad conceptualization may be acceptable for conducting initial research in this area; however, scholars will need to continue building up knowledge about this construct and eventually narrow this definition. Maintaining the current definition of authentic leaders would pose measurement challenges. They suggest that:

> "a refined definition must include a specification of the nature of the dimension (e.g., trait, behavior, attribution, etc.), the observer/perspective of the person(s) providing the report (e.g., self, subordinate, peer, etc.), the level(s) of analysis involved (for example: individual, dyad, group and organization), the response category measurement units to be employed (such as frequency, magnitude, extent of agreement, etc.), and the dimension's content domain (including whether there are sub-dimensions involved)."
>
> (ibid., p. 478)

Authentic leadership

There have been a number of perspectives describing authentic leadership which have largely helped to establish a broader understanding of the authentic leadership construct and its possible scope. Luthans and Avolio (2003) define authentic leadership as a process that draws from both positive psychological capacities and a highly developed organizational context, which results in both greater self-awareness and self-regulated positive behaviors on the part of leaders and associates, fostering positive self-development. They describe authentic leadership as a positive construct and relate to it descriptions such as genuine, reliable, trustworthy, real, and veritable. Others note that knowing oneself and being true to oneself are essential qualities of authentic leadership, although there have been assertions that there may be much more to authentic leadership than just being true to oneself (see Ilies et al., 2005; Walumbwa et al., 2008). Avolio and Gardner (2005) also argue that this definition was adopted to avoid prior criticisms of leadership constructs for not adequately recognizing the complexity of the phenomenon, including ignoring the context in which leadership is embedded (Bass, 1990; Rost, 1991; Yukl, 2002).

Kellett et al. (2006) describe authentic leadership as developing an honest, open, and transparent relationship between leaders and followers. Scholars assert that compassion ("heart") and relationships with followers are elements in the authenticity of leadership (see George et al., 2007; George and Sims, 2007). Terry (1993) establishes that leadership is essentially an authentic action, a unique and honorific mode of engagement in life. Bass and Steidlmeier (1999) related authenticity and inauthenticity with the behavior of leaders—including their moral character, values, and programs. To Toor and Ofori (2008a), authentic leadership is about authenticity of character, faithfulness of the individual, genuineness in beliefs, practicality of ideas, veracity of vision, sincerity in actions, and openness to feedback.

To define authentic leadership, Shamir and Eilam (2005) introduce the construct of authentic followership, which is achieved by "followers who follow leaders for authentic reasons and have an authentic relationship with the leader" (pp. 400–401). Gardner et al. (2005)

argued that authentic followership: "mirrors the developmental processes of authentic leadership" (p. 346) and is characterized by "heightened levels of followers' self-awareness and self-regulation leading to positive follower development and outcomes" (p. 346).

The current definition of authentic leadership portrays it as a multidimensional and multilevel construct as suggested by Luthans and Avolio (2003) and Cooper et al. (2005). To Cooper et al. (2005), it contains elements from diverse domains – traits, states, behaviors, contexts, and attributions. Moreover, the observers or perspectives involved vary from the leader, to followers (at various distances), and possibly to additional observers. Following the advice in the early work on authentic leadership, Walumbwa et al. (2008) presented a modified definition of authentic leadership that is:

> a pattern of leader behavior that draws upon and promotes both positive psychological capacities and a positive ethical climate, to foster greater self-awareness, an internalized moral perspective, balanced processing of information, and relational transparency on the part of leaders working with followers, fostering positive self-development.
>
> (ibid, p. 94)

Sidani and Rowe (2018) address some challenges that face authentic leadership scholarship, especially how the construct is understood and measured. Their conceptualization of authentic leadership is not about the leadership style but an outcome of a legitimation process under which the overlap between the value systems of leaders and followers leads to impressions of authenticity. Sidani and Rowe (2018) argue that being authentic is not enough for a leader unless the leadership is embraced by a follower who grants moral legitimacy to the leader. Steffens et al. (2016) also show that authentic leadership, in fact, enhances followership, if the leaders advance the interest of the groups they lead.

The proponents of authentic leadership label it the most prized organizational and individual asset (Goffee and Jones, 2005). Others note that authentic leadership is a "root construct" of all positive and effective forms of leadership, such as transformational leadership, ethical leadership, and servant leadership (Avolio and Gardner, 2005; Avolio et al., 2004; Gardner et al., 2005). This root construct transcends other theories of leadership and helps one to understand what is and what is not a "genuine" form of leadership (Avolio et al., 2005).

Although most researchers have studied authentic leadership as a standalone construct, others have found connections with, and critical differences from, other forms of leadership. For example, Toor and Ofori (2009) identify "authentic transformational leaders" as those who possess a moral character, have a strong concern for self and others, and exhibit ethical values which are deeply embedded in their vision. Van Dierendonck (2011) posits that servant leadership is demonstrated by expressing authenticity, in addition to empowering and developing people, expressing humility and providing direction. In his categorization of various values-based leadership theories and styles, Van Wart (2014) believes that authentic leadership is closely associated with ethical leadership.

Cooper et al. (2005) argue that ignoring the important differences among multiple conceptualizations and measures of the same constructs can potentially lead to the development of distinct literatures whose interpretations are not comparable or additive across independent (for example, Schriesheim et al., 1977) or dependent variables (for example, Lowe et al., 1996). Cooper et al. (2005) provide recommendations to further refine the definition of authentic leadership in the early stages of theory-building. They suggest that authentic leadership researchers will need to identify the key dimensions of authentic leadership and then create a theoretically based definition of the construct. They recommend that qualitative methods might be a useful way of identifying these specific dimensions.

Despite the popularity that authentic leadership has gained over the years, some scholars have also provided critiques of the construct. Ladkin and Spiller (2013) are of the view that much of the literature on authentic leadership focuses on the individual, the leader. This may be at the cost of the fact that leadership is a relational phenomenon which involves the leader, the followers and the context in which the leader decides to take the leadership role. Another difficulty with authentic leadership is that it advocates "being oneself", whereas "self" is not necessarily well-bounded or clearly defined; instead, the "self" is theorized as being more fluid and formed through relationships and engagements. If one accepts that the "self" is fluid, and "work in process," as Ladkin and Spiller (2013) put it, then what are the implications of changing or evolving the "self" on authentic leadership?

Hopkins and O'Neil (2015) also contend that the current view of authentic leadership is not gender-neutral and is especially challenging for women, especially for those women leaders who want to enact authentic leadership. They argue that expected leadership role behaviors make it difficult for women leaders to act as authentic. Organizations themselves are not gender-neutral, requiring women to adjust their leadership styles to conform to male-dominated environments. A meta-analysis by Miao et al. (2018) shows that emotional intelligence is significantly and positively related to authentic leadership and this relationship does not differ between male-dominated and female-dominated studies.

Benefits of authentic leadership

As posited by its proponents, authentic leadership is beneficial for the leader-follower relationship. It also produces positive outcomes for the organization. Several empirical studies have been undertaken on authentic leadership and its impact on followers and organizations.

Benefits for organizations and teams

Research shows that ethical leadership is positively associated with all the four dimensions of authentic leadership (self-awareness, relational transparency, internalized moral perspective, and balanced processing), organizational citizenship behavior, organizational commitment, and satisfaction with the supervisors (Walumbwa et al., 2008).

Hmieleski et al. (2012), in a study of top management teams of new ventures and the performance of their firms, found that authentic leadership behavior has a positive indirect effect on a firm's performance. They argued that authentic leadership may be particularly beneficial for team performance when shared among team members. Guerrero et. al. (2015) studied the relationship between the chairperson's authentic leadership on boards and the motivation and commitment of non-executives sitting on the boards of a Canadian credit union. This study showed that chairs with an authentic leadership style favor motivation and commitment through the emergence of a participative safety climate based on transparency and the sharing of ideas. Stander et al. (2015), in a study of public health employees in public hospitals and clinics in South Africa, found that authentic leadership is a significant predictor of optimism and trust in the organization and that optimism and trust in the organization mediated the relationship between authentic leadership and work engagement. In another study, Iqbal et al. (2018) found that authentic leadership positively predicts organizational citizenship behavior in the banking sector in Pakistan. They also found that corporate social responsibility positively mediated the effect of authentic leadership on organizational citizenship behavior.

In a study of authentic leadership and its influence on team performance, Lyubovnikova et al. (2017) found a significant fully mediated relationship between authentic leadership and team performance. Their research found that authentic leadership shapes team behavior, which in turn positively predicts team effectiveness and productivity.

Benefits for leaders

Jensen and Luthans (2006) studied the psychological strengths of entrepreneurs as business leaders and founders of relatively new and small organizations, and found empirical support for the relationship between psychological capital and self-perceptions of authentic leadership among these entrepreneurs.

Toor and Ofori (2009), in their study of construction leaders in Singapore, found that authenticity is significantly correlated with psychological well-being and negatively correlated with contingent self-esteem. These findings indicate that authenticity results in healthy psychological functioning of leaders and may lead to several positive work-related outcomes which may include meaningful work, positive functioning and happiness, better work aspirations and achievements, and resilience.

Benefits for followers

Woolley et al. (2011) found a positive relationship between authentic leadership and followers' psychological capital, partially mediated by a positive work climate, and with a significant moderating effect from gender. Liu et al. (2018) examined authentic leadership influence on employees' workplace behaviors in a large health organization. They found that authentic leadership is positively associated with subordinates' proactive behavior and negatively related to subordinates' workplace deviance behavior. The analysis also showed that three factors: supervisor identification, psychological safety, and job engagement, have mediating effects on this relationship between authentic leadership and subordinates' proactive behavior.

Rego et al. (2012) studied the psychological capital of employees and their supervisors' authentic leadership and found that authentic leadership predicts employees' creativity. The study empirically validated the finding that both authentic leadership and psychological capital may foster employees' creativity, which in turn gives organizations a competitive advantage.

How authentic leadership differs from other forms of leadership

Cooper et al. (2005) emphasize the discriminant validity of the authentic leadership construct to ascertain whether the construct is similar to, or different from, other leadership constructs. As authentic leadership is a new construct, the difference between it and other forms of leadership is a pertinent question to consider. As discussed above, proponents of authentic leadership conceptualize it as a "root construct" underlying all positive approaches to leadership (Avolio et al., 2004; May et al., 2003). However, in contrast with transformational leadership, in particular, authentic leadership may or may not be charismatic, as noted by George (2003). Also, authentic leadership theory includes an in-depth focus on leader and follower self-awareness, self-regulation, positive psychological capital, and the moderating role of a positive organizational climate (Adler and Kown, 2002). This feature is not a prime consideration in other leadership theories.

Avolio and Gardner (2005) argue that, to be viewed as transformational by both the definitions of Burns (1978) and Bass (1985), a leader must be authentic. However, being an authentic leader does not necessarily mean that the leader is transformational. For example, authentic leaders may or may not actively or proactively focus on developing followers into leaders, even though they have a positive impact on them through role modeling (Avolio and Gardner, 2005).

Distinguishing between authentic and charismatic leaders, the proponents of the authentic leadership construct expect that authentic leaders will influence followers' self-awareness of values/moral perspective, more based on their individual character, personal example, and dedication, than on inspirational appeals, dramatic presentations, or other forms of impression management (Gardner and Avolio, 1998, cited in Avolio and Gardner, 2005). Like authentic leadership, both servant and spiritual leadership include either explicit or implicit recognition of the role of leader's self-awareness and regulation. The servant leadership theories encompass the discussion of leader awareness, empathy, conceptualization, and foresight (vision). However, by contrast, the authentic leadership development perspective is drawn from the clinical, positive, and social psychology fields for the discussion of self-awareness and self-regulation.

The discussions of these constructs within servant leadership theory have been largely atheoretical and not grounded in, or supported by, empirical research. Also, servant leadership theory lacks explicit recognition of the mediating role of followers' self-awareness and regulation, as well as positive psychological capital, and a positive organizational context. Finally, contributions of servant leadership to sustainable and veritable performance have not been articulated (Avolio and Gardner, 2005). Areas of overlap between the authentic and spiritual leadership theories include their focus on integrity, trust, courage, hope, and resilience. However, the discussion of these factors is not well integrated into the existing theory of spiritual leadership, and research on the self-systems of leaders and followers or positive psychology, and consideration of self-regulation and the moderating role of the organizational context are missing (ibid.).

Despite these differences between the two seemingly similar theories presented by the authentic leadership scholars, Cooper et al. (2005) argue that there is a need to create a reliable and valid measure of authentic leadership which would further help to differentiate it from other forms of leadership. If the scale items of authentic leadership are not well discriminated from the items of other measures (such as the Multifactor Leadership Questionnaire (MLQ), the Multicultural Leadership Questionnaire (MCLQ), the Leadership Behavior Description Questionnaire (LBDQ), and Least Preferred Co-worker (LPC)), this would imply that there is no difference between authentic leadership and some other form of leadership in the literature. If this discrimination is proven theoretically as well as empirically, it would then be possible to attain further developments in the construct. Therefore, Cooper et al. (2005) encouraged researchers to focus on this subject to increase the credibility of the construct and help to garner support from authentic leadership skeptics.

Measuring the authenticity of leadership

There have been attempts to further articulate the construct of authentic leadership. However, measuring authenticity remains a challenge. Walumbwa et al. (2008) note that the current low volume of empirical work on authentic leadership is essentially because of the inherent complexity involved in measuring the authenticity of leadership behavior. In one of the early attempts to measure authenticity, Terry (1993) suggests these seven criteria to judge the reliability of the knowledge of authenticity: correspondence, consistency, coherence, concealment, conveyance, comprehensiveness, and convergence. The first six of these criteria are benchmarks for judging authentic action. Terry believes that these six criteria point toward a seventh measure, "convergence," which, he suggests, is the end and means to authenticity.

Cooper et al. (2005) suggest that studies can be undertaken on leaders who meet the current broad criteria for authenticity. Also, people who appear to be authentic leaders could be interviewed, and grounded theory methodology (Glaser and Strauss, 1967) employed to map the dimensions of authentic leadership and their nomological networks. Case studies and interview-based studies could also be complemented with the "life story" (Shamir et al., 2004; Shamir and Eilam, 2005) or "narrative" (Sparrowe, 2005) approaches in which leaders' biographies (written or oral) are analyzed.

Moreover, Toor and Ofori (2006a) propose that the authenticity of behavior may also be determined by questionnaire surveys and interviews which ask respondents to differentiate famous authentic leaders from inauthentic leaders, authentic behaviors from inauthentic behaviors, authentic decisions from inauthentic decisions, and other authentic choices from inauthentic choices. Individuals can also be asked to rate their close relationships (such as those with parents, siblings, romantic partners, close friends, teachers, mentors, and role models) on an authenticity scale.

Furthermore, existing psychometric tests for appraising emotional and social intelligence, attachment style, dimensions of personality, and one's knowledge and awareness about the self can also help in judging authentic leadership behavior. Some measures can also be prepared which differentiate authentic motives from inauthentic motives of individuals (Toor and Ofori, 2006c). However, authenticity still remains a subjective term as people can be considered to be inauthentic, less authentic or more authentic. To overcome bias and social desirability in studies, Toor and Ofori (2006a) propose a 360-degree measurement approach when studying authenticity.

To measure authenticity of individual behavior, Goldman and Kernis (2004) have proposed a 45-item inventory in which they measure four components of authenticity: awareness; unbiased processing; behavior; and relational orientation. This measure has been found to help to produce satisfactory results in studies. Walumbwa et al. (2008) also proposed a 16-item higher-order and multi-dimensional measure of authentic leadership which they refer to as the Authentic Leadership Questionnaire (ALQ). Their initial studies in China, Kenya, and the United States produced encouraging results and statistical models have demonstrated the predictive validity of the ALQ measure for important work-related attitudes and behaviors (ibid.). Neider and Schriesheim (2011) developed and tested the Authentic Leadership Inventory (ALI); and it has been applied in several other studies as an instrument to measure authentic leadership (Stander et al., 2015; Fusco et al., 2016; Davidson et al., 2018), although ALQ remains a more popular and widely used tool.

Authentic leadership development

Although a rich body of knowledge on conceptualization and empirical research on authentic leadership has emerged in recent years, what seems slow to follow is a scientific approach to authentic leadership development (Fusco et al., 2016). Many researchers on leadership believe that "leadership development" remains an underdeveloped area of research and needs a lot more attention (Zaccaro, 2007; Avolio, 2010). Some would even say that "the mismatch between leadership development as it exists and what leaders actually need is enormous and widening" (Rowland, 2016). Some pertinent questions may be asked at this juncture. Do researchers and practitioners know how to help leaders to become more authentic? Is it possible to teach someone to be more authentic?

Different perspectives for authentic leadership development have been proposed (see, for example, Luthans and Avolio, 2003; May et al., 2003; Gardner et al., 2005; Ilies et al., 2005; Shamir and Eilam, 2005; and Michie and Gooty, 2005; Goffee and Jones, 2005; Walumbwa et al., 2008). Based on positive organizational scholarship and the transformational/full-range leadership theory, Luthans and Avolio (2003) propose a developmental model for authentic leadership which considers authentic leadership development as a dynamic lifelong process. Various trigger events during different stages of a person's life help to shape this leadership development over time. In authentic leaders, these experiences in life would lead to the development of positive psychological capacities (confidence, hope, optimism, and resilience).

In a positive organizational context, the trigger events and life challenges result in positive self-development which provides self-awareness and self-regulation, eventually resulting in the development of authentic leadership in individuals. Luthans and Avolio describe the authentic leader as being: confident, hopeful, optimistic, resilient, transparent, moral/ethical, future-oriented, and associate building. They assert: "to create authentic leadership requires the development of the individual and the context in which he or she is embedded over time" (ibid., p. 258).

Gardner et al. (2005) present a self-based multi-level model of authentic leader and follower development. The theoretical foundation for their model is the literature on the self and identity. They believe that the developmental processes of leader and follower self-awareness (values, identity, emotions, goals, and motives) and self-regulation (balanced processing, authentic behavior, and relational transparency) are vital for authentic leader and follower development. In this model, the influence of the leader's and followers' personal histories and trigger events are considered to be antecedents of authentic leadership and followership, as well as the reciprocal effects with an inclusive, ethical, caring, and strength-based organizational climate. The proponents of the model view positive modeling as a primary means whereby leaders develop authentic followers. Posited outcomes of authentic leader–follower relationships include heightened levels of follower trust in the leader; engagement; workplace well-being; and veritable, sustainable performance.

Ilies et al. (2005) present their multi-component model of authentic leadership. To explore the specific links between authentic leadership and both leaders' and followers' eudaemonic well-being, Ilies et al. developed their model, starting from the nascent multi-component conceptualization of authenticity. They propose a four-component model of authentic leadership that comprises: self-awareness; unbiased processing; authentic behavior; and authentic relational orientation. Proponents of this model argue that, from a developmental perspective, these dimensions can be roughly mapped onto the six aspects of human wellness proposed by Ryff and Keyes (1995) to reflect human self-actualization (self-acceptance, environmental mastery, purpose in life, positive relationships, personal growth, and self-determination). In other words, self-awareness and unbiased processing should lead to increased self-acceptance and environmental mastery, and also help one to define one's purpose in life; authentic relational orientation should lead to positive relationships; self-awareness and unbiased processing should enhance one's personal growth through self-development; and authentic behaviors and actions are, by definition, self-determined (Ilies et al., 2005).

According to Shamir and Eilam (2005), the development of an authentic leader has four components: (1) development of a leader's identity as a central component of the person's self-concept; (2) development of self-knowledge and self-concept clarity, including clarity about values and convictions; (3) development of goals that are concordant with the self-concept; and (4) increasing self-expressive behavior, namely consistency between leader behaviors and the leader's self-concept. They also present a life-story approach to leadership development which comprises four themes: (1) leadership development as a natural process; (2) leadership development out of struggle; (3) leadership development as finding a cause; and (4) leadership development as learning from experience. This approach suggests that self-knowledge, self-concept clarity, and the internalization of the leader's role into the self-concept are achieved through the construction of life stories (Shamir and Eilam, 2005).

Michie and Gooty (2005) base their conception of an authentic leader and authentic leadership on emotion and the positive psychology literature to present an alternative approach to the role of emotions in leadership development. Their approach suggests that frequent experiences of positive other-directed emotions motivate leaders to act on their other-regarding values. They introduce an interactive approach to the cognitive and emotional processes that motivate authentic leaders to act in ways that are consistent with their self-transcendent values. They argue that consistency between a leader's self-transcendent values and actions is strengthened by the leader's capacity to experience positive other-directed emotions.

In the model presented by Michie and Gooty (2005), self-transcendent values comprise universal values (social justice, equality, and broadmindedness) and benevolent values (honesty, loyalty, and responsibility). They describe the self-transcendent behaviors of authentic leaders as: treat others fairly; treat others with respect; be open to the ideas and opinions of others; be transparent; and transcend self-interest for the common good. Positive other-directed emotions include: gratitude, appreciation, goodwill, and concern for others.

Goffee and Jones (2005) present a simpler version of authenticity of leadership. They argue that establishing one's authenticity of leadership requires two developments. First, the leader must make sure that his or her words are fully consistent with what the leader does. They note that great leaders remain consistent in their behavior and deeds. Second, authentic leaders should be able to find common ground with the people they lead. To manage the expectations of different people, authentic leaders are able to present to them different faces and adapt to the situation as needed. That does not mean that they manipulate people or pretend to them. Authentic leaders, as Goffee and Jones put it, are flexible and have learnt to manage their authenticity in front of others, however ironical that may sound.

A framework for authentic leadership development

After reviewing mainstream leadership literature, and works from construction, and authentic leadership, a framework for authentic leadership development was developed. The aim of the framework was to integrate the likely antecedents of leadership development, in particular, those which are relevant to business leaders. This framework sought to do the following: discuss the antecedents of leadership development, presented in the current literature, that affect the internal leadership schema of an individual; and formulate a model of leadership development that includes most of the often highlighted antecedents, such as significant individuals, significant social institutions, and significant experiences.

The model primarily emphasizes the importance of the developmental context or antecedents of leadership that play a significant role in developing authentic leadership in individuals. Antecedents are usually perceived as dramatic and high-profile events (Cooper et al., 2005) that immediately precede a certain behavior. These stimuli may or may not serve as being discriminative for a specific behavior. These events have conventionally been viewed as negative events although a modern approach emphasizes positive growth events (see Luthans et al., 2007; Avolio and Luthans, 2006). Bennis and Thomas (2002) refer to such vital episodes of life as "crucibles" which they define as "a transformative experience through which an individual comes to a new or an altered sense of identity."[2]

It is therefore pertinent to ask: "Why should one be interested in exploring the leadership antecedents, trigger events or crucibles?" Gardner et al. (2005) posit that influential persons and positive trigger events serve as positive forces in developing leaders' self-awareness that determine psychological well-being and help to generate more self-clarity and self-certainty. Cooper et al. (2005) believe that such trigger events, if replicated appropriately through an intervention, can help to generate leadership authenticity. This is because the event may remind the individual of a similar past experience, create more self-awareness, generate disturbance in the conscience, and remind the person of something that is crucial. If employed with proven techniques, such as chaining and shaping (devised by B.F. Skinner, cited in Cooper et al., 2005), the effect of antecedents can be augmented to accelerate leadership development in individuals.

Luthans et al. (2006) argue that positive reframing of trigger events helps to develop positive psychological capacities (or what they refer to as "PsyCap"). When individuals come across such planned or unplanned trigger events, they are in a better position to enjoy and fully benefit from these episodes (Avolio and Luthans, 2006). In other words, it is a cyclic or iterative process in which trigger events result in the development of PsyCap; and with improved PsyCap, one is in a better position to identify and comprehend a certain trigger event. In this framing and reframing process, individuals can learn from their everyday experience and thus develop their leadership potential. Luthans et al. (2007) contend that unplanned and unpredictable trigger events can carry the risk of exposing the individuals to negative perceptions. However, it is more beneficial for organizations to be proactive and expose the leaders and future leaders to planned trigger events. This will challenge the leaders to employ their strengths and talents to positively react to such situations and will eventually result in the development of resilience (Tugade and Fredrickson, 2004; Luthans et al., 2007). Planned leadership interventions can be particularly effective for young, promising talent as they help them to hone their leadership potential.

The above discussion supports the view that antecedents play a significant role in shaping authentic leaders' personality, niches, behavior, values, motivations, and leadership styles – these may be referred to here as "significant characteristics." These "significant characteristics" from leadership antecedents have an impact on leadership outcomes. Hence, it is relevant to find possible correlations among the antecedents and leadership outcomes. Toor and Ofori (2006b) suggest that leadership antecedents can be categorized into three groups: (1) significant individuals; (2) significant social institutions; and (3) significant experiences. Possible links among these categories of antecedents and leadership characteristics suggest a case for research to address the subject of leadership development. In the following sections, an antecedental model of leadership development is presented.

The Integrated Antecedental Model of Authentic Leadership Development

Elder (1974, cited in Lopata and Levy, 2003, p. 3) argues that:

> Overall the life can be viewed as a multilevel phenomenon, ranging from structural pathways through social institutions and organizations to the social trajectories of individuals and their developmental pathways. In concept, the life course generally refers to the interweave of age-graded trajectories, such as work careers and family pathways, that are subject to changing conditions and future options, and to short-term transitions … multiple … interdependent … pathways.

Therefore, to understand life changes and their consequences, one must not only study the individual and those aspects of self that are more or less changeable, but one must also study the physical and social environment with an eye toward the pressures they exert and the affordances they present (Tesser, 2002).

In order to fully explain how leadership antecedents affect leadership outcomes and shape leadership behaviors, the Integrated Antecedental Model of Authentic Leadership Development (IAMALD), shown in Figure 4.1, is proposed. This model comprises four significant aspects of a leader's life. Leadership antecedents are categorized into significant individuals, significant social institutions, and significant experiences as suggested by Toor and Ofori (2006b). These lead to the development of significant leader characteristics.

Significant individuals include biological relations – those who are inherently present in a person's life without the person having any choice about them; accidental – those whom the person comes across by chance; chosen – those whom a person chooses to make significant in the person's life; and forced or unnatural – those who are forced to become a part of a person's life.

Significant social institutions include the home –such as parents, siblings, immediate family; academic institutions – school, college, university; work and voluntary organizations; religion and religious institutions; society, culture, and community; and economy and

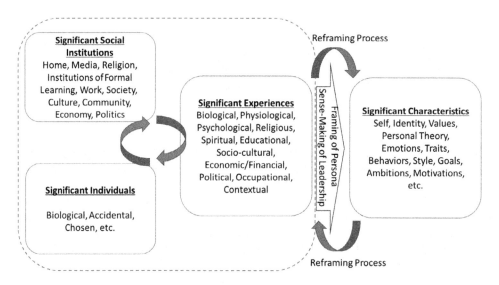

Figure 4.1 The Integrated Antecedental Model of Authentic Leadership Development

politics. Significant experiences include biological, physiological, economic, psychological, socio-cultural, educational, and occupational experiences.

Significant characteristics or leadership outcomes can be categorized as self and identity, values, emotions and personal theory, traits, behaviors and style, goals, and motivations.

The notion underlying the proposed antecedental model is that leadership development is a social phenomenon. Leaders are developed through a social process where they interact with various individuals in different capacities and different contexts. At home during childhood, at school during early adolescence, at college or university during young adulthood, and during working life in organizations, individuals deal with numerous other characters in a variety of situations and circumstances. These characters can be parents and siblings at home (Toor and Ofori, 2006d), teachers and friends at school, and line managers and mentors in organizations. They instigate some significant episodes or events in the life of a person, which are referred to as "significant experiences" (Toor and Ofori, 2006b). Under the effect of these events, individuals develop certain significant leadership characteristics, such as values, traits, motives, behavior. and styles.

Figure 4.1 illustrates the taxonomies of significant individuals, significant social institutions, and significant experiences which lead to the development of significant characteristics in individuals who emerge as leaders in social settings.

The antecedental model of leadership development asserts that the life of a leader comprises various episodes in which significant individuals in different social institutions play their role in different capacities. These episodes can have both positive and negative influences on the personality of a leader, or "leader-self." The outcomes of these episodes eventually build the profile of the actual leader-self or leader's persona. This leader-self characterizes the self-concept, self-knowledge, personal values and norms, personality, nature of social relationships, motives, leadership style, and various other dimensions which are attributed to leadership.

The reframing process

The literature suggests that the "reframing process" is a vital feature of authentic leadership development. Authentic leaders structure their life-stories according to their self-belief, self-identity, and self-perception. However, these stories are subjected to what happens later in their lives. Avolio and Luthans (2006) argue that leaders continuously frame and reframe their life stories depending on the nature of the experiences they go through. A leader's life story is shaped by the past leadership antecedents and it then shapes how the leader experiences the future antecedents that arise in the leader's life (ibid.). These experiences give meaning as to how the leader develops particular behaviors by framing and reframing the leader's life-story. This positive or negative reframing occurs as the leader adapts the leader's actual self with the leader's possible self in the course of the leader's life. With the occurrence of positive events, self-image and possible self of the leader are likely to develop, resulting in the development of positive psychological capacities (Luthans et al., 2007). The development of positive psychological capacities or PsyCap (Luthans et al., 2006) helps to reframe the past and perceive future events more positively. As a result of this reinforcement, the leader is now in a better position to fully benefit from these episodes (Avolio and Luthans, 2006).

Social forces and mediating variables

Granovetter (1985) argues that human behaviour is influenced by social realities. Living in their social settings, humans interact with each other and develop their social networks.

Since leadership has been perceived as the relational extension of the authenticity of a person who is embedded in a network of social relationships (Avolio and Gardner, 2005), the social learning process helps the leaders to form a global self-concept of themselves as leaders (Lord and Brown, 2004; Kempster, 2006) and specific dimensions of their views which relate to their own morality (Hannah et al., 2005). Over the course of their social interactions, leaders form their own sociological concept which is known as "social capital" (Varella et al., 2005). This is defined as "the ability of actors to secure benefits by virtue of membership in social networks or other social structures" (Portes, 1998, cited in Varella et al., 2005, p. 404).

Social capital engenders "prosocial behaviors" and is an indicator of positive inter-personal relationships (Cameron et al., 2003). Social interactions and relationships instill much of how an individual construes the world around him or her. Kan and Parry (2004) argue that leadership is a dynamic, situation-based social process that is contingent upon culture and context. Therefore, from a leadership development perspective, it is related to how different social forces join together to trigger leadership behavior in individuals. A deeper insight into these social forces is not only useful from the developmental perspective but it is also vital for designing interventions which can help to develop individuals as leaders.

There are mediating variables that interact with leader-self and leadership antecedents over the person's lifetime. An important variable is positive psychological capital that has been much highlighted in the literature on authentic leadership (see Avolio and Luthans, 2006; Luthans and Avolio, 2003; Avolio et al., 2004). Luthans and colleagues describe psychological capital or PsyCap as the positive psychological state of development that is characterized by confidence, hope, optimism, and resilience of an individual (see Luthans and Avolio, 2003; Luthans et al., 2006; Luthans et al., 2007). According to Luthans et al. (2006), PsyCap can be reliably and validly measured, and can also be developed by micro-interventions, which has a positive impact on performance and satisfaction.

Other researchers have highlighted a positive moral perspective and positive self-development as related variables which mediate leadership development. Toor and Ofori (2006b) propose leadership quotients, such as social, emotional, human, spiritual, ethical, innovative, cultural, and contextual, as important variables for leadership development. Toor (2006) argues that the positive mediation of leadership quotients with positive psychological states, positive moral perspective, positive self-development, and leadership antecedents results in positive leadership development under favorable organizational and positive social forces. On the other hand, a negative mediation of those variables results in less positive leadership development or even no development. This mediation is iterative and occurs over the life of the leader to develop his or her personal leadership concept and leadership style. Another pertinent factor is the perceived congruence of followers around the motives and methods of the leader. Whittington et al. (2005) posit that when the leader's motives and methods are seen as congruent, the leader becomes a valid example of a role model for the followers. Thus, the leader is better able to influence others when others see him or her as a genuine individual, with no divergence between motives and actions.

Shamir and Eilam's (2005) life story approach suggests that leaders' life stories should be approached as "depositories of meaning" and further analyzed to discover those meanings. The data for such analyses can come from biographies and autobiographies of leaders, interviews in the media, interviews of leaders for research purposes, interviews of leaders' colleagues and followers, leaders' speeches, and leaders' sharing of their life stories with others. In this regard, Shamir and Eilam suggest several directions of further research.

They propose that leaders' life stories can be compared to those of others such as artists, scientists, or ordinary people, to examine the proposition that they contain specific leadership-related contents.

To test whether leaders' life stories are selectively constructed by the leaders, such life stories should be compared with the stories that others – family members, colleagues, and followers – tell about the leaders' lives. Another line of inquiry that Shamir and Eilam suggest is to focus on the process of constructing life stories by leaders. From the life story perspective, leadership development is, to a great extent, the development of self-knowledge and clarity through reflection, interpretation, and revision of life stories. Therefore, the construction of life stories is what studies of authentic leader development should focus on. Shamir and Eilam also suggest that there should be further work to develop methods which can distinguish authentic stories from inauthentic stories and authentic leadership from inauthentic leadership. Finally, they propose that the followers' responses to leaders' life stories, and the effects of these stories on the followers should also be studied.

Research areas of authentic leadership

Several areas for research are highlighted in the emerging theoretical frameworks on authentic leadership. However, as the research on authentic leadership is still at a nascent stage and only conceptual and theoretical perspectives have been presented (see Luthans and Avolio, 2003; Gardner et al., 2005; Ilies et al., 2005) in addition to some empirical work on the construct (Walumbwa et al., 2008), it is essential to explore various dimensions that lie under the authentic leadership construct using qualitative or inductive methods. Cooper et al. (2005) observe that qualitative methods "are appropriate (perhaps even necessary) when there is little extant research on which to base hypotheses" (p. 479). They further observe that qualitative methods can be instrumental to explore "conceptual frameworks that relate authentic leadership to its key antecedent, moderating, mediating, and dependent variables" (ibid., p. 479).

It is argued that a lot more qualitative empirical evidence is required before the field becomes mature enough to be able to embark on quantitative or hypotheses-driven research. Hence, it is essential to explore some areas such as: the antecedents of authentic leadership development (Luthans and Avolio, 2003; Gardner et al., 2005; Avolio, 2007); strategies that authentic leaders use to influence their followers and organizations (Avolio and Luthans, 2006); and basic social processes that yield authentic leadership influence. This research is focused on the discovery or development of theory rather than testing previously set hypotheses for the purpose of the study.

Summary

The chapter presents a summary of the research which has been undertaken on the authentic leadership construct so far. First, several perspectives of authors on authenticity, authentic leaders, and authentic leadership are presented. The broad claims made for the construct are examined. The limitations of the existing field are considered. Second, ideas in the literature on the development of authentic leaders are considered. As suggested in the literature, it is agreed that it is possible to take systematic steps to develop leaders, with appropriate interventions. For example, Luthans and Avolio (2003) stress the need to construct "taxonomies of trigger events" that promote positive leadership development. Rothstein et al. (1990) suggest that a leader's personal history, trigger events, experiences at work, and personal and organizational factors may be potential antecedents to their emergence as leaders and their effectiveness in this role. Third, a tentative model for authentic leadership development, IAMALD, is developed from the review of the literature. This model will be applied in the analysis and discussion of the empirical data. Fourth, a discussion of future directions of research on authentic leadership and its development is presented.

Notes

1 See www.etymonline.com/word/authentic
2 See https://hbr.org/2002/09/crucibles-of-leadership

5 Research design and the grounded theory approach

Introduction

The field of leadership research has changed in terms of how one thinks about, studies, and defines leadership (Hunt and Ropo, 1995; Bryman, 2004). Bryman (2004) suggests that this is because of the greater optimism about the field and greater methodological diversity being employed by researchers to study leadership.

There has been a significant rise in the use of qualitative research approaches to study leadership. Interpretivist or qualitative approaches (such as case studies, phenomenology, ethnography, grounded theory, and historeometry) are increasingly being applied to explain the complex phenomenon of leadership. It is claimed that qualitative methods provide greater richness and comprehensiveness (Antwi and Hamza, 2015).

This chapter explains the method used to accomplish the objectives of the study outlined in Chapter 1. This research focuses on "authentic leadership development and influence in the construction industry" and the key output of this research is to develop a "grounded theory framework of authentic leadership development and influence" in the construction industry. This chapter discusses the choice of the method used. The discussion has three sections. First, why a qualitative approach was used as the principal method. The discussion establishes the epistemological and ontological grounds for the use of the grounded theory methodology for leadership research. Second, how the data was collected. This considers the data collection approach, the instruments for data collection, the target population, the sampling method, the sample size, and other related issues. Third, how the data was analyzed and validated using the grounded theory framework. This chapter also highlights the issue of validity and reliability of the application of the resulting grounded theory.

Choosing a research method

An examination of possible methods and methodologies available to examine the research question is imperative for a quality research study (Blismas and Dainty, 2003). Goulding (2002) acknowledges that choosing a research method is time-consuming, laborious, and difficult; it is a personal and reflective process. Goulding views research as part of an integrated process involving researchers, their beliefs and experiences, the cooperation of various stakeholders of the research, and the suitability and implementation of a chosen methodology which results in an answer that is a single perspective and not an absolute explanation of the problem.

In this process, Guba and Lincoln (1994) suggest that researchers should address questions on their own philosophy, including: (1) the paradigmatic question – "What is the basic belief system or world-view that defines the nature of the world, the individual's place within it and the range of possible relationship to that world?"; (2) the ontological question – "What is the relationship between the researcher as the would-be knower and what can be known?"; and (3) the methodological question – "How can the enquirer go about finding out what he or she believes can be known?"

Bryman (2007) highlights the importance of the research question in the selection of the method. He explains that the research question provides a point of departure for finding the solution to a particular problem; and that the nature of the research question guides decisions about research design and methods. Bryman's interviews with researchers reveal that other disciplinary requirements – what should pass as acceptable knowledge – policy issues, expectations concerning the kind of knowledge they require for policy, and expectations of funding bodies all play a role in choosing the research methodology.

Yet another possible factor is the personal skills of the researcher to conduct a particular kind of research. The researcher's methodological preferences, such as a commitment to a quantitative or qualitative research approach, lead to research questions being formulated in such a way that can help the researcher to choose a particular research method. However, it is essential that the researcher substantiates the method chosen and provides a justification for the choice made (Blismas and Dainty, 2003).

A number of other issues play an important role in the choice of the research method. Buchanan and Bryman (2007) highlight many examples of these issues: the aims of the research, epistemological concerns, and norms of practice, which are also influenced by organizational, historical, political, ethical, evidential, and personal factors. Trauth (2001) suggests that the factors that influence the choice of qualitative methods include: the nature of the research problem, the researcher's theoretical lens, and the degree of uncertainty surrounding the phenomenon. The research method is an integral component of a wider, iterative, and coherent research system in which a number of influences need to be accommodated in decisions concerning the choice of method. Buchanan and Bryman (2007) suggest that the design of organizational research work and the choice of data collection methods are partly a creative process, despite a number of constraints and influences. Therefore, it is important to recognize the role of personal interests, preferences, biases, prejudices, and creativity in addition to technical skills, knowledge, and the competence of the researcher. To Buchanan and Bryman, this must encompass the ability to address, systematically and coherently, the organizational, historical, political, ethical, evidential, and personal influences on the choice of the research methods.

Research traditions for studies on leadership

Reviewing a decade of works in the influential journal, *The Leadership Quarterly*, Gardner et al. (2010) note that leadership is a critical field of inquiry in the organizational sciences, one that captures the curiosity of a large number of micro-, meso-, and macro-organizational scholars. Some of the approaches which have been adopted in research on leadership are discussed in this section.

Quantitative approach for leadership research

Quantitative research methods are characterized by the assumption that human behavior can be explained by social facts. Such methods employ the deductive logic of the natural sciences (Horna, 1994). Due to the turn of leadership research toward psychology during the 1960s and the 1970s, the use of positivism and quantitative methods associated with research in psychology was common, particularly in the United States (Parry, 1998). House and Aditya (1997) observe that about 98 percent of the empirical evidence in leadership research is American in character. Other researchers confirm that the vast majority of empirical research on leadership originates in North America (see Avolio et al., 2003; Lowe and Gardner, 2000; Conger, 1998). Reviewing 10 years of publications in *The Leadership Quarterly*, Lowe and Gardner (2000) found that 64 percent of studies employed a questionnaire-based method of collecting data. They also reported that just around one-third of all the articles were based on a qualitative methodological approach. Many studies (see Conger, 1998; Lowe and Gardner, 2000; Bryman, 2004) showed that the number of qualitative studies on leadership is significantly lower than those using quantitative methods. Takahashi et al. (2012) found relatively few qualitative studies of leadership in international contexts.

There have been many criticisms of the application of quantitative approaches to study leadership. Bryman (2004) observes that leadership research has been dominated by a single kind of data-gathering instrument – the self-administered questionnaire. Most studies on leadership employed questionnaires in various contexts, including experimental settings, cross-sectional designs, and longitudinal investigations. A few standard sets of questionnaires have been used in a large number of studies. Some researchers suggest that surveys and questionnaires mostly measure attitudes toward behaviors and not the actual behaviors, mainly due to social desirability (Phillips, 1973). Such surveys also mostly measure the static situations and do not explain the processes behind them. Descriptions of leadership from such studies are not very helpful in understanding the deeper structures (Conger, 1998) and dynamism of leadership phenomena. Avolio and Bass (1995) note that the inability of quantitative research to draw effective links across the multiple levels to explain leadership events and outcomes has been a major shortcoming. Yukl (1994) shares this perspective and argues that quantitative approaches mostly focus on a single level of analysis and hence ignore several other mediating factors. Alvesson (1996) asserts that the inadequacies of quantitative and hypothesis-driven approaches have encouraged researchers to look for alternative qualitative methods. Quantitative methods focus on objectivity and attempt to capture the reality, but are unable to explain the subjective and ever-shifting realities of the leadership process. Other criticisms of the quantitative approach in the social sciences is that it is pseudo-scientific, inflexible, myopic, mechanistic, and limited to the realm of testing existing theories (Goulding, 2002).

Qualitative approach to leadership research

Jones (1997) observes that the qualitative methodologies are strong in areas that have been identified as potential weaknesses in the quantitative approach. Interviews and observations provide deep knowledge about a particular phenomenon. Bryman (2004) selected leadership studies using qualitative methods published in major social sciences journals between 1979 and 2003. He found that only 10 such studies were published during the first half, between 1979 and 1991, whereas 56 studies were published during the second half, from 1991 to 2003. Bryman observes that the upward trend in qualitative research on leadership did not begin until 10 years after the influential issue of *Administrative Science Quarterly* (on qualitative methods) in 1979.

In terms of data collection method, Bryman found that 56 of the studies (more than 80 percent) employed qualitative interviews. Twenty-five studies used interviewing as the sole technique of data collection. It is ironic that qualitative researchers tend to use only interviewing when other techniques of data gathering (such as document analysis, observation and anthropology, discourse and conversation analysis, visual data and case studies) can be used to complement interviewing. However, there are a number of issues here. For example, observational techniques are hard to use for several reasons, such as: the need for greater investment of time for relatively little return in terms of data; observing acts of leadership is a complex matter; it might be difficult to find leaders who are willing to allow the observation; the potential of contamination of observation due to the presence of the participant observer(s); ethical and methodological dilemmas associated with entering the field; and disengagement from being an observer (Labaree, 2002).

In terms of location, Bryman (2004) found that 61 percent of all the articles were based on US participants or materials, and 20 percent were conducted in the UK or based on UK materials. This suggests that, unlike quantitative research on leadership, qualitative research on the subject is less focused on the USA. Qualitative research on leadership has also been conducted in a variety of situations and contexts. Bryman argues that qualitative studies on leadership are predominantly concerned with how leaders and their styles of leadership promote change and how leadership styles themselves change in response to particular circumstances.

The reviews by Lowe and Gardner (2000) and Bryman (2004) showed that there are relatively fewer studies on leadership using qualitative methods as compared to those using quantitative methods. Conger (1998) also notes the dearth of qualitative studies in leadership research. He observes that qualitative approaches may be intensive, complex, expensive, and time-consuming. However, they are rich in detail and illuminating in new ways to explain the complex phenomenon of leadership. Conger also notes that qualitative research studies are particularly important during the exploratory phases of researching a new area. However, he argues that, in leadership research, qualitative research plays an important role at all stages of investigation because of the complexity of leadership itself (ibid., p. 109). Parry (1998) shares this sentiment and observes that qualitative approaches, due to their painstaking data collection and analysis techniques, have been under-utilized by leadership researchers. Parry (2004) observes that greater use of qualitative approaches in the recent past is resulting in a fuller and richer understanding of the nature of leadership.

Martin and Turner (1986) argue that qualitative approaches allow richer descriptions, increased likelihood of developing empirically supported new ideas with practical relevance, and an increase in the interest of practitioners. Qualitative approaches help in discovering new ideas and phenomena rather than verifying old and existing theories (Bryman, 1984). Conger (1998) advocates that qualitative methods offer many benefits including: greater opportunity to examine the process in depth; the flexibility to discern other contextual factors; and greater effectiveness in investigating symbolic dimensions.

However, qualitative research is not without its shortcomings. To Alvesson (1996), qualitative research can be as superficial as the quantitative form. One of the most often mentioned limitations of qualitative research is its over-reliance on the interview as a principal method (Conger, 1998). The qualitative approach as a research method has been criticized as exploratory, filled with conjecture, unscientific, and involving a distortion of the canons of "good" science (Goulding, 2002).

Some authors view qualitative research and the resulting findings as soft-hearted, and not to be taken seriously (Gherardi and Turner, 2002). In the opinion of Morse (1998), qualitative research leads nowhere and predominantly relies on inference, insight, logic, and luck. Qualitative research is also accused of being novelistic, descriptive, not sufficiently rigorous, incapable of explaining why things happen, and having no hard and fast rules of procedures (Goulding, 2002). The qualitative approach to research, in general, has been undergoing refinement. Morgan and Nica (2020) introduce the Iterative Thematic Inquiry, a new method for the analysis of qualitative data, based on a search for themes that not only begins the analysis process but continues throughout. Eakin and Gladstone (2020) note that much qualitative research produces little new knowledge because of deficits of analysis; researchers seldom venture beyond cataloguing data into pre-existing concepts and scouting for "themes". Eakin and Gladstone propose a "value-adding" approach to qualitative analysis to extend and enrich researchers' analytic interpretive practices.

The processual nature of leadership

It is necessary to discuss some aspects of leadership from the methodological point of view. The first issue which needs to be addressed is whether leadership is solely about qualities, behaviors, and attributes of the leader, as advocated in many classical theories, or whether it is actually a social process comprising the leader, followers, and situations. Most studies on leadership are predominantly centered on the leader and too few studies investigate the process of leadership (Parry, 2006). Meindl et al. (1985) share the same sentiment and argue that outcomes are often linked to the leader while forgetting that many other factors also play a significant role. It is noted that this excessive focus on the leader is a characteristic of North American research (Bryman, 2004; Chan, 2005; Parry, 2006).

These views indicate that leadership is a social process of influence (Bass, 1998; Conger, 1998; Parry, 1998; 2004; Yukl, 2002) that engages everyone in the community (Drath and Palus, 1994; Barker, 1997; Wenger and Snyder, 2000). Therefore, leadership is a function of the social resources that are embedded in relationships (Day, 2000), the environment, the structure and technology of organizations (Osborn et al., 2002). As noted by George and Sims (2007), it is about bringing people together around a shared purpose and empowering them to step up and lead in order to create value for all stakeholders. This definition shows that leadership involves a number of social agents – leaders, followers, and many other stakeholders, who receive the impact of leadership.

Parry (1998) explains that a process involves change which occurs over time and links interactional sequences. Researchers note that change is inherent in the leadership process (Osborn et al., 2002; Yukl, 2002 Kan and Parry, 2004). This implies that the relationship between leaders and followers is also a process which involves changes in the beliefs and motivations of followers that occur within organizations, which act as societies. Many others consider leadership to be about the future (Sarros, 1992), about creating and coping with dramatic change in organizations (Kotter, 1982; Bennis and Nanus, 1985; DuBrin, 1995). Change incidents may form the basis for investigating the leadership process. Kotter (1990) argued that leadership can be differentiated from management due to the inherent notion of change in leadership.

Finally, leadership occurs within a group context as well as within a dyadic relationship (Hackman and Johnson, 1996; Zaccaro et al., 2001). The effectiveness of leadership is largely dependent upon the context. Osborn et al. (2002) argue that leadership "is socially constructed in and from a context where patterns over time must be considered and where history matters" (p. 798). The concept of context is similar to what contingency theories of leadership have advocated, that leadership style cannot be separated from the prevailing conditions under which leadership is being exercised (Fiedler, 1967; House, 1971; Hersey and Blanchard, 1982). Osborn et al. (2002) suggest that contextual macro views need greater recognition than they receive currently. They suggest four contexts of leadership in organizations – stability, crisis, dynamic equilibrium, and edge of chaos – to explain the contextual of leadership in organizations. They consider leadership as a series of attempts, over time, to alter human actions and organizational systems. Others also note the importance that context – historical, proximal, and distal—plays in the effectiveness and sustenance of leadership (see Avolio, 2007).

From the above discussion, there are conceptual bases which support the notion that leadership is a social and processual construct (Parry, 1998). Therefore, researchers have purported that processual analysis can enhance one's understanding of leadership (Hunt and Ropo, 1995; Parry, 1998). This conception of leadership is a departure from "role" to "process". Kan and Parry (2004) observe that defining leadership as a process supports the notion that leadership is not a linear phenomenon; it is rather multi-directional, and involves formal leaders, informal leaders, and followers. Thus, Parry (1998) argues that leadership research should focus on studying the social processes that occur between people and which have a leadership impact. The reason for this is that irrespective of what behavior the leader employs, there are many other variables that influence the impact which these leadership behaviors have upon followers and upon the context of the work.

In a review of empirical studies of leadership in the construction industry, it is shown that most of the empirical studies applied quantitative methodologies using survey questionnaires to collect information. Few of the studies used qualitative methodologies employing interviews and case studies. However, such qualitative studies did not extend beyond interviewing the subjects or analyzing some official documentation to develop rigorous analyses. For future research on leadership in construction, Toor and Ofori (2008b) suggest the use of more qualitative techniques, such as the grounded theory approach. They also suggest the use of observational techniques (Conger, 1998), narratives, personal writing, stories, or biographies (Shamir and Eilam, 2005), ethnographic studies (Noordegraaf and Stewart, 2000), and psychometric neuro-scientific methods (Rock, 2006) for leadership research.

Grounded theory as a qualitative research approach

Grounded theory is an established method well suited to enhancing knowledge of leadership. Originally proposed by Glaser and Strauss (1967), the method has evolved and gone through many developments over the years (see Strauss and Corbin, 1990; Glaser, 1992; Corbin and Strauss, 2008; Ralph et al., 2015). Bryant (2017) provides a contemporary guide on the approach. Charmaz and Bryant (2010) note that constructivists have re-envisioned and revised grounded theory in ways that make the method more flexible and widely adoptable. They also note that whereas, in the past, grounded theory had often been viewed as separate from other methods, the constructivist version makes the usefulness of combining grounded theory with other approaches clearer.

Grounded theory uses qualitative research methods with the aim of generating theory which is grounded in the data, rather than testing existing theories (Glaser, 1978; 1992; Strauss and Corbin, 1990). Glaser (1992) noted that grounded theory is useful for research related to human behavior in organizations, groups, and other social configurations. Parry (1998) suggests that, as leadership is a process of social influence, this makes grounded theory a pertinent method of analysis which emphasizes theory development rather than testing an existing theory. To Hunt and Ropo (1995), grounded theory discovers the underlying social processes and forces that result in a particular activity or phenomenon. Given the comprehensive guidelines that the grounded theory method offers for qualitative research, it has emerged as a popular and effective method for research in management studies (Locke, 2007). The grounded theory approach now has its own journal: *The Grounded Theory Review*, which is an interdisciplinary, online academic journal for the advancement of classic grounded theory and scholarship (Nathaniel, 2020). In recent decades, there has been increasing interest among researchers in using the grounded theory approach to study leadership. This development is now discussed.

Support for the use of grounded theory for leadership research

Mainstream leadership researchers have employed, and called for the application of, qualitative methods to analyze leadership style and behavior (see Parry, 1998; Takahashi et al., 2012). Others have proposed a mixture of quantitative and qualitative methods (Bernerth et al., 2017). Many researchers have also proposed the application of the "grounded theory approach" (Strauss and Corbin, 1990; Komives et al., 2005; Kempster and Parry, 2011), to theorize leadership (see Parry, 1998; 2002; Kan and Parry, 2004). Toor and Ofori (2008c) argue that grounded theory is an appropriate method for leadership research in construction owing to the social nature of both leadership and construction.

The main argument here is that leadership is a basic social process (Bass, 1998; Conger, 1998; Parry, 1998; 2004; Yukl, 2002; Gardner et al., 2010) and processual in nature (Hunt 1991; Alvesson, 1996; Bryman, 2004). It is further compounded by a number of intervening variables. As explained above, this constitutes a move from "leadership as a role" to "leadership as a process" and therefore calls for more grounded qualitative approach that can dig into social realities and can uncover the intervening variables and forces that influence leadership (Bryman et al., 1988; Parry, 1998). It needs to be examined through the study of leadership incidents in various organizational contexts (Kan and Parry, 2004; Cooper et al., 2005; Kempster, 2006). Others note that a grounded theory approach will be helpful to refresh and complement the existing works on leadership (Goulding, 2002; Locke, 2007; Kempster and Parry, 2011).

Kan and Parry (2004) suggest that the grounded theory approach is well suited to enhance knowledge of leadership, and capable of capturing the complexities of the leadership process without discarding, ignoring, or assuming away relevant variables. While using the grounded theory approach, the richness of the data ensures that the resulting theory is fully able to elaborate upon the leadership process for participants and researchers. Many researchers agree that a full grounded theory approach can provide valid and reliable findings to explain the leadership phenomena in a given context (Hunt, 1991; Hunt and Ropo, 1995; Conger, 1998).

Parry (1998) underlines a number of aspects one should consider while using the grounded theory approach for leadership research. He observes that one must observe or interview in depth concerning the process of social influence. To do this, one must have a definition of leadership to ensure that the phenomenon under consideration is really leadership. Also, such research should be more concerned about the leadership process rather than leaders themselves. The interview subjects should be statistically random. Therefore, the interviewees can come from a range of levels in the hierarchy, various functional areas, and from different stages of the change process. Interviewing people across the hierarchy would help to capture perceptions of the leadership process at different levels. During this process, until the grounded theory has been generated, it is not appropriate to consider any existing leadership theories. However, after the generation of the grounded theory, it is useful to compare it with existing theories.

Despite its strength, the grounded theory approach for leadership research has various weaknesses and limitations. The weaknesses mostly pertain to the validity and reliability of the generated theory. Parry (1998) suggests that the application of multiple sources of data collection is useful to improve the validity of the findings, but an interviewing sub-strategy should be the core of the data-gathering strategy for grounded theory research on leadership.

Application of grounded theory in leadership research

A number of studies have been conducted in leadership research using the grounded theory approach. These include Irurita (1990; 1996), Hunt and Ropo (1995), Harchar and Hyle (1996), Parry (1999), Kan and Parry (2004), Larsson et al. (2005), Fernando and Jackson (2006), Hay and Hodgkinson (2006), Jones and Kriflik (2006), Kempster (2006), Sarros et al. (2006), Sjoberg et al. (2006); Lakshman (2007) and Kempster and Parry (2011). These studies have been conducted across a wide spectrum of organizations, contexts, and leadership situations. They explore a number of topics related to leadership, such as administrative instructional leadership, leadership in turbulent change, indirect leadership, the role of spirituality in leadership, leadership in a cleaned-up bureaucracy, leadership learning, and leadership in stressful and complex rescue operations.

Studies have been conducted in various sectors of industry and internationally. Unlike quantitative research, which is predominantly North American in nature, qualitative grounded theory approach has been used in America, Europe, Asia, and the Asia-Pacific region. Qualitative interviewing has been used as the primary source of data collection while some studies have used informal interviewing, observational techniques – participant and non-participant, and document analysis. Kan and Parry (2004), however, used both qualitative and quantitative data for triangulation purposes. They used the Multi-factor Leadership Questionnaire (MLQ) to bring quantitative psychometric data into grounded theory analysis. Such an innovative way of using quantitative data into grounded theory analysis is not common in leadership studies.

Kan and Parry observe that leadership research traditionally tends to control for variables, such as hierarchy or groups, to comply with the positivist nomothetic tradition. On the other hand, the grounded theory method attempts to find the emergence of new theories and propositions rather than testing any existing theories. However, adequately mixing quantitative data and qualitative data helps to better understand the leadership phenomenon in greater depth. Takahashi et al. (2012) proposed that, to understand leadership phenomena in a global context, it is necessary to take a triangulation approach, employing surveys, experimental manipulations, company records, and qualitative interviews.

Grounded theory in construction research

The grounded theory approach has not gained much popularity in research in construction although a number of calls have been made for its wider use (see Toor and Ofori, 2008c). Although there have been instances in which researchers have employed grounded theory, these examples are rare, considering the scope and potential of grounded theory for construction management problems. When online databases of leading journals in construction and project management were searched using the keyword "grounded theory", very few articles were found. Also, a search within the International Construction Database (ICONDA) produced few papers on grounded theory that were published in the proceedings of various major conferences. A list of some studies using the grounded theory methodology in construction management research is presented in Table 5.1. A review of these

Table 5.1 Examples of Grounded theory studies in construction

Journal and sources	Total	Nature of study
Building Research and Information	2	1. Factors facilitating construction industry development (Fox and Skitmore, 2007) 2. Post-occupancy evaluation in architecture: experiences and perspectives from UK practice (Hay et al., 2018)
Construction Management and Economics	7	1. Construction crisis management (Loosemore, 1999) 2. Development of design (MacPherson et al., 1993) 3. Women's career underachievement in construction companies (Dainty et al., 2000) 4. Modelling the determinants of multi-firm project success: a grounded exploration of differing participant perspectives (Phua, 2004) 5. Competitive strategy revisited: contested concepts and dynamic capabilities (Green et al., 2008) 6. Institutions and institutional logics in construction safety management: the case of climatic heat stress (Jia et. al., 2017) 7. From Finnish AEC knowledge ecosystem to business ecosystem: lessons learned from the national deployment of BIM (Aksenova et. al., 2018)
Journal of Management in Engineering	2	1. Organization's capabilities in managing facility capital projects (Dettbarn et al., 2005) 2. Using grounded theory methodology to explore the information of precursors based on subway construction incidents (Zhou et al., 2015)
Journal of Computing in Civil Engineering	1	1. Building process modeling for design management (Platt, 1996)
International Journal of Project Management	3	1. Strategizing for anticipated risks and turbulence in large-scale engineering projects (Floricel and Miller, 2004) 2. How can the benefits of PM training programs be improved? (Thiry, 2004) 3. A grounded theory examination of project managers' accountability (MacDonald, 2020)
The Electronic Journal of Business Research Methodology	1	1. Knowledge and value management in construction (Hunter et al., 2005)
CIB conferences	5	1. Career dynamics of professional women (Dainty et al., 1999) 2. Construction industry development (Fox, 1999) 3. Project member behaviour and client service expectations (Hall, 2006) 4. Role of cultural capital towards the development of sustainable business model for design firm internationalization (London and Chen, 2007) 5. Strategic management practice and tendency in Vietnamese small- and medium-size construction firms (Nguyen et al., 2006)
Unpublished PhD dissertations	7	1. A grounded theory of the determinants of women's underachievement in large construction companies (Dainty, 1998) 2. Construction industry development: analysis and synthesis of contributing factors (Fox, 2003) 3. Risk allocation in contracts: how to improve the process (Zaghloul, 2005) 4. Women in construction management: creating a theory of career choice and development (Moore, 2006) 5. Facilitating knowledge management initiatives for improved competitiveness in small and medium enterprises in construction (Hari, 2006) 6. A link between project value management and best value in the public sector (Hunter, 2006) 7. A grounded theory study of younger and older construction workers' perceptions of each other in the work place (Oluokun, 2008)

studies found that most of them tend to employ a partial grounded theory approach. Many of them also tend to limit the illustration of procedures of data collection, analysis technique, triangulation methods, and so on. No evidence was found in this search of the use of the grounded theory approach for leadership research in construction.

This rare use of grounded theory methodology in construction research reveals the narrowness of the ontological and epistemological standpoints of the construction management research community (Dainty, 2007). A pluralistic attitude toward research methodology will be helpful in developing a better, richer, contextually adequate, and nuanced elucidation of construction management problems (Love et al., 2002; Blismas and Dainty, 2003 ; Dainty, 2007).

The research design for this study

The research design for this study was based on the established guidelines of grounded theory methodology (see Glaser and Strauss, 1967; Strauss and Corbin, 1990; Corbin and Strauss, 2008). These guidelines can be used to analyze social processes that are present in human interactions within a certain system. Application of the theory results in explanations of social processes or structures that are grounded in the qualitative data. Strauss and Corbin (1990) define grounded theory as the:

> "one that is inductively derived from the study of the phenomenon it represents. That is, it is discovered, developed, and provisionally verified through systematic data collection and analysis of data pertaining to that phenomenon. Therefore, one does not begin with a theory and then prove it. Rather, one begins with an area of study and what is relevant to that area is allowed to emerge."
>
> (Strauss and Corbin, 1990, p. 23)

McCaslin (1993) asserts that the grounded theory approach is best to apply to phenomena about which a theory does not already exist. Therefore, the grounded theory approach is not used to verify an existing theory; it is the generation of a theory (Glaser and Strauss, 1967; Strauss and Corbin, 1990) or a "discovery" of a theory from the grounding of data. The application of grounded theory methodology to explore a certain phenomenon can be broken down into four phases that are further broken down into eight different steps, summarized with their given rationale in Table 5.2. Each of the phases noted in Table 5.2 is elaborated upon below.

Figure 5.1 explains the flow of the research in this study, the development of the design, data collection, and data analysis.

Research design phase

The process of the review of the literature, the identification of the research problem, the establishment of the study objectives, the definition of the research delimitations, the choice of the research methodology to address the study objectives, and the development of a research design technically fall under the research design phase and have been elaborated upon in Chapters 1–4 and the current chapter on methodology. During the research design phase, an understanding was developed on the usefulness of the methodology to address the study objectives, target cases or subjects to be interviewed in the beginning were identified, and an interview protocol (see Appendix 1) was designed. Completion of the research design phase led to the data collection phase.

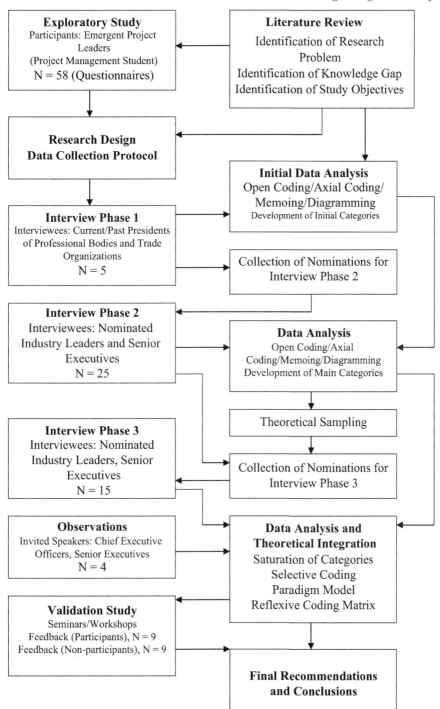

Figure 5.1 Flow chart of research: data collection and analysis

Table 5.2 Process of grounded theory research

Phase		Activity	Rationale
Phase 1: Research Design Phase			
Step 1	Review of technical literature	Definition of research question Definition of *a priori* constructs	Focuses efforts Constrains irrelevant variation and sharpens external validity
Step 2	Selecting cases	Theoretical, not random, sampling	Focuses efforts on theoretically useful cases (e.g., those that test and/or extend the theory)
Phase 2: Data Collection Phase			
Step 3	Developing rigorous data collection protocol	Create case study database Employ multiple data collection methods Qualitative and quantitative data	Increases reliability; increases construct validity Strengthens grounding of theory by triangulation of evidence Enhances internal validity Synergistic view of evidence
Step 4	Entering the field	Overlap of data collection and analysis Flexible and opportunistic data collection methods	Speeds analysis and reveals helpful adjustments to data collection Allows investigators to take advantage of emergent themes and unique case features
Phase 3: Data Analysis Phase			
Step 5	Analyzing data relating to the first case	Use open coding Use axial coding Use selective coding	Develops concepts, categories and properties Develops connections between a category and its sub-categories Integrates categories to build a theoretical framework All forms of coding enhance internal validity
Step 6	Theoretical sampling	Literal and theoretical replication across cases (go to step 2 until theoretical saturation)	Confirms, extends, and sharpens theoretical framework
Step 7	Reaching closure	Theoretical saturation when possible	Ends process when marginal improvement becomes small
Phase 4: Literature Comparison Phase			
Step 8	Comparing emergent theory with extant literature	Comparisons with conflicting frameworks Comparisons with similar frameworks	Improves construct definitions, and therefore internal validity Also improves external validity by establishing the domain to which the study's findings can be generalized

Source: Pandit (1996).

The findings and lessons from the exploratory study conducted on graduate students provided deep insights for the development of interview protocols and data collection with professional leaders in the field. The aim of the exploratory study was to capture the leadership development influences of emergent project leaders who were enrolled in a graduate degree program at the National University of Singapore. The exploratory study sought to develop a picture of possible leadership antecedents (divided into two main categories:

significant individuals and significant experiences) based on the theoretical model of leadership development presented by Toor and Ofori (2006b) and Toor (2006). The exploratory study was helpful in refining the research approach, especially the interview structure for field research.

In May to July 2020, the researchers sought written answers to a few pertinent questions from some of the selected leaders who had been interviewed in the main study, especially those who were still holding leadership positions. Requests were sent to 10 leaders; responses were received from five of them because many of the leaders had retired. Four of the leaders provided written answers to questions which mainly focused on how their leadership had evolved over these years and some of the lessons they had learned. The list of questions is shown in Appendix 9.

Data collection phase

> Our grounded theory adventure starts as we enter the field where we gather data. We step forward from our disciplinary perspectives with a few tools and provisional concepts. A grounded theory journey may take several varied routes, depending on where we want to go and where our analysis takes us. Ethnographic methods, intensive interviewing, and textual analysis provide tools for gathering data as we traverse these routes.
>
> (Charmaz, 2006, p. 13)

In order to address the objectives of the study, qualitative data was collected through interviewing, in addition to some quantitative data that was collected during the exploratory study phase. For grounded theory analysis, interviewing enables a rich perspective of processes to be captured, giving rise to leadership development (Bryman, 1996; Conger, 1998; Egri and Herman, 2000).

Interview sampling and selection method

The sampling strategy has a profound effect on the research quality (Groulding, 2002). Bryman (2004) notes that there is still a tendency to put emphasis only on the role and leadership practices of formally designated leaders. However, leadership is not necessarily about position; many people who are not in designated leadership positions still perform a leadership role, and are perceived as leaders by their peers and followers. Parry (2006) notes that a leader is not necessarily a person in senior management but someone who has an impact upon followers.

In this study, an attempt was made to address the criticism in the earlier studies about sampling of leaders. Therefore, attempts were made to study those who were perceived as leaders by others – industry peers, colleagues, superiors, and subordinates. Therefore, a peer nomination process – or snowball sampling technique (O'Leary, 2004) – was adopted for interviewee selection. This procedure has been used by many who have used the grounded theory approach (see Komives et al., 2005; 2006). George and Sims (2007) interviewed 125 leaders they identified through a peer nomination process in their study of authentic leaders in the United States. Harchar and Hyle (1996), in their study of teaching leadership, also selected their respondents through nomination by other educators and administrators. Parry (1999) used this sampling approach in order to achieve the saturation of various emerging categories and variables. He interviewed individuals who were mentioned in prior interviews as having contributed to the leadership process.

In the current study, interviews were conducted across the whole spectrum of the development industry to capture the process of leadership. Although the interviewees were busy senior executives within their organizations, most of them accepted the invitations and showed a keen interest in the subject of leadership. The interviewees were purposely chosen through a peer nomination process. In the first phase, five past or current presidents of professional institutions and trade associations in the Singapore construction industry were interviewed, considering that they were leaders in their respective areas as they had been elected by their peers to lead the organizations. Nominations were also gathered from senior academics about those who were close to the portrayal of authentic leaders by George (2003), that is, authentic leaders understand their purpose, practice solid values, lead with the heart, establish connected relationships, and demonstrate self-discipline.

The leaders interviewed in the first phase were asked to nominate professionals whom they thought could be considered to be close to the definition of authentic leaders proposed by George. Since the interviewed leaders were nominated and perceived as "leaders" by their peers, a certain level of reduction in social desirability and personal bias was achieved.

In the second phase, invitations to interviews were sent to those nominated by the past or current presidents of trade associations and professional bodies. The invitation letters gave a brief introduction to the study and assured the invited participants of their anonymity and the use of the information and responses they provide solely for the study. During this phase, more nominations of possible interviewees (meeting the George (2003) definition of authentic leadership) were collected from the leaders for further interviews. The process of seeking nominations continued until the names of most of the nominees were either repeated by many former interviewees or they had already been interviewed.

In the second phase, 25 leaders were interviewed. Further nominations for third phase interviewing were driven by theoretical sampling, explained in the section on sampling. Fifteen interviews were conducted in the third phase. This yielded a total of 45 interviews. Although it was not the intention to choose formally designated executives or managers, the interviewees were mostly presidents or chairpersons, chief executive officers, managing directors, managing partners, general managers or deputy general managers, and directors of various construction-related businesses and organizations including developers, architects, engineers, contractors, and quantity surveyors in Singapore. The list of interviewees is shown in Appendix 2.

A typical interview was designed for one hour. However, most interviews took 75–100 minutes to complete. Interviews were digitally recorded. Following each interview, an independent person transcribed the audio recording of the session. The researchers reviewed the transcripts for readability and consistency with the recorded version. In addition to the recording of the interviews, notes were made during the interviews. The interview questions are presented in Appendix 1. The questions asked related to a range of subjects including:

1   Introduction, covering the interviewees' history and career
2   The interviewees' leadership philosophy
3   The influence of significant individuals and events on their leadership development
4   Key turning points in their leadership development
5   The personal and professional challenges they had faced
6   Their personal ambitions and aspirations as leaders
7   Their approach to developing followers
8   What they wanted to achieve as leaders
9   The legacy the interviewees hoped to leave behind as leaders.

Besides the semi-structured interviews, the researchers also attended and made observations in four lecture sessions of chief executive officers invited for a special series of seminars on leadership, as part of a module on the MSc (Project Management) programme at the National University of Singapore, which were also included as data for this study. These CEOs were well-known leaders in the construction industry and were perceived as "authentic" by the academics who invited them to deliver lectures on professional practice to the masters students. The authors recorded the sessions and noted observations. These CEOs typically discussed their personal profile, their company profile, how they rose to executive positions, how they overcame the major challenges they had faced in their careers, and how they envisioned the future of their companies as CEOs.

Some of the participants in the study also provided documents which were useful in the data analysis. Examples of the documents were personal curriculum vitae, articles they had authored, company brochures, presentations on company philosophy, and values. Data collection spanned almost two years, from January 2006 to February 2008. The follow-up small study was undertaken in May to August 2020.

Data analysis phase

Data immersion

Grounded theory methodology provides comprehensive guidelines for analyzing the qualitative data. The first step is immersion in the data which refers to the transcription of the interviews, a thorough reading of the transcription and undertaking necessary editing to achieve accuracy. During the interviews and transcribing exercises, it is useful if the researcher is able to note when and how the interviewees wanted to convey some messages through emotional cues, when they tried to explain things through examples, laughter, emphasis, and high or low pitch of voice. In this study, the transcription and reading process helped the researchers to note some ideas to establish an initial coding scheme as suggested by Strauss and Corbin (1990).

Internal triangulation of qualitative interviewing

Kempster (2006) presented the concept of "interview triangulation" through which the reliability of information collected is enhanced. In his triangulation procedure, he compared the responses of interviewees during different stages and cross-checked whether the information within a single interview was coherent or whether interviewees showed some social desirability or made-up answers. Kempster's (2006) concept was used in this research to enhance the internal reliability of the data. Internal triangulation was carried out by comparing the responses of interviewees during various stages of each interview when interviewees spoke about their biography, leadership philosophy, personal ambitions and aspirations from their leadership, leadership practices, and the legacies they sought to leave behind. It was noted that all the leaders interviewed in this study were consistent in their responses throughout the interviews. This assured reduced social desirability on the part of the interviewees who were already perceived as "authentic" by their nominators.

Data coding

In the grounded theory approach, data analysis takes place in parallel with the data collection until the saturation of the concepts and categories is reached. The challenge is how the researcher deals with the massive amounts of empirical data, text, observations, and multiple possible meanings behind a particular conversation, at both the individual and social levels (Walker and Myrick, 2006). The analytical process involves coding strategies: the process of breaking down interviews, observations, conversations, field notes, documents, and other forms of data collected in the study into distinct units of meaning which are labeled to generate concepts, theories, descriptions, and models (Goulding, 2002). Proponents of the grounded theory approach describe coding as an essential aspect of transforming raw data into theoretical constructions of social processes (see Glaser, 1978; 1992; Strauss and Corbin, 1990). In grounded theory analysis, concepts and categories emerge as the researcher becomes immersed in the data and analyzes some of it. Goulding (1999, p. 868) notes that: "Coding is an ongoing process moving through a series of stages, from open or NVivo codes, through to more abstract conceptual codes which indicate a series of relationships which offer a plausible explanation of the behavior under study."

Analysis or development of categories in the grounded theory approach was driven by comparison of interviewee responses across the cases. Strauss and Corbin (1990) suggest the use of the constant comparative method and questioning (what, when, where, why, how, and so on) to analyze the qualitative data. In constant comparison, the data was examined for all possible meanings. Comparison occurred in tandem with naming. As Strauss and Corbin recommend, an initial coding scheme was constructed. This scheme developed further and changed substantially as the analysis proceeded. Initial coding helped to structure the coding process in the initial phase when random codes were evolving through the open coding process (explained in the next section). However, after analysis of a few interviews, the coding scheme matured into lower-order, middle-order, and higher-order categories as the axial coding was carried out.

Open coding

Open coding is the initial step in the coding process. Strauss and Corbin define open coding as "the process of breaking down, examining, comparing, conceptualizing, and categorizing data" (ibid., p. 61). In other words, open coding is "breaking data apart and delineating concepts to stand for blocks of raw data" (Corbin and Strauss, 2008, p. 195) and then rearranging the data with a fresh description (Dey, 1999). Strauss and Corbin (1990) also suggest the use of various techniques to obtain theoretical sensitivity in the open coding process. These techniques include extensive questioning, analysis of words, phrases, sentences, the flip-flop technique, making close-in and far-out comparison, and waving the red flag. Open coding in the current research was conducted by examining the data (the interview transcripts and post-interview notes) for similarities and differences to identify discrete categories and subcategories (Strauss and Corbin, 1998). Open coding was undertaken to achieve two basic goals: (1) to ask questions about the data; and (2) to compare various concepts across the cases.

In the open coding process, concepts were identified and their properties and dimensions were discovered in the data. As Strauss and Corbin (ibid.) recommend, "microanalysis" or "word-by-word" analysis was carried out to code the meaning found in words or groups of words. It was performed by giving different incidents a conceptual label, by grouping together similar conceptual labels under a common conceptual category, and then developing each category in terms of its properties, sub-properties, and dimensions (Jones and Noble, 2007). In effect, this gives each occurrence of a category a "dimensional profile" that represents the specific properties of a phenomenon under a given set of conditions (Strauss and Corbin, 1990, p. 70). In order to enhance theoretical sensitivity during open coding, a number of techniques suggested by Strauss and Corbin and Corbin and Strauss (2008) were used. These included: extensive questioning; intense analysis of a word, phrase or sentence; comparison techniques; and waving the red flag.

Open coding in this research started during the transcription of the interviews conducted during the first phase of interviewing. Coding time was considerably high during the first few interviews. However, it reduced as the coding proceeded. Constant comparisons of incidents were made among the cases. As a result, categories started emerging at an early stage of the study.

Axial coding

Next to open coding, a complex and higher level of coding is axial coding. It is defined as "a set of procedures whereby data are put back together in new ways after open coding, by making connections between categories. This is done by using a coding paradigm involving conditions, context, action/interactional strategies, and consequences" (Strauss and Corbin, 1990, p. 96). Corbin and Strauss (2008) define axial coding as "crosscutting or relating concepts to each other" (p. 195). Whereas open coding breaks the data into categories, axial coding puts the data back together by making connections between the categories and subcategories. It focuses on the conditions that give rise to a category (phenomenon), the context (the specific set of properties) in which it is embedded, the action/interactional strategies by which the processes are carried out, and the consequences of the strategies (Kendall, 1999). Therefore, axial coding primarily links a category to its subcategories in a set of relationships that Strauss and Corbin (1990) call the paradigm model. These authors admit that the lines between open coding, axial coding, and selective coding are somewhat artificial.

Various concepts which emerged in the open coding were integrated into multiple categories during the axial coding process. Axial coding helped put the fractured data back together by making connections among various sets of the data. Categories were put together based on causal conditions, intervening conditions, contextual conditions, and consequences. While open coding helped to develop a wide range of concepts related to leadership development, manifestation, and functioning within the organizations, axial coding helped the researchers to understand – at a higher level of abstraction – how leaders develop and what processes take place during various developmental phases as well as what strategies the leaders adopt to effectively function within organizations. Search tools and query options available in NVivo7 software package were used during the axial coding process. These tools were useful in the analysis process in addition to memoing and diagramming (explained in the subsequent sections).

The paradigm model

The paradigm model is an organizing scheme that connects subcategories of data to a central idea or phenomenon, to help the researcher think systematically about the data and pose questions about how categories of data relate to each other (ibid.). There are six predetermined subcategories that guide data collection and analysis: (1) conditions; (2) phenomena; (3) context; (4) intervening conditions; (5) actions/strategies; and (6) consequences. The development of a paradigm model was useful for the researchers to think about how various cognitive and social processes were related to authentic leadership development and influence. The paradigm model was completed by asking questions and employing the constant comparison technique (Glaser and Strauss, 1967; Strauss and Corbin, 1990; Merriam and Associates, 2002) during the analysis process. In developing the paradigm model, the researchers shifted between inductive and deductive styles of thinking which rendered the emergence of categories related to leadership development and influence in a more systematic and methodical way. It also made it easier for the researchers to establish connections among various categories and to understand what was happening within the data.

Selective coding: selecting the core category

Selective coding is a process of integration and further refinement of the theory by bringing together all the categories that have been developed to form the initial theoretical framework. It is aimed at integrating the various categories to "form a larger theoretical scheme" (Strauss and Corbin, 1998, p. 143) and is the "final leap between creating a list of concepts and producing a theory" (Strauss and Corbin, 1990, p. 117). Selective coding was done by relating all the major categories – leadership development, self-leadership, self-transcendent leadership, and sustainable leadership – to the core category through the paradigm model. The core category that emerged in this research was labeled: "dynamic and creative reconciliation of self and social realities." As Corbin and Strauss (2008) observe, this core category represented the main theme of the research and was related to all the other major categories within the research. The core category was chosen from the main categories and was found to possess the "greatest explanatory relevance and highest potential for linking all of the other categories together" (ibid., p. 104). The core category explained, at the highest level of abstraction, how authentic leaders develop as well as exert their influence on their followers. It was developed by fulfilling the criteria noted by Corbin and Strauss as follows:

- it must be abstract so that all other categories can be related to it;
- it must appear frequently in the data and should be indicated within all or almost all cases;
- it must be logical and consistent with the data;
- it can be used to do research in other substantive areas, leading to the development of a more general theory;
- it should grow in depth and explanatory power as each of the other categories is related to it through statements of relationships.

Other techniques which were used to integrate the categories and eventually select the core category included:

- filling up the core category with sub-categories within the data;
- writing up an initial story line;
- moving from the descriptive story to the theoretical explanation;
- making use of integrative diagrams;
- reviewing and sorting through memos.

Memos and diagrams

Memos are a specialized type of the written records of the analysis (Corbin and Strauss, 2008). Developing memos about the data, the emerging patterns of concepts, and the linkages of categories was a continuous exercise that remained central to the grounded theory data analysis in this research. Memoing and diagramming started from the beginning of the data analysis until the final theory emerged. Memoing helped the researchers to keep a track of all the categories, properties, hypotheses, and generative questions that emerged from the analysis. Memos were useful in data exploration, identifying and developing the properties and dimensions of concepts and categories, making comparisons and asking questions across the cases, and elaborating upon the paradigm model – the relationships between conditions, actions/interactions, and consequences, and finally building up a story line and then explaining it. Strauss and Corbin (1990) note that memos can be of three types: code memos, theoretical memos, and operational memos. The researchers made use of all three types of memos in order to gain analytical distance from the data. Over 50 memos were written during the course of analysis in addition to several hand-written notes.

In order to visually comprehend the relationship between various categories, diagrams or models played a key role. They were helpful in illustrating the density and complexity of the theory and synthesize the relationships between general categories and the core category. Through diagrams and models, it was much easier to locate relationships that were significant and processes that were taking place within the data. Sorting of memos and diagrams assisted the researchers to undertake the integration of categories.

The reflexive coding matrix

A conditional matrix is the highest level of analysis possible in the grounded theory approach of Strauss and Corbin (1990). The conditional matrix generated from the selective coding procedures increases the theoretical sensitivity to the range conditions and potential consequences, hence providing the means for relating these factors to a particular social phenomenon (McCaslin, 1993). A conditional or consequential matrix can be regarded as a device to stimulate thought regarding the possible conditions and consequences that can enter into the context (Corbin and Strauss, 2008).

As McCaslin (1993) suggested, developing a reflective coding matrix was a useful exercise in this research. It helped to contextualize the core category, the central phenomenon to which all other major and minor categories related. Once the core category was determined, all other categories became sub-categories – or "nearly core categories." The sub-categories in the relational hierarchy became the core category descriptors: the properties, processes, dimensions, contexts, and modes for understanding the consequences. The reflexive coding matrix was developed with the help of conditional matrices drawn to enrich the analysis by sorting through the range of conditions and consequences in which events were located.

Theoretical sampling

An essential step in grounded theory research is theoretical sampling that provides guidance about the sample size sufficient to attain a given research objective. There are established methods and guidelines to determine the sample size according to the population size in quantitative research. Conger (1998) expresses the criticism that statistical analysis reinforces the "long-standing belief" that scientific studies should examine large sample sizes to uncover the truth, and contends that this thesis is debatable in qualitative research.

There are no established requirements on sample size; or the number of interviews, observations, or documents for theoretical sampling. Conger cites Mintzberg (1979, p. 583): "What … is wrong with the samples of one? Why should researchers have to apologize for them? Should Piaget apologize for studying his own children, a physicist for splitting only one atom?" (ibid., p. 116). Kelle and Laurie (1995) also argue that an in-depth analysis in even a single interview can yield deep insights into the distribution of organizational power. Proponents of grounded theory argue that there is no definite sample size for qualitative research. One needs to continue to sample until saturation of the categories is achieved, no matter how small the resulting sample size (see Strauss and Corbin, 1990; Glaser, 1992).

The rigor of qualitative research is not driven by the number of cases but by the sufficiency, completeness, and quality of the ideas these cases generate. To achieve rigor, one crucial task in the grounded theory approach is to achieve the saturation of categories (Charmaz, 2006). Hence, the number of interviews is not an indicator of the "closure of research" in the grounded theory approach; instead, the researcher closes the data collection when he or she thinks that the categories in the data analysis process have been saturated and there are no further pertinent issues emerging in the data (Strauss and Corbin, 1990). This is achieved through "theoretical sampling" which is a method of data collection based on concepts/themes derived from the data (ibid.; Dey, 1999; Goulding, 2003; Locke, 2007; Strauss and Corbin, 2008). The purpose of theoretical sampling is to collect data from places, people, and events that will maximize the opportunities to develop concepts in terms of their properties and dimensions, uncover variations, and identify relationships between concepts (Corbin and Strauss, 2008, p. 143). Theoretical sampling is "data collection based on concepts that appear to be relevant to the evolving story line" (ibid., p. 195) and is driven by the saturation of concepts and categories –in terms of their properties and dimensions, including variation, and if theory building, the delineating of relationship between concepts (ibid.).

In the current research, theoretical sampling procedures played a key role in generating and saturating the categories and concepts. After the first phase of interviews, while the analysis was already underway, few notable trends were seen to be emerging. In the subsequent interviews during the second phase, questions related to such emerging trends were emphasized and a few follow-up questions were added to the interview protocol (see Appendix 1). He significance of theoretical sampling was more evident at the end of second phase of interviews when it was clear that certain issues needed more discussion and deeper understanding. These included: social issues and pressures for women leaders; perceived and practical difference between leadership and management; self-transcendence in leadership manifestation; leadership during times of change; and leadership succession. These issues were followed up during the interviews in the third phase. This process went on until nothing new was occurring in the data and most of the major categories had reached saturation.

Grounding of findings in the existing literature

In a key stage in the grounded theory approach, the researcher grounds the emerging theory in the existing literature, in order to examine the relevance of social processes with the existing body of knowledge. It also shows how grounded theory contributes new knowledge or interprets the existing knowledge in new ways. Grounding in the literature shows any existing gaps that the emerging theory covers. In the current research, as "dynamic and creative reconciliation of self and social realities" emerged, its properties were identified and then a literature review was carried out to see how the existing body of knowledge interprets these properties. It was found that the emergent theory was fully explicable through the existing literature, and that it was able to explain a number of concepts in new ways.

Use of computer software

Despite several benefits that computer-aided qualitative data analysis software (CAQDAS) offer, Dainty et al. (2000) concluded that CAQDAS does not provide a definitive solution to the wider acceptance of qualitative research methods in construction management research. Various researchers advocate the use of computer software, although they also advise the user to be cautious in using it during the analysis process (see Blismas and Dainty, 2003; Corbin and Strauss, 2008). Fielding and Lee (1998) note that computer programs can lead the researcher to a point where the researcher has either to succumb to technological determination or let the computer program drive the analysis. Blismas and Dainty (2003) also note that computers can restrict the diversity of approaches available to qualitative researchers. Corbin and Strauss (2008) suggest that computer programs can stifle creativity or mechanize the analytic process, making the researcher obsessed with the technical details of analysis rather than focusing on concepts, ideas, complex thinking, and linkages among various concepts and categories. If researchers do not effectively use the computer software, they might distance themselves from the data or rely upon coding and retrieval tasks at the cost of other analytical activities.

Use of CAQDAS does not replace the necessary and continuous intuitive input of the researcher. Despite the reservations that researchers have about CAQDAS, it is pertinent to note that the large volumes of unstructured data associated with qualitative methods are difficult to manage and analyze. Analysis of qualitative data involves several tasks (Brent, 1984) such as: (1) recording data; (2) storing data; (3) developing and constructing concepts and theoretical categories; (4) classification of data by categories, that is, giving conceptual labels to data and assigning data items to categories; (5) querying and retrieval of data; and (6) summarizing the data. Undertaking these tasks manually would be tedious and time-consuming.

Computer programs can enhance the creativity of analysis by providing various options to analyze and manage large amounts of data in different ways. They offer innovative ways of performing coding. They also offer choices for linking various documents – such as interview transcripts, memos and diagrams – so that the researcher can establish creative linkages in the data. Computer programs for qualitative data analysis assist to manage data, query data, graphically model the concepts, and report from within the data (Bazeley, 2006). The latest generation packages can facilitate the time-consuming processes necessary to analyze qualitative data and can provide a methodological framework on which the processes are carried out. Computer-aided methods also assist in the management, manipulation, and exploration of larger and wider data sets. Parry (1998) suggests the use of computer-aided techniques to assist in the recording and analysis of data and to enhance the rigor of the analysis. Appropriate computer software also assists in generating a valid and reliable grounded theory (Parry, 1999).

To facilitate the data analysis in this research, NVivo7 was used. Nvivo7 aided the data analysis in data management, allowing text or discourse to be edited, visually coded, contextually annotated, hyperlinked to other texts or multimedia data, and searched according to parameters specified by the user. Nvivo7 offers features such as drawing diagrams, preparing memos, linking concepts and categories, preparing sets (comprising concepts, memos, diagrams, and so on), and classifying cases through attributes. It also offers features for retrieval of data. Queries which can be performed under Nvivo7 include text search, coding, matrix coding, word frequency, and compounded queries. The software was useful for data management and handling during the analysis stage.

Issues of validity and reliability

LeCompte and Goetz (1982) argue that the reliability and validity of findings are important in all fields that engage in scientific inquiry. They also note that the value of scientific research partially depends on the individual researchers to demonstrate the credibility of their findings. However, qualitative research does not necessarily have the same concern with validity and reliability as quantitative research. In qualitative research, validity and reliability have different meanings and are achieved through different means. Parry (1998) emphasizes that the evaluation criteria of objectivist or positivist research are not appropriate for qualitative research. Reliability and validity in qualitative research are redefined with the concepts of credibility, neutrality, consistency (Lincoln and Guba, 1985), trustworthiness (Seale, 1999), truthfulness (Emden and Sandelowski, 1999), rigor (Mays and Pope, 1995), reliability and transferability (Miles and Huberman, 1994), integrity (Watson and Girad, 2004), narrative probability (Fisher, 1987), confidence in the research, and making the findings defensible (Mishler, 1986).

Eisner and Peshkin (1990) note that validity of qualitative research pertains to the degree to which claims of research about knowledge correspond to reality or the research participants' constructions of reality. Parry (1998) observes that issues pertinent to validity and reliability of the data are central to discussions about weaknesses and strengths of the grounded theory approach. He argues that if grounded theory research can be shown to be both valid and reliable, it will make the subjective research seem objective (ibid., p. 95), although objectivity is also important in grounded theory research and must be maintained all the time.

Others argue that validity and reliability in qualitative research can be achieved through representative case selection (Rubin and Rubin, 2005), open-ended narrative interviews (Elliott, 2005), and focused research questions (Alvesson, 1996). Morse et al. (2002) argue that reliability and validity should be actively achieved by the researcher during the research process by the active responsiveness of the researcher to what happens during the research process, ensuring methodological coherence and adequacy of the theoretical sampling, and taking an active analytic stance on saturation. Aguinaldo (2004) suggests that

> qualitative researchers should not be constrained within a "methodological straitjacket" and must be allowed to use whatever methods necessary to explore the social phenomenon under consideration. However, it is not just the choices, but the reasons for those choices ... that need to be made explicit and held up to scrutiny. This demands a certain degree of reflexivity from researchers.
>
> (p. 133)

To Eisner and Peshkin (1990), validity in qualitative research is about determining the degree to which researchers' claims about knowledge correspond to the reality (or research participants' constructions of reality) that is being studied. Miles and Huberman (1994) suggest that maximizing the credibility of the research and minimizing the threats involves five standards:

1   *Objectivity/confirmability*: Conclusions represent researcher neutrality and freedom from unacknowledged bias.
2   *Reliability/dependability/auditability*: The research process is consistent, and stable over time and across methods used.

3   *Internal validity/credibility/authenticity*: The results make sense and are credible to the research participants.
4   *External validity/transferability/fittingness*: Conclusions can be transferred to other contexts.
5   *Utilization/application/action orientation*: Conclusions have pragmatic validity, including usable knowledge providing the basis for action.

Cho and Trent (2006), in their discourse on the validity of qualitative research, divide validity into two categories: transactional validity and transformational validity. They define transactional validity "an interactive process between the researcher, the researched, and the collected data that is aimed at achieving a relatively higher level of accuracy and consensus by means of revisiting facts, feelings, experiences, and values or beliefs collected and interpreted" (p. 321). They note that the use of transactional validity depends on the researcher's belief as to the extent to which it achieves the level of certainty. They define transformational validity as "a progressive, emancipatory process leading toward social change that is to be achieved by the research endeavor itself" (ibid., p. 322). From this perspective, research involves a deeper, self-reflective, empathetic understanding of the researcher while working with the researched.

Parry (1998) suggests that validity in qualitative research on leadership can be enhanced by employing multiple sources of data and collecting the maximum variety of data to ensure the saturation of categories which emerge from the analysis. The researcher should also apply multiple perspectives on a given critical event of leadership. From the perspective of reliability, Parry (1998) argues that appropriate and rigorous use of sampling frame, critical incidents, and introductory questions can take the researcher closer to a replicative grounded theory study. Moreover, grounded theory should be able to fit with the existing body of knowledge on leadership, or able to explain the variation in the situation of research.

Other authors prefer to use the terms "pragmatic usefulness" and "credibility," "originality," and "resonance" to illustrate the goodness of the composed theory (see Locke, 2007). Pragmatic usefulness means that a good theory is one that will be practically useful to both social scientists and laymen. Charmaz (2006) considers pragmatic usefulness to refer to the emergence of generic processes, potential for further research, and contribution to knowledge (p. 183). Credibility is usually discussed in terms of: the analytic practices of the researcher, the rhetorical issues involved in producing a credible publication, and the relationship between the composed concepts and experience of readers as well as the beliefs of the researcher (ibid., p. 59). Credibility also means sufficiency of data to explain the process, the saturation of categories which emerged, and the link between the data, argument, and analysis (ibid.). Originality means freshness of new categories and insights, social and theoretical significance of the theory, and refinement or extension of existing frameworks. Resonance refers to the fullness of the studied experience, possible linkages between larger collections or institutions and individual lives, and persuasion of participants about the framework developed.

Evaluating grounded theory

The problem of how to assess qualitative research remains unsolved (Flick, 2002). Corbin and Strauss (2008) note that "everyone agrees evaluation is necessary but there is little consensus about what that evaluation should consist of" (p. 297). Strauss and Corbin (1990) note that "the reader should be able to make judgments about some of the components of the research process that led to the publication" (p. 252). Therefore, in qualitative research, it is essential to provide readers with the necessary information so that they can judge the adequacy of the research process. This information includes how the original sample was selected; the major categories which emerged in the analysis; some of the events, incidents, actions, and so on that pointed to some of these major categories; the categories which formed the basis of theoretical sampling; the result of the theoretical sampling; some of the hypothesis pertaining to conceptual relations among categories and how they were tested; how and why the core category was selected; whether the conditions and consequences are built into the study and explained; whether the theoretical findings seem significant; and whether the findings become part of the discussion and ideas exchanged among relevant social and professional groups.

Corbin and Strauss (2008) note the following criteria for judging qualitative research:

- fit (do the findings resonate with the experience of the target population and participants?);
- applicability (usefulness of the findings in terms of new insights and explanation);
- concepts (organization of findings around the concepts);
- contextualization of the concepts (inclusion of the context in emergent concepts);
- logic (logical flow or sense of ideas);
- depth (descriptive richness and detail of findings);
- variation (examples of cases that do not fit with findings);
- creativity (newness of research);
- sensitivity (was the research driven by preconceived ideas or did analysis drive the research?);
- evidence of memos (discussion of memos in the final report).

Validation in this research study

In order to achieve validity and maximize reliability in the current research, rigorous procedures were adopted. Data was collected from multiple sources across the industry (such as developers, contractors, designers, engineers, quantity surveyors) and across the hierarchy in the organizations (senior executives and middle-level managers). The use of theoretical sampling, internal triangulation in the interviews, and the inclusion of quantitative data obtained in an exploratory study helped to achieve validity and the internal reliability of the data. Moreover, analysis of some secondary documents and observations was carried out to enhance the validity and reliability. Table 5.3 shows the measures that were taken to improve methodological rigor.

Table 5.3 Achievement of reliability and validity in the current research

Issue	*Remedial measure*
Researcher's bias in sampling procedures	Rigorous selection criteria of authentic leaders
	Leaders chosen through peer-nomination process
Researcher's bias in analysis	Data collection and analysis were driven by theoretical sampling
	Internal triangulation of interviews was carried out to ensure the internal consistency of responses
The findings may not make sense or seem credible	Through grounding of resulting theory in the literature
	A validation study was carried out by involving research participants
	Validation was also carried out with non-participant leaders in the industry

To enhance the validity and credibility of the current research, a validation exercise was carried out. For this purpose, opinions were sought from both study participants and non-participants on the resulting grounded theory model. Briefing sessions were conducted with nine research participants which typically lasted for one hour during which the research design, general findings, and resulting grounded theory model were explained to the executives who had previously been interviewed. Following the presentation, participants were asked to rate various questions on a Likert scale from 1–5 where 1 = Least Likely and 5 = Most Likely. These questions sought the participants' opinion on three major aspects: (1) do research participants see themselves in the theory and relate to the findings?; (2) do research participants think that the findings are comprehensive?; and (3) do research participants think that the findings are useful from a practical perspective and that their colleagues and subordinates can benefit from the resulting frameworks? In order to achieve external validity of the findings, nine non-participant leaders were also approached. These non-participant leaders were nominated as authentic leaders by the participant leaders. The non-participant leaders were also given briefings similar to the ones given to the leaders as mentioned above. Non-participants also provided ratings on the three major aspects noted above. A comparison was made between the ratings given by participants and non-participants (see Chapter 6).

Many leaders who participated in the validation study also showed an interest in open seminars and workshops on the results of the study in their firms so that their followers and colleagues could also learn from the resulting framework. Five such seminars and workshops were carried out, which were attended by young professionals, management trainees, operational managers, and senior managers. Discussion sessions in these seminars and workshops were lively and attendees frequently mentioned that they could relate their own life story and leadership development to the findings of this research. Although these seminars and workshops did not contribute to the study directly, feedback obtained from the participants in the sessions was useful to make recommendations for future research.

A note on the construction industry in Singapore

A short note on the construction industry in Singapore is presented in Box 5.1. It provides basic economic data on Singapore, and some background information on the size, importance, and structure of the construction industry, the practices it adopts, the challenges it is facing, and the efforts being made to address them. Thus, Box 5.1 provides readers with the context of the field study and its results.

Box 5.1 A note on Singapore construction

Basic socio-economic information

Some recent basic social and economic data on Singapore is published by Department of Statistics (2018a; 2018b). The total land area of Singapore is approximately 720km². The total population was 5,612,253 with an annual growth rate of 0.1 percent. The literacy rate among residents was 97.7 percent. Life expectancy at birth in Singapore was 80.7 years for male residents; 85.2 years for female residents; and 83.1 years for total residents. The Gross Domestic Product (GDP) of Singapore at current market prices in 2017 was S$447,284 million (the average exchange rate in 2017 was US$1.00 = S$1.34) (Monetary Authority of Singapore, 2019). The GDP per capita was S$76,863. The contribution of construction to GDP in 2017 was S$17,809 million (or

4.0 percent). The home ownership rate among resident households is 90.7 percent. Of those who live in public housing apartments, the ownership rate is 92.1 percent.

Singapore has undergone a remarkable economic and social transformation since the 1960s. The Economic Development Board (2017) provides an outline of the history: the GDP of Singapore in 1960 was S$2.16 billion, with per capita income at S$1,310. In the 1960s, the country faced high unemployment as a result of high population growth and also had a low skill base. The country focused on labor-intensive export-oriented industries. By the 1970s, Singapore had attained full employment. It set up programs to develop workers' skills and raised wages. In the 1980s, the country was poised for capital-intensive industrial activities. It set up the first petrochemical complex in Asia and factories for wafer fabrication, industrial machinery, machine tools, and automation equipment. Singapore's economy went through a technology-intensive stage in the 1990s; it strengthened its capabilities in chemicals, electronics, and engineering, and entered the pharmaceutical, biotechnology, and medical technology segments of industry. In the early 2000s, the focus was on knowledge- and innovation-intensive activities. Singapore's economic development has been well researched (see, for example, Lim, 2016). The construction industry has played a significant role in Singapore's economic and social transformation. It was systematically developed to enable it to undertake the widening range and increasingly more sophisticated tasks it was required to undertake. The industry is being further developed to enhance its capacity and capability.

Construction in the economy

The gross fixed capital formation in Construction and Works in 2017 was S$46,953 million, or 42 percent of the total of S$110,972 million (Department of Statistics, 2018a). The construction industry employed about 101,000 Singapore residents in 2017 (this does not include the foreign workers who constitute the bulk of the construction site workforce; the data shows that the total labor force in Singapore is 3,657,000, of whom residents were 2,269,700). The 2017 resident construction workforce comprised, in percentage terms:

- Managers and Administrators (14);
- Working Proprietors (10);
- Professionals (11);
- Associate Professionals and Technicians (18);
- Clerical Support Workers (14);
- Service and Sales Workers (1);
- Craftsmen and Related Trades Workers (19);
- Plant and Machine Operators and Assemblers (7);
- Cleaners, Labourers and Related Workers (6).

This data indicates that residents take up the positions in the industry requiring higher expertise.

The volume of construction contracts awarded in the period, 2011–2017 ranged from a peak of SS39 billion in 2014 and SS25 billion in 2017 (ibid.). This is an indication of the volume of demand for construction, a figure which is closely monitored by both practitioners and administrators. The total value of certificates certified by the responsible project consultants for payment by clients also ranged from S$36 billion in 2014 to S$28 billion in 2017. This is an indication of the overall output of the contractors, which forms almost all of the value of the gross output of the construction industry.

The construction industry in Singapore retains the features of the industry in most countries of being the most demanding segment of the economy in terms of the number of hours worked in a week. The average for construction ranged from the highest of 53.2 hours in 2013 to 50.8 hours in 2017 (ibid.). For the economy as a whole, the corresponding figures were 46.2 in 2013 and 45.1 hours respectively in 2017.

Managing the industry's development

The government of Singapore shows an understanding of the role of the construction industry and of the industry's needs, which is quite rare. The government has led efforts in the development of the industry (Centre for Liveable Cities, 2015). The Construction Industry Development Board (CIDB) was established in 1984 under the Construction Industry Development Board Act Cap. 51, 1984, to spearhead the improvement and expansion of the industry. In 1999, after a restructuring, the CIDB was succeeded by the Building and Construction Authority (BCA) under the Building and Construction Authority Act Cap. 30A of 1999.[1] The BCA describes itself as an agency "championing the development of an excellent built environment for Singapore."[2] It considers that it has influence in the following areas: Safety, Quality, Sustainability, and User-Friendliness; and that these distinguish Singapore's built environment from those of other cities. The current mission of the BCA is: "We shape a safe, high quality, sustainable and friendly built environment."

The functions of the BCA are:[3]

1 to promote the development, improvement and expansion of construction to be implemented;
2 to provide a search service for searches on building records and plans in the construction industry, including the use of advanced technology in the construction industry;
3 to advise and make recommendations to the government on matters affecting or connected with the construction industry and on the control of building works and the safety of buildings;
4 to raise standards and efficiency in the construction industry by encouraging the standardization and improvement of construction processes, techniques, products, and materials;
5 to promote good procurement methods and practices in the industry and advise and assist the government in the procurement of construction works and services;
6 to provide consultancy and advisory services related to the construction industry;
7 to promote the advancement of skills and expertise of persons in the construction industry;
8 to raise the professionalism and capabilities of firms in the construction industry;
9 to promote the adoption of internationally recognized quality management systems in the industry;
10 to facilitate the supply of essential construction materials and secure and manage land and facilities related to their import and production;

11 to promote and carry out research for the development and improvement of the construction industry and in respect of matters including the building control system, building codes and regulations, building maintenance and management, energy usage in buildings;

12 to promote the efficient use of energy in buildings and to advise the government on the measures and regulations.

Roles and responsibilities of practitioners on a typical project

The construction industry in Singapore is broadly similar to that of the United Kingdom in terms of its structure (with regard to the roles and responsibilities of the participants in a typical project) and the main elements of practice and procedures. The architect is usually the first consultant to be appointed by the client. The architect's role includes: (1) preparation of the sketch design and detailed design; (2) advising the client on the most preferable tender; (3) supervising construction; and (4) issuing interim and final payment certificates.

Engineers from many branches of knowledge are involved in construction projects in Singapore. The structural engineer and mechanical and electrical (M&E) engineers undertake the design and supervision functions on the relevant aspects of the project. With environmental performance of buildings assuming an important role in Singapore, especially in the area of energy conservation and management, the M&E engineers have become even more important players in construction projects. Other specialists include landscape architects, interior designers, acoustics engineers, and architectural designers who specialize in particular building types, such as performance halls, hospitals, and airports.

The quantity surveyor provides advice to both the design and construction teams on the cost and financial aspects, as well as relevant issues in the contract. The tasks include: preparation of bills of quantities; compilation of tender documents; preparation of interim payment certificates; provision of advice on cost implications of variations; and preparation of the final accounts for the project. The land surveyor undertakes the preparation of site plans, the setting out for building work, and the establishment of levels and dimensions as the work proceeds.

In Singapore, construction contracts are awarded under general contracting arrangements. The client enters into the contract with one entity, the main contractor, to deliver the entire project. The client may appoint nominated subcontractors for some specialist works, such as the ventilation and air-conditioning; and vertical transportation installations. On large construction projects, the substructure and superstructure are awarded separately, and are often undertaken by different organizations. Another major feature of the construction industry in Singapore is the predominance of labor subcontracting. Very few construction companies employ more than a few essential personnel. Multi-level subcontracting is also often evident on some projects. The subcontracting arrangement is blamed for many of the weaknesses in the industry, and its poor performance, such as in quality, health and safety (Construction 21 Steering Committee, 1999).

Summary

The research design for the study is discussed in this chapter in some detail for the benefit of existing and aspiring researchers on leadership. In the methodology, the merits and weaknesses of various approaches to the study of leadership are compared. The grounded theory approach is proposed and justified as the preferred research method for the study. Relevant aspects of grounded theory are explained in some detail. Details of the sampling techniques, data collection, and analysis were presented. Issues pertinent to analysis software selection and validation of the study were also explained. The approach to the validation of the grounded theory was explained. As the field study was undertaken in Singapore, a note on Singapore and its construction industry is included to provide a background to readers in Box 5.1. It includes basic socio-economic data on Singapore; the role of construction in the economy; the structure and practices of the construction industry; and measures for continuously developing the construction industry and enhancing its performance.

Notes

1 See www.bca.gov.sg/AboutUs/about_bca.html
2 Ibid.
3 Ibid.

6 Leadership development
A lifelong journey

Introduction

Qualitative analysis of the information obtained in this study using a grounded theory approach resulted in categories that pertained to leadership development and leadership influence. This chapter presents the analysis and discussion of the concepts and categories that reflect the process of authentic leadership development over a person's life-span. The chapter begins with a consideration of the exploratory study which was undertaken prior to the in-depth field interviews. It then explains the concepts and categories that emerged as a result of open and axial coding of interviews with authentic leaders. A framework of authentic leadership development that emerged through the integration of these categories is presented. The framework comprises the four phases of various socio-cognitive processes that occur during authentic leadership development.

Analysis of the exploratory study

The exploratory study sought to capture the leadership developmental influences of emergent project leaders who were enrolled in a graduate degree program at the National University of Singapore (NUS). It was an attempt to identify possible leadership antecedents based on the theoretical model of leadership development presented by Toor (2006), and Toor and Ofori (2006b). The exploratory study also sought to further explain the Integrated Antecedental Model of Authentic Leadership Development (IAMALD) that is presented in Figure 4.1 in Chapter 4.

A set of questionnaires was developed from the literature review. Leadership developmental antecedents were divided into two major categories: "significant individuals" and "significant experiences." A 7-point Likert scale was used: 1 = "extremely negative influence" up to 7 = "extremely positive influence." Respondents could indicate N/A = "not applicable" when the relevant significant individual or experience was not applicable to them. In the literature, there is a debate on whether one should use a 5-point or a 7-point Likert scale. Dawes (2008) found that the 5- and 7-point scales produced the same mean score, once they were rescaled. Dawes further argues that it is "good news" for research departments or agencies who ponder whether changing the scale format will destroy the comparability of historical data. It is quite easy to rescale 5- and 7-point scales and the resultant data is quite comparable. A straightforward rescaling and arithmetical adjustment also make it possible to compare data from a 5- or 7-point scale to those from a 10-point scale. Other authors also show that 5- or 7-point Likert scales are most commonly used and have proven to generate reliable results (Malhotra and Peterson, 2006).

The questionnaire was distributed to 90 students on the MSc (Project Management) program at the NUS. It is essential to note that all students under this program were required to have at least two years of relevant practical experience after obtaining their first degrees. Owing to the rigorous admission criteria, most students had previously had management or leadership roles, or possessed leadership potential from the excellent academic background, strong references, and performance in the interviews of the candidates accepted onto the program. Thus, the students were deemed to be emergent or potential leaders with considerable promise in their professions.

A total of 58 questionnaires were returned, yielding a response rate of about 65 percent. An internal reliability test of taxonomies of significant individuals and experiences was undertaken; it produced a Cronbach Alpha of 0.823, indicating high internal consistency of the items. Of the 58 respondents, 33 (57 percent) were males, while 25 (43 percent) were females. Some 80 percent of the respondents were aged over 25 years. They had significant experience. Some 43 (76 percent) had 1–5 years' experience; 8 (14 percent) had 6–10 years' experience; another 8 (14 percent) had 11–20 years' experience; and 3 (5 percent) had more than 20 years' experience. Another feature of the sample is their diversity of nationalities; the respondents encompassed 10 different nationalities: China, India, Indonesia, Malaysia, Myanmar, Nigeria, Pakistan, Singapore, Sri Lanka and the UK.

Analysis of variance (ANOVA)

The respondents were asked to rate the importance of various significant individuals (SI), people who had significantly influenced the respondent's life; and significant experiences (SE), experiences that had significantly shaped the respondent's world-view. In order to ascertain whether the respondents had a similar opinion on SI and SE when they were divided into different groups – based on gender, age, working experience, experience as a leader, and nationality – an analysis of variance (ANOVA) was carried out. It was hypothesized that respondents had a similar opinion about significant individuals and significant experiences when they were divided into groups based on various demographic features. This was the null hypothesis or H°. The null hypothesis or Ho was to be rejected if the test significance level was less than 0.05 at the 95 percent confidence interval. The results for all the variables are presented in Appendix 4.

Out of 140 computations, only seven computations (5 percent) showed significant difference ($p<.05$). Also, none of the variables showed significant difference at the confidence interval of 90 percent ($p<.1$). Based on these results, H_o was rejected and the alternate hypothesis or H_1 was accepted which means that the respondents, when divided into groups based on the features, had a similar opinion about the different variables shown in Appendix 4. This indicates that those perceptions about significant individuals and significant experiences remain largely similar among respondents, regardless of their gender, age group, working experience, experience as a leader, or nationality.

Significant individuals

The frequencies of the ratings accorded various significant individuals are shown in Appendix 5. Fathers, mothers, and teachers were most frequently rated for having an "extremely positive influence." Most of the significant individuals were positively rated for their influence on the leadership development of the respondents. Siblings, peers, teachers, and colleagues were negatively rated in two instances. One respondent rated their spouse as having had an "extremely negative influence." It is also observable that significant percentages of respondents rated all the significant individuals "neutral" in terms of their influence on the respondents' leadership development.

The mean rating scores given to each significant individual are presented in Appendix 6. They show that teachers, fathers, mothers, mentors, and spouses were rated as the top five significant individuals having an influence on the leadership development of the respondents. Thus, respondents do not perceive remote individuals such as political figures or social figures as important to their leadership development. Among the family members, "father" was rated higher than "mother" and "siblings"; and the mean score of "father" is second only to that of "teachers." This appears to suggest that many people choose their fathers as their ideal persons or role models (such as political, religious, entertainment or media figures). These findings are similar to those of Avolio and Gibbons (1988), who found that parental factors play a central role in the development of transformational or charismatic leaders; and Brown and Gardner (2005) who noted that the most influential individuals in their respondents' lives were friends, family members, teachers, coaches, ministers, and mentors. Arvey et al. (2007) also found that the significant individuals who were triggers of leadership were (in rank order): parents, peers, siblings, mentors, role models, and other family members.

The political leaders among the "ideal personalities" identified by the respondents were: Mahatma Gandhi, Subhash Chandra Bose, Rajiv Gandhi, Allama Iqbal, Nelson Mandela, Mao Zedong, André Malraux, Tony Blair, A.P.J. Abulkalam, and Atal Bihari Vajpai. The businessmen mentioned by the respondents included Bill Gates, Warren Buffet, Jack Welch, Azim Premji, and M.M. Kothari. Some respondents mentioned authors such as Anthony Robbins and Steven Covey, the chief executive officers (CEOs) of their own organizations, as well as religious figures, and well-known personalities in their respective societies. Whereas many of the personalities mentioned by the respondents are region-specific and may not be well known around the world, it shows that people tend to get inspiration from heroes and leaders of their own societies, because of their socio-cultural congruence with them. However, notably, the mean scores suggest that the influence of these ideal personalities or role models is not as high when compared to those of teachers, parents, and mentors. One explanation for this low mean is that such personalities only act as role models but do not provide direct emotional support or guidance to individuals. On the other hand, teachers and parents act as close social educators and play an important role in leadership development.

Significant experiences

The respondents were asked to rate the influence of a taxonomy of experiences on the 7-point Likert scale, from 1 = "extremely negative influence" to 7 = "extremely positive influence." The frequency distribution of various significant experiences is shown in Appendix 7. The results indicate that organizational experiences, early childhood experiences within the family, early childhood experiences outside the family, and educational experiences at university were most frequently rated as having had an "extremely positive influence." Most frequently mentioned experiences which had an "extremely negative influence" included "experience of a comfortable life," some special incident, and "loss or death of someone." These findings are similar to those of Arvey et al. (2006) who found that "training and developmental experiences," "educational experiences," and 'prior challenges in life" were experiences which were most frequently indicated by the respondents as having had an influence on their leadership development.

In the open-ended questions, some respondents mentioned "marriage," "natural calamity," "personal illness," "loss of a close friend," and "loss of a sibling" as significant experiences in their lives, which had a negative influence on their personalities. Some respondents also mentioned other significant experiences such as: participation in sports; reading articles by Subhash Chandra Bose (a political figure in Indian history); meeting Rajiv Gandhi (a former Prime Minister of India); meeting a senior figure in the person's company; receiving advice from one's husband; meeting some friends of the opposite sex; and watching some movies, as experiences that influenced their leadership development positively.

Similarly, research has shown that various experiences during a person's education play a significant role in the development of leadership identity among individuals (see Komives et al., 2005; 2006). With reference to experiences during education, particularly during university education, studies have shown that individuals get opportunities to take up leadership positions. Evidence on leadership development also shows experiential learning in organizations as an important part of leadership development (see Cheetham and Chivers, 2001; Hirst et al., 2004; McCall, 2004; Kempster, 2006). These findings are also in line with studies which underscore the importance of familial influences, particularly that of parents (Keller, 1999; Zacharatos et al. 2000; Keller, 2003; Popper and Mayseless, 2003) and birth order within the family (Sulloway, 1996; Andeweg and Van Den Berg, 2003).

Discussion of the exploratory study

The respondents perceive their "teachers" and parents to be influential individuals in their leadership development. The higher rating for "father" as compared to "mother" suggests that individuals tend to gain professional inspiration from fathers rather than from mothers. The importance that respondents accorded to their parents in the development of their leadership capabilities is in line with the findings of earlier researchers. Bronfenbrenner (1961) noted that leadership was more likely in families in which fathers were more educated and in which both parents were less rejecting, less punitive, and less overprotective. In turn, parent-child interactions reflecting these more positive qualities predisposed children to leadership behaviors. Hartman and Harris (1992) also found that college students modeled their management style on the leadership style of their parents. These findings underscore the importance of the role of parents in developing their children as leaders.

The descriptive analysis of the responses in the exploratory study is shown in Appendix 8. Among significant experiences, educational and occupational experiences were rated high with an equal mean value (Mean = 5.91). These findings are in line with previous studies which suggest that experiences during one's university education and professional career play a key role in a person's identity construction (self-awareness) (see Avolio and Luthans, 2006). Bennis and Thomas (2002) also found vital developmental experiences, turning points in life, familiarity with failure, and learning from failures as antecedents to the leadership development of their subjects. In relation to organizational experiences, Scully et al. (1996) noted the reasons for the development of rigid behavior in CEOs. They observed that the CEOs of poor-performing firms were "tougher" in their leadership behavior toward the members of their top management team who reported directly to them than the CEOs of higher-performing companies. Although the findings in the current study do reveal that educational and organizational experiences are important for leadership learning, they do not show the nature of experiences (for example, failure, success, frustration, hardship, opportunity of leadership, influence of bosses and colleagues, and so on) which result in leadership development. Most professionals attend university, followed by a professional life; so, what experiences turn certain individuals into leaders and others into followers? To answer this question, it is useful to further expand these categories in order to explain the causal conditions, intervening conditions, specific context, and consequences attached to such significant experiences.

With reference to early childhood experiences, O'Connor et al. (1995) found that events in childhood were vital in shaping the vision and world-view of charismatic leaders. The current exploratory study also found that experiences during a person's childhood – both within the family and outside the family – significantly influence the person's leadership development. Childhood is the time when individuals construct their personal identities and implicit leadership constructs. Various familial and socio-cultural influences play important roles in the development of the identity and implicit leadership theories of the individual. It is essential to explore which specific experiences lead to a certain kind of identity development which eventually motivates the individual to exhibit leadership behaviors at later stages in life. Investigations of such childhood experiences can make useful contributions to the understanding of the processual nature of leadership development during various phases of a person's life.

In general, the findings of the exploratory study revealed that emergent leaders relate their leadership development to various significant individuals and significant experiences in their lives. Although some studies have shown that genetics do play an important role in leadership emergence (see, for example, Ilies et al., 2004; Arvey et al., 2006; Avolio and Luthans, 2006; Arvey et al., 2007), others argue that leadership development is largely a social construction in nature (Sooklal, 1991; Conger 1998; Chan, 2005). Osborn et al. (2002) also argue that leadership "is socially constructed in and from a context where patterns over time must be considered and where history matters" (p. 798). They consider leadership as a series of attempts, over time, to alter human actions and organizational systems. The findings in this study so far also support the view that leadership development is heavily influenced by various developmental events during different phases of leaders' lives. Some critical events result in the construction or reconstruction of a leadership identity which largely mediates the leadership that the individual will exhibit. Therefore, the influence of such events deserves further research.

The results of the exploratory study showed the importance of significant individuals and significant experiences for leadership development. However, the questionnaire-based quantitative approach was unable to unearth the socio-cognitive processes that occurred due to the interaction of significant individuals and significant experiences. It also could not explain how leadership development was related to how leaders exercised their leadership. The results of the exploratory study point to a need for a deeper investigation of "understanding of life events" to determine the nature and influence of leadership trigger events on organizational leaders. The findings of the exploratory study showed that "leadership development and influence" can better be captured by a qualitative approach. Therefore, following the exploratory study, a qualitative research approach was adopted to explore leadership development and influence. Analysis of the qualitative information collected during the in-depth study using grounded theory methodology is now presented.

Analysis of the main field study

A full qualitative grounded theory research approach was adopted in the detailed study. As recommended in the grounded theory guidelines (Strauss and Corbin, 1990; Corbin and Strauss, 2008), data analysis was undertaken in parallel with data collection. Analysis began after the first phase of interviews in which presidents or vice presidents of professional institutions and trade associations in the construction industry of Singapore were interviewed. After the first phase of the interviews, full transcripts were prepared and analyzed as explained in Chapter 5. More interviews were conducted in the second and third phases of data collection, which followed the guidelines for theoretical sampling. As a result of theoretical sampling, the analysis of subsequent interviews resulted in the emergence of more categories and saturation of earlier categories. A total of 49 cases (45 interviews and observations made in four CEO lectures) were included in the current study. Demographic details of the interviewees are shown in Table 6.1.

The "consultants" were from firms whose principal expertise was engineering design (civil, structural, and mechanical and electrical engineering). Some interviewees worked in firms of architects, whereas some worked for firms engaged in both architectural and engineering design. Three interviewees were former practitioners who had left construction to work in other sectors, but they were very senior executives who had earned a reputation for being authentic leaders in the industry. Qualitative data obtained from these cases was analyzed in accordance with the process explained in Chapter 5. Open coding and axial coding led to the emergence of lower-, middle-, and higher-level categories that are summarized in Table 6.2.

The leadership support group

During the interviews, the leaders were asked about the people who had had the greatest impact on them during their early years (before university), late years (university life, early professional career), and late years of professional careers. They were also asked about those who inspired them as their leadership role models. Mentions were made of numerous individuals who were influential in inspiring leadership in the interviewees. Categories pertinent to significant individuals that emerged from the coding process include: family members, close social educators, and distant social educators. These categories were integrated under the broader category of "leadership support group" which refers to various individuals who help, support, or assist the leader in various phases of leadership development. Although there were individuals who negatively influenced leaders – such as a tyrannical boss or an unhelpful colleague – interactions with such individuals are rather considered as experiences. These individuals are not included in "leadership support group" through they are largely the part of "people" who influence authentic leaders.

Family members

Based on the responses obtained on the influential figures, all family members were grouped under the bigger category of "Family Members." This category includes: parents (father and mother), grandparents (grandfather and grandmother), uncles and aunts, siblings, intimate partners, spouse, children. More than half of the interviewees mentioned their parents as the most influential figures in their lives. The CEO of a developer's organization shared:

> My parents had the greatest impact on me. My father, to us, is a paragon of hard work and diligence. I do not know anybody else as hard working as him … My mother also had a strong impact on me … She was really the person who guided us most and set goals for us.

Table 6.1 Demographic details of the subjects

Attribute	Range/properties	No. of cases
Gender	Male	42
	Female	7
Age group	30–40	9
	40–50	14
	50–60	23
	60–70	3
Company type	Architects	8
	Consultants (engineers, designers)	9
	Contractors	7
	Developers	11
	Quantity surveyors	7
	Architects + engineers	4
	Others	3
Education	PhD	3
	Post-graduate degree	18
	Graduate degree	27
	Polytechnic graduate	1
Experience as a leader	5–10 years	5
	10–15 years	12
	15–20 years	12
	20–25 years	15
	25–30 years	5
Experience in the industry	5–10 years	3
	10–15 years	4
	15–20 years	6
	20–25 years	12
	25–30 years	10
	30–35 years	12
	35–40 years	2
Position in the organization	Manager and senior manager	7
	General/deputy general manager	2
	Director/executive director	20
	Managing director	2
	CEO/deputy CEO	10
	Managing partner	2
	President/vice president	4
	Chairman/group chairman	2

Table 6.2 Categories and processes under leadership development

Categories and processes	Leadership development					
Higher level			Construction of identity Social learning Recruitment of role models and support group	Expansion of support group Development of higher-order attributes Action-based reflection Transformation of identity	Positive reframing Self-regulation Self-transcendence Systemic perspective	Meta-reflection Role modeling for followers
Middle level	Leadership support group	Social institutions	Preparation	Polishing and practicing	Performing	Passing
Lower level	Family members Close social educators Distant social educators	Familial institutions Educational institutions Professional institutions Other social institutions	Influence within familial settings Struggling and making my own way in the world Academic inspiration Exposure to new things in life Extracurricular activities and early brushes with leadership Other social experiences	Facing frustrations Struggling with failures and challenges Learning by doing many things Learning from seniors, bosses, and mentors Molding the self over time Personal initiatives for self-development Various turning points in professional life Facing social challenges	Getting confidence from success and achievement Leading through intense organizational change Handling professional challenges	Developing followers Sustained succession planning

It was notable that the interviewees also mentioned why their parents were so influential. Some got inspiration because of the values their parents imparted to them. Others were inspired because of the guidance they got from both of their parents. The CEO of a quantity surveying firm related:

> I would say the person who had impact on me was my father ... I took a lot of guidance from him. He was also a very philosophical man. Quite often, we would meet with him to talk about aspects of the businesses including the parts that he had already entrusted to us ... He wouldn't impose his style, but he would tell you the good and bad and let you get on with it.

A senior leader mentioned that her mother had a very strong personality, and was full of wisdom and strength. She stated:

> My Mom! She is just very strong ... independent ... full of drive and motivation ... Compared to a lot of women, she is much stronger ... She has not had much education. But her old ways of planning and managing things are remarkable. I think I have her strength.

Such responses not only revealed that parents had a significant influence on many of the leaders, but also showed that parents were vital characters who taught them how to live a better life and how to grow further.

The interviewees also mentioned other family members having influenced them during their early days. These individuals included grandparents, uncles and aunts, siblings, spouses, and children. For example, a director in an architectural practice mentioned that one of her uncles was very influential in building her confidence. She recounted: "A person who shook my life, apart from my Mom, was this gentleman, Mr. Lim. He was an uncle of mine ... He gave me public speaking confidence and polished my leadership [ability]."

The CEO of a large development company mentioned his late sister who had a great influence on him:

> My sister (she left us in 1978) ... always guided me well in values, filial piety, and the love for education. I owe lots and lots to her. She taught us about life values ... to value hard work in order to achieve our ambitions, and not to look for short-cuts in whatever we do.

Close social educators

When asked about influential figures in their lives, the interviewees mentioned a number of other individuals. School and university teachers, friends, peers, colleagues, coaches, and mentors had influenced the leaders in several ways. They attributed inspiration, emotional support, intellectual stimulation, and guidance to such people. For example, the CEO of a large consulting firm shared:

> One form teacher, who was also my badminton coach in my secondary school days influenced me a lot ... He was very encouraging and always asked us to look at the positive side of things. [He was] nurturing and engaging.

A director of an architectural practice mentioned her university tutors who had influenced her significantly. One particular tutor on her architecture course was "very encouraging":

> He would take us out to Chinatown to draw the buildings, people, anything we liked. He taught us freehand drawing. He had a strong impact on me. He taught us how to use our eyes when drawing, to gauge distances and dimensions, and to sketch buildings and other figures to scale.

A senior executive reflected on the fact that his peers had an influence on him:

> My peers! In architectural school, we spent 5 years together, and we were only a few people. We were only 25 in all, attending the classes, working in the studio, sharing

sleepless nights, etc. You would actually follow closely how others are developing the skills and we would all learn together ... In other courses, you may learn from your peers, but I think, the impact is really a lot in the architecture school.

The CEO of a consulting firm also mentioned that his childhood peers influenced him in many ways. He stated:

I have always studied in boarding schools and that has done me a lot of good. I first left home to be a boarder when I was 12 years old. One thing I learnt is how to ... survive in the face of competition because you are competing with all the boys for space, for food, for everything, you know ... The other thing is that you are living and working with the best students, so I think that way you are constantly upgrading your knowledge.

A sizeable majority of leaders interviewed in this study mentioned their superiors and colleagues who had made a strong impact on their lives. One senior manager in a contracting firm related:

I was assisting my boss when I was working in Shanghai with a Japanese company. He was very hands-on. I benefited a lot from him ... The special thing about him was his leadership. The way he actually dealt with the staff and his attitude in dealing with problems really inspired me. That is actually left deep inside my mind.

The chairman of a quantity surveying consultancy mentioned several leaders in the local industry who had influenced him in different ways. He noted:

I would say, in my working life, I would keep a particular look-out for certain industry leaders [names omitted] ... I would actually ... look at the things they do, listen to, or read their speeches, look at their actions. And I would try to get their company prospectus every year, look at the report and how well they have done, and look at the targets they set for themselves.

A senior director in an architectural practice considered one of his colleagues to be the most influential figure in the profession; he was professionally highly competent, and also empathetic:

Within the office, my biggest influence was probably [name omitted] ... I don't think he's the best designer in the office, but he has really good eyes for getting things to work well ... He also is the partner who knows more people in the office by their first names ... There was a time when he knew every single person even when we are about 300 in the firm.

A vice president of a developer's firm shared similar sentiments about his boss:

My boss is a tremendous leader who holds the values. And even in her presence, you know that you can speak your mind about important things even if it contradicts her point of view and still not be punished. You need that kind of assurance so that you can do things.

Distant social educators

Some interviewees mentioned role models, political leaders, social leaders, famous professionals, and other individuals who had influenced them. Although such individuals had never come in direct contact with the interviewees, their influence was significant and it had a positive impact on the leaders in this study. One of the leaders mentioned a famous project manager. She had never met him but had read his book. She mentioned that this project manager was her friend's boss and had a very positive influence on his followers. She reflected:

> Strangely, the most influential person is the one I have not worked with or ever met. He was considered the father of project management in our times … At that time, he wrote a book. [Previously] I thought the project manager should be technically sound and an expert; and he had to know everything. But the book said that project management was actually about people management. I enjoyed reading the book, and I read it again and again. It was interesting and it was all true what it said. So, to say that, he is my mentor!

A large majority of the interviewees mentioned various political figures as their role models. The person most frequently referred to was Lee Kuan Yew, the first Prime Minister of Singapore. One CEO paid tribute to Mr Lee:

> My leadership role model has always been Lee Kuan Yew. I just admire his hands-on approach, his leadership style and the fact that he is able to stand up to the rest of the world … [He is] absolutely brilliant … Making Singapore such a big economy out of nothing is not an easy job.

Another leader reflected upon Lee Kuan Yew in the following way: "Another person who really inspired me and whom I always admire the most is Lee Kuan Yew. He built up the country and brought it from the third world to the first world in a single generation." The CEO of an engineering firm mentioned a number of role models:

> I admire Gandhi a lot. As a young man, I liked Lee Kuan Yew and Nelson Mandela. For an ideal, they can sacrifice their family … just on one principle. It is not easy to make such a huge sacrifice for a principle.

While admiring the role models, the interviewees mentioned what they liked about their role models. These attributes usually included the ability of the role model, values that the role model holds, and actions that the role model takes in order to serve the people. In a few cases, the leaders did not mention any particular person's name but they did say that there were a number of people who had inspired them in various ways and had brought positive changes to their lives.

Events, incidents, and turning points

The interviews in this study were focused on stories and narratives of leaders about how they came to be the people they were and what events had really changed them as individuals. The interviewees were asked questions related to events that had significantly changed or transformed their lives positively. The questions related to the interviewees' life narratives form part of the set presented in Appendix 1. Examples of these questions are:

1 How would you briefly describe your early life? (Was it comfortable, easy-going, diffi-cult, struggling, challenging, or fearful?)
2 From your early life, can you remember two events when you felt you were really at your best and it boosted your leadership development?
3 Can you remember a single most important event/incident/turning point that trans-formed you as a person and inspired your leadership style? And do you consider this moment as the defining moment for your leadership development?
4 What has been your greatest professional challenge in your career and how did you overcome it?

The interviewees related numerous stories explaining how their early life had influenced them. Many of the interviewees knew what had positively contributed to their leadership development. Some took a while to reflect on their past to remember the incidents and events that shaped their lives. A few of them could not remember any event that had shaped them. However, such interviewees did mention that there were a lot of events and they were not able to remember the most crucial one(s). The stories that the interviewees narrated mostly pertained to their childhood experiences at home and school. Adolescence and adulthood experiences were mostly about their university life, professional challenges, and various other social challenges.

The incidents and events are categorized into four phases: (1) preparation for lea-dership (events, incidents, and experiences during childhood and early adolescence); (2) polishing and practicing of leadership (events, incidents, and experiences during adult-hood); (3) performing leadership (events, incidents, and experiences when leaders actu-ally gained a sustained success by performing leadership); and (4) passing leadership (events, incidents, and experiences when leaders were passing the reins of leadership on to the next generation). In the following sections, these categories and their sub-cate-gories are discussed.

Preparing for leadership

Preparing for leadership relates to the events, incidents, and experiences during inter-viewees' childhood, adolescence, and early adulthood. Such experiences took place in var-ious social settings including: in the family, at school, at university, in a religious group, in the army, and in a foreign country. The following basic level categories emerged under "preparation for leadership":

- influence within familial settings;
- academic inspiration;
- struggling and making my own way in the world;
- exposure to new things;
- extracurricular activities and early brushes with leadership;
- other social experiences.

Influence within familial settings

Describing the impact of familial settings during early childhood, a director in an archi-tectural firm said:

It is the way we have been brought up. We are a very open family. My parents respect you, no matter who you are, even us, while we were children. It was passed down well to us. It has been a big part of my life. My parents were very influential.

The CEO of an engineering firm related her story in a very emotional way:

Oh, my early life! I would never want to go through it a second time. I come from a large family. My parents were rubber tappers. They could hardly make ends meet. We used to wake up at 5 a.m. and help our parents in tapping the rubber trees. Then we would go to school, and help with the work again after we come back from school. It was a very hard life. And money was not enough. That sort of childhood life molded my character. Three meals a day were not guaranteed. Sometimes, we would go hungry because there was no food in the house.

Academic inspiration

Many interviewees mentioned that they were very good in academic work during their childhood and school significantly influenced them in several ways. One of the interviewees described her excellent academic performance at school which contributed to her development as a leader. She noted:

My father lost his job when I was in pre-university. I knew I had to study hard. When I was very young, I always did well in school. At the end of each year, my father would ask a single question, "What position did you get?" I was always the top student in my class. This trust of my parents made me very independent. Also, when I got high marks, I would also become the pet of the teachers. It was childhood motivation to excel in learning.

A senior manager recounted that the legacy of his prestigious school was an influential factor that inspired him and left indelible impressions on him:

From primary school, I went to the Raffles Institution. In terms of its own character, it must have influenced me a lot ... The institution was 100 years old at that time. The school was well known as the premier school in the society. It had produced the professionals, doctors, and the people needed for the administrative and civil services. The first Prime Minister, many ministers, army generals, ambassadors ... This school inculcated in me that, when you walk in there, you are entering an institution with 100 years of educating the scholars who built the nation. It was a sense of perfection. And that you have feeling that several thousand better, and far better scholars have passed through there. You are now the flag bearer for this illustrious group ... You and your cohort ... it was a tradition. We were told that ... there was history behind us, the history of great people. This built into us the sense of striving to achieve ... That helped us in developing personality. The sense of achievement and getting things done was inculcated in us. It becomes your basic character at later stages.

Struggling and making my own way in the world

There were several instances when the leaders related the hard times they had gone through during their early years. They talked about the tough financial conditions they had to face. The CEO of a consulting firm revealed that she had very little money at university. She noted:

> When I came to Singapore [to study at university], my father gave me 20 [Malaysian] Ringgit and told me that it was all he had. He told me I had to find my own way. I started to give tuition. I borrowed money from my friends ... The first year was very difficult. I had to pick up English ... Time went on and by the end of the year, I found that I had not paid the examination fee for the year. I went to the administration office and told them the situation. I also told them that I had no money but I was willing to work. They were helpful and they arranged some part-time work for me in a bank. They also applied to the Lee Foundation to ask for a bursary for me.

Another executive mentioned about the hardship he faced and how this hardship had molded his character:

> I came from a large family of four brothers and two sisters. I lived for over 20 years in a village. Out of these more than 20 years, I think more than 10 years were without electricity, without tap water and without modern sewers ... the long and short of it is ... that's how I view things ... I have a baseline that is pretty hard ... Through my years, whatever it changed, that is always at the back of your mind; that's where I came from.

Exposure to new things

"Exposure to new things" refers to novel experiences that significantly influence the individual's outlook on life. Several interviewees mentioned that their early lives changed because of exposure to certain new things. These ranged from doing the mandatory National Service (which involves military training and service in the armed forces) to traveling abroad. One of the interviewees shared:

> I had the opportunity to go around the whole of USA ... I managed through the university to go there [for an internship]. I went ... on my own ... worked there for three weeks, and then went around the whole of the USA for about three months. I think that was a very valuable personal experience in developing what I consider to be a vivacious personality.

The managing partner of a large quantity surveying firm reflected that his National Service in the Army had deeply influenced him:

> One event that I think is fundamental is that during National Service, because I was in the artillery conversion course ... it was a period when one was subjected to a lot of physical as well as mental pressure and challenges. And so that is where I had a test of my will. But having gone through it, it ... made me a better person.

The executive director of a quantity surveying firm mentioned an event from his youth when he first participated in the National Youth Achievement Award (NYAA). He shared this:

> My teacher in the polytechnic approached me and encouraged me to join the National Youth Achievement Award. Probably I won't be living half the life I am living right

now … If I had not been approached by my teacher; I would not be where I am now. Probably, I would not have had the kind of experience I have had.

Extracurricular activities and early brushes with leadership

Many interviewees mentioned their early leadership experiences at school. One of the leaders actually said that he "had the yearning to become a leader" and that he "was involved in everything under the sun." Another interviewee said that he was involved in many extracurricular activities during his early life:

> I think from my school days, even when I was in the Boys Scouts, I was a leader of the squad. In secondary school, I was also president of a social club. In the army, I was selected to be an officer and then I became a company commander. There are probably some leadership traits which others saw in me. Some intangible things made me a leader.

The executive director of a quantity surveying group shared his story:

> Ever since secondary school, I have volunteered to participate in youth groups … I used to be very active in the National Youth Achievement Award. I was the chairman there. I won gold medals. I organized trips for the handicapped … I organized archery lessons for wheelchair-bound people. But the underlying thing is that … you are actually working with a group of volunteers with no commitment to you, with no contract … you are not paying them. How do you spur them on, encourage them? How do you motivate them to work together, and with you? … That is where, since the age of 13, I have been learning in terms of the leadership side.

Other social experiences

In the category of "other social experiences," most of the interviewees mentioned a wide range of experiences they had gone through, including the death of a loved one, spiritual influences, and some minor incidents in childhood. Some mentioned that they developed as leaders gradually, and there were no critical episodes in their lives. For example, one CEO noted that: "I don't think there is any particular turning point in my life … it was very much the case of a growth, building up slowly. There was not much of a particular upward or a down moment as such."

Another executive noted that:

> I can't remember any two events or a single event in these terms. I can't think of any specific events which shaped me into what I am. There have been probably a series of events combining and leading to how things have just panned out. So, no specific event has made me change my course.

Processes during the preparatory phase

It is pertinent to analyze what is actually going on during the "preparing for leadership" phase. From the analysis of the interviews, emergent concepts and categories, and memos, three processes were identified: social learning, construction of identity, and recruitment of role models and support group.

Construction of identity

"Construction of identity" refers to the development of implicit perceptions of one's persona. During this phase, when a person performs a leadership role within the family, in school, or in other social settings, the person also develops their individual leadership construct. Identity development is an on-going process; however, the early phases of life play a vital role in the self-concept of a person. Individuals make sense of the world around them and differentiate their own roles from others. They synthesize past experiences and in the light of these experiences they anticipate their future possible selves. For example, "influences within familial settings," "academic inspiration," and "struggling making my way in the world" are the categories which explain how individuals develop a certain identity during childhood and adolescence. During "preparation for leadership," leaders also learn and understand values, develop ambitions, and set goals for their lives.

Social learning

Social learning is a generic process in human development. Social learning can be both vicarious and observational in nature. Categories such as "exposure to new things," "extracurricular activities and early brushes with leadership," and "other social experiences refer to how individuals learn new things in the early stages of their life. Social learning is a process that goes on throughout life. as individuals continue to learn from various experiences in their social settings. However, social learning during childhood and the early stages of life is vital as it leaves deeper impressions on a person's identity.

Recruitment of role models and support group

During the preparation phase of leadership development, individuals recruit their support group, both consciously and unconsciously. They seek help from others and learn from them. The support group during this phase usually includes family members (parents, siblings, and close relationships), teachers, and friends. In the earlier section on "leadership support group," it has been explained that individuals are heavily influenced by people around them. They learn values, develop ambitions and goals of life, positive attributes, and an understanding of their roles in society. During this process, individuals recruit role models who can range from parents, siblings, and teachers to political figures, celebrities, and so on. They learn to regulate their behaviors and emotions.

Polishing and practicing leadership

The category of "polishing and practicing leadership" emerged as a result of the integration of sub-categories that pointed to those events, incidents, and experiences that significantly changed the interviewees' life course. Such events could be minor or major frustrations, intense organizational and environmental changes, major failures or achievements, critical turning points, decisive episodes, or transformative experiences. These events, incidents, or experiences usually resulted in learning by the leaders and provided them the opportunity of "polishing and practicing" their leadership. These lower-level subcategories emerged under the "polishing and practicing" phase of leadership development:

- facing frustrations;
- struggling with failures and challenges;
- learning by doing many things;
- learning from seniors, bosses, and mentors;
- molding the self over time;
- personal initiatives for self-development;
- various turning points in professional life;
- facing social challenges.

Facing frustrations

The interviewees mentioned some events when they felt frustrated and disappointed for certain reasons. In most cases, frustrations occurred because of the attitude of the superior, colleagues, or subordinates. For example, an interviewee mentioned an incident in which she had to deal with a staff member who was older than she was. She had difficulty in dealing with this staff member due to his attitude toward work. She noted:

> In my previous work, sometimes dealing with people frustrated me. This gentleman was 20 years older than me. I found it difficult getting anything across to him … He would send letters out … to the client. This is not the practice in today's work … the man left the company when I tried to take corrective measures. I sat down and thought about whether there is a way that I can improve my leadership …That was a time when I was quite disappointed with myself … I thought I can be a little more open.

The CEO of a consulting firm related an incident in which he had fired a staff member on the spot. This was a frustrating event for him and his leadership:

> There was one occasion when one of our directors on a major project was supposed to have delivered some drawings with our proposal to the project architects. I think he spent the whole night before at a night club so he just gave the package to our office attendant to go and deliver it. This young man probably had a small accident along the way and lost all the drawings and these drawings were the original tracings. At that time, we used to do drawings in tracings; there were no computer files. The whole design was lost and basically the architect had to start to prepare the design from scratch and obviously that was a big blow for us because of him. And I basically sacked that man, and this is the only time I have sacked someone on the spot. I thought what he did was a very irresponsible act. And unfortunately he took it quite badly and he actually fainted. So that's where I felt pretty bad actually. Maybe I should have had more patience, … [and] done things differently.

Struggling with failures and challenges

Have you ever faced any failures? Do failures constrain you? The interviewees related stories of the challenges and failures they had come across during their early life. However, each of them was adamant that no matter how big the failure or challenge was, they knew they could face it and find a way to address it. A senior vice president told the following story:

I was a new project manager in my previous company … I had a difficult boss who thought I could do planning although I had never done it before. It was a great challenge. I wanted to prove myself and I was very driven … I took over a new project which was about halfway toward completion. In the first meeting, everybody was pushing tasks and responsibilities away … "This is not my job, this is your job." They all looked miserable and unhappy, very unhappy! What I did was to tell everyone, "Let's work together." Still today, the people of that team are very close friends of mine. We all became close friends. It was a terrible job before that incident. But at the end of the day, everyone was very happy. It won a few awards. It was very satisfying.

The CEO of a large conglomerate said that failures were part of a leader's life. If one has not failed, one does not really appreciate success. He shared:

Actually, even for any successful person, very rarely can you say that they have never fallen flat on their faces. I suppose everyone has the ability to get up and stand again. And make it even bigger. That is most satisfying. In 1987, I lost a lot of money. I was launching a public company in Australia. I got the telex that all the investors had backed off … The question was whether to commit suicide or go back to your family. Build again … As a true businessman, you have to take up the challenge and try to make a success out of it.

A senior executive in a quantity surveying firm also shared that he had faced failures many times. However, each time he failed, he took it positively and learned from it. He explained:

To me, failure is a positive development … I would take this as a learning curve and think about how to do things better. I am very glad that all our partners have this thinking. They are not easily defeated; rather, they would look at it and say, "Where have we gone wrong?" or "What did we not do correctly?" or "What should we have done?" … and they will set about addressing the problematic issue.

Learning by doing many things

The interviewees frequently mentioned that they had learnt leadership by doing many different things through business ventures, working on many projects, and so on. A director interviewed mentioned that she had learnt a great deal by working on a diverse range of projects. She explained:

I think in my entire career over more than 15 years, my last project was the toughest. After I finished polytechnic, in 1993, there was a boom and there was so much work. I was very junior and I had a lot of opportunity to do everything within two years. I learnt through the projects. On average, I was working on almost 10 projects [at a time] … I had real-life training.

The CEO of a large firm shared that he had polished his leadership skills by learning through working in many different organizational settings. He explained: "I believe that my training with the various companies where I have been placed in leadership positions and have had to make important decisions independently, certainly helped me to learn how to make the right decisions".

The CEO of a design firm shared the challenges of business which he encountered in his initial days:

> Getting projects was a problem in the beginning … I could not get government projects as I had no record in the beginning … they told me that I had to prove myself to get government's contracts. So, I had to work solely on private-sector jobs and build up my record. Initially, I was helping many people without any money. Now, I tell people that one should very well be prepared to take risks in business. But there is a lot of satisfaction in seeing the company grow. When you build up a company, there is a sense of achievement.

Learning from seniors, bosses, and mentors

The interviewees consistently mentioned their bosses, mentors, and coaches who had played pivotal roles in training them. They had found such supporting figures at various stages of their careers but they were grateful to them for the mentorship, guidance, coaching, and direction they had provided to them. One of the interviewees, a senior director in a consulting firm, shared this about his mentor:

> I had a good mentor, an architect by training. He was the executive director of the firm. He was a very loyal executive. He mentored me and the training was very, very good. We embarked on many huge projects at that time together. It was a huge amount of work and sheer volume and a large amount of repetition. It gave me great opportunities to learn.

The CEO of a large contracting firm was also appreciative of his mentor. He shared: "So far, I have come across one very major mentor who saw it as a personal duty to mentor me, who volunteered to mentor me." The vice president of a developer's firm also mentioned her boss who had a great influence on her: "Actually, my former boss, in my previous company, is the one who inspired me … He was very down-to-earth. He was a very warm person and a very different one." A senior manager also praised his boss for the mentorship he provided him: "It was learning with my boss. I learnt how he dealt with time; he taught me leadership and management."

Molding the self over time

Many of the interviewees referred to taking up a bigger role in society. They mentioned the times when they realized that they should think beyond themselves and achieve something for others, for the country, and for society. A CEO of a development company shared that he had changed significantly after he realized his role in life. This happened after he became a committed Christian. He shared:

> My leadership style has changed over the last few years … I see myself as a steward and a leader for future generations. I don't see things in the present. I do things so that the successive generations can be at peace.

Another senior executive noted:

When I was younger, even until I was out of the army, I was so sure of myself. I remember I was really arrogant and had this enormous self-belief that one can overcome any challenges. [It was] really enormous. To a large extent I am more tempered and more realistic now.

Personal initiatives for self-development

Most of the interviewees mentioned the occasions when they took personal initiatives for their self-development. Attending a leadership course, reading books and articles, and participating in leading professional institutions were often mentioned under this sub-category. A director in a quantity surveying firm mentioned an event when she realized her leadership capability. She reflected:

Leadership has been part of my life. I attended a leadership course in the UK. There was a game where I had to lead the group and solve the problem which had been set to our group. I was given the opportunity to lead very senior people. After the … course, I realized that, after all, although I am young, I can be a good leader too. Otherwise, leadership is day-to-day life. Since I was young, leading in the family, leading during my education, taking care of my siblings' education, planning the food for the family on a daily basis, and so on … that is all leadership.

A senior executive in a quantity surveying firm stated:

My involvement in the professional bodies … helped me to hone my leadership skills … I was made the Vice President of the Quantity Surveying Division of the Singapore Institute of Surveyors and Valuers, and from there, I became the first Vice Chairman of the Pacific Association of Quantity Surveyors.

A senior vice president in a developer's firm reflected that she was very interested in bringing about changes in rules and regulations on buildings while she was a junior architect. Therefore, apart from her own job, she took an interest in proposing the possible changes to the higher management:

I did manage to have some guidelines changed, saying that they are made by people, just like us, and so can be changed. Finally, the bosses accepted my proposals. So they changed the rules … I always challenge the rules.

The CEO of a consulting firm mentioned reading a book during his early career which significantly changed his perspective and greatly influenced him in the management of his company during later years: "I read the book, *Built to Last* by Jim Collins. I read the book in my early days, and I took a lot of wisdom from there. I am now reading his second book, *Good to Great*."

Various turning points in professional life

Although the interviewees highlighted several points in their professional careers which could be regarded as defining moments, they emphasized some events which they considered to have had a major influence on them. The CEO of a contracting firm shared one such experience; he decided to leave a government agency for the private sector:

> Moving into the private sector was a major milestone decision. But I was always able to take a decision at various points in time. I went to join a developer first which was like a safe route before I jumped into the world of contracting. Making a decision to join a contracting firm subsequently was a brave one. There must be a lot of belief in my own self to be able to jump into it … it was almost a quantum jump for me.

Another senior executive shared the turning point in her life. She left her professional practice, and she had two children; she rejoined the profession after a long gap:

> It was tough coming [back] after the eight years break. There were many things different when I came back. I was a computer illiterate. I still can't type properly, even today. I was not good in any aspect of the computer. It would take me a longer time to do the work. It was miserable. I worked hard and I wanted to do it well. Due to my commitment … In two years, I caught up.

Another senior executive shared that her emotional quotient (EQ) was not very good until she suffered a blow in her career:

> My EQ level at that time was very low, may be minus 1 [laughing]. My style was very direct. I was straightforward. But this style was not fine with many people … After I faced the crisis, I changed. I never anticipated in my life that I would face retrenchment, which I did. When you face retrenchment, you think about things from a different perspective. When the career path hits something drastic, there you look back. I did not know that some people are too slow by nature; that is how they are. Previously, I was not sympathetic, and not considerate.

The president of a large consulting firm expressed how she realized her role when she was given an opportunity to bring about a change; she came to know that she could contribute:

> There was a change in management in this company … previously you needed to go through the hierarchy and you needed to be extremely good at every level to be promoted. Then there was restructuring and I was suddenly given a chance; and I attended a higher-level meeting. It was through this meeting that I understood what the whole organization was all about. Then I could do better myself. I could set a proper direction for those I was leading. That was the first corporate planning exercise for me.

Facing social challenges

The category of facing social challenges contains issues that female leaders faced during their professional careers. Social challenges mostly pertained to gender-based overt and covert discrimination that female leaders faced. A leader described her personal experience

> I must say, as a woman, this is not an easy industry to be in. When I ran for president of [my professional body], I was warned by many friends that I would not have a

chance. But I persevered, and I got through. I do not play golf where people say many important decisions are made ... so it was a difficult thing.

Another female leader often had difficulties with some clients initially:

When I approach some clients, they call my bosses and say ... "I don't want to see a lady engineer"... Well, I do face this situation sometimes. Another "weakness," I suppose I have, is that I won't entertain my client in night clubs.

Other female leaders also shared similar views about social attitudes they had to face. One senior vice president related her story which involved cultural attitudes to gender:

I remember that as a woman, it was very difficult working with them as Korean contractors will not take any instructions from a woman. I remember going to the meeting and I was very tough; in that situation, I had to insist on things being done the proper way. He [the contractor's Project Manager] was very unhappy. Eventually, when we got to know each other, we got along well.

Another female leader described issues in her own firm affecting her career development:

The issue of my promotion was a challenge. My own boss often said that "If I were to retrench people, I will retrench the women first." I was upset about this as I knew that one of my subordinates was her family's breadwinner. Her husband was not working. I was upset that he could be so presumptuous. Indeed, I could see that the men in our group were promoted more quickly than the women. I had to work harder, much harder to prove myself. At the end of the day, he gave me the promotion.

Some female leaders also noted that the construction industry is a physically demanding one to be in. One of them explained:

I have tried project management. Indeed, it is tougher for a woman to lead in managing a construction project. Being a woman, there is an issue when you have to be there with architects and other people on site. It is a lot more difficult to achieve in terms of physical ability. I had this vivid experience when I was pregnant. I had to climb up to the roof. It is physically challenging. You sometimes need men to pull you up some steps.

Processes during the polishing and practicing phase

Processes that were identified during the "polishing and practicing" phase of leadership development include:

- expansion of the support group;
- development of higher-order positive attributes;
- action-based reflection;
- transformation of identity (leading to self-awareness).

"Expansion of support group" is more pertinent to social development of the individual whereas the other three processes concern cognitive development. Details pertaining to "expansion of the support group" and "development of higher-order positive attributes" are discussed further in Chapters 7 and 8.

Expansion of the support group

During the "polishing and practicing" phase, individuals expand their support group, build relationships, and establish connections and form social networks with people around them, within and outside their organizations. They recruit more people into their support group and seek help from close as well as distant social educators. Leaders need to expand their support groups to achieve organizational success. They focus on developing relationships: within their own organizations, across various professions within the industry, and with the clients they serve. These relationships serve personal as well as organizational purposes for leaders.

Development of higher-order positive attributes

During the "polishing and practicing" phase of leadership development, as leaders go through substantial developmental experiences, an advanced form of complex social learning takes place, resulting in the development of complex and superior skills that are essential for leadership. These "higher-order positive attributes" include: social intelligence, self-confidence, positive self-image, self-drive, optimism, hope, courage, resilience, benevolence, creativity, and wisdom. Categories such as "learning by doing many things," "learning from seniors, bosses, and mentors," "molding self over time," and "personal initiatives for self-development" form part of the higher-order learning although the development of higher-order positive attributes is a more complex process which is intertwined with other categories such as "struggling with failures and challenges" and "various turning points."

Action-based reflection

During the analysis of the interviews, a number of memos were written that helped to perform higher-order abstraction of concepts and ideas. In one memo, it was noted:

> Reflection is a major process that takes place during all phases of leadership development. Leaders reflect and see how they went through different phases of life. They do not necessarily live their past. But they do remember the events and stories which formed their personal and professional lives. They also remember the lessons they learnt from such events. They relate to those people who were significant characters during the critical episodes of their early life. Such people eventually become a part of the support system of leaders and these episodes become guideposts for the leaders to follow. They learn the best ideas and necessary cautions of leadership from such critical episodes.
>
> A related issue about reflection is that the leaders always reflect upon their past to get more knowledge about themselves. They compare their circumstances with the critical episodes which serve as thresholds for them. Reflection also brings them back to conform to the values they stand for and believe in. Reflection about critical episodes reminds them of what is important in their lives.
>
> Each leader has a story about a critical episode in his or her life. Each of these episodes has dimensions such as: why it happened, who were significant players, what the individual did to overcome the situation, what were the intervening conditions and

what was the context, and finally, what were the succeeding happenings or consequences of such episodes. All these dimensions have a cognitive side as well. Leaders tend to imagine what works best, what should be done, what should not be done, who can be a source of guidance, and who can be trusted if similar or nearly similar episodes occur again. Authentic leaders do not just engage in reflection, they learn lessons and take appropriate actions to improve in the future. Therefore, it is "action-based reflection" that plays a key role in developing authentic leaders.

Successful leaders continually engage in reflection. They reflect upon their successes and failures and learn lessons. This part of tacit knowledge can only be used when leaders want to unlock it. They also transfer this tacit knowledge to their followers by telling stories, describing events, and relating to their past in many ways. This knowledge sometimes becomes explicit knowledge in the form of organizational stories, written metaphors, and so on.

During the reflection exercise, leaders analyze the reasons of both failures and successes. They analyze what failed them, what were the underlying reasons, how they can improve in the future, and what lessons are learnt. They also transfer these lessons to their followers. Similarly, they reflect on the successes and consider how things could be done better. What brought them success? What factors were missing and what risks were taken during the process? This constant reflection results in increased self-awareness leading to increased self-regulation. It is a key for developing and sustaining self-leadership.

Finally, most people do not do much reflection until there is a need. Leaders, when struck by critical episodes and trigger events, get the chance to transform the challenge into an opportunity for personal growth. Reflecting on an event also brings an altogether new perspective to the event. Some negative moments or events can eventually bring up to (or teach) the leader some life-long lessons which otherwise would not have been possible. Leaders can also make mistakes, but they learn from them.

The above memo was written as many of the leaders mentioned that they had a habit of reflecting. The general manager of a developer's firm noted: "All of us have a conscience. Sitting down and reflecting is the key. You have to undertake a continuous struggle to improve, and reflecting from time to time is a good practice."

A director in an architectural practice shared:

When something goes wrong, you then reflect. You reflect back on how things could have been done better. This is particularly relevant when people leave the firm. In the current industry, this is happening so much. It's peculiar to construction. When it happens, it is good to reflect and then you might think: "We could have done more to retain the person."

The CEO of a developer's company also discussed reflection:

There are a lot more events, success, failure. Things that happen to people around us and make us reflect how we could have done better to help them. I still think about whether I am adopting the right policies for our company and our people, and doing the right things. Right by people, right by God. It's a continuous thing.

These excerpts from the interviews show that action-based reflection is a valuable process in authentic leadership development. However, as noted above, leaders typically engage in such reflection when they face critical dilemmas or come across a defining episode in their lives. Such episodes can create an altered sense of identity that leads to a higher level of self-awareness.

Transformation of identity (leading to self-awareness)

"Transformation of identity" is another related process that takes place in a person's authentic leadership development. Crucibles, significant events, transformative events, critical episodes, and turning points stimulate the transformation of the leader's identity. These are followed by critical events and reflection on events in the person's early life which result in the transformation of personal identity, eventually leading to higher levels of self-awareness. Transformative events provide an opportunity for reflection and as individuals look back on such events, they realize their role in life and the goals they should be achieving. As noted above, "facing frustrations," "struggling with failures and challenges," "molding self over time," "various turning points in professional life," and "facing social challenges" are categories that mention events, incidents, and experiences that result in reflection and transformation of identity.

A senior manager related how his boss transformed his life:

> It was [through] learning with my boss. He actually changed me. I learnt from him the way he dealt with time, leadership and management. He had great problem-solving skills. To me, as a young engineer, that really affected me deeply.

A senior director described his experience which he considered a turning point:

> It was quite a good experience [abroad], having the opportunity to break out to do your own thing. It was like being thrown up there, alone, and growing up. I think that was a kind of turning point for me in life where I was thrown in at the deep end and I had to learn on my own.

A director in an architectural firm noted that the start of his professional career was a turning point: "When I started to work, I thought I had to take a responsible role. It was a turning point. It came when I started working … I thought I had to be responsible, much more responsible."

These excerpts show that leaders were able to identify certain points in their lives when they went through transformations of their identity. These turning points gave them a fresh look at life and an opportunity to become better individuals. As a result of the identity transformation, leaders are able to engage in more self-regulation and develop higher-order attributes during the "performing" phase of leadership that is discussed next.

Performing leadership

During the "performing" phase of leadership, authentic leaders are usually in a position where they are well settled in their career and are playing leadership roles in teams, in organizations, or in their professional communities. The "performing" phase is different from "polishing and practicing" in a sense that leaders are usually more mature and established in their character and leadership position. Sub-categories that emerged under the performing phase include:

- gaining confidence from success and achievements;
- leading through intense organizational and environmental changes;
- handling professional challenges.

Processes that take place during this phase of leadership development include:

- positive reframing of past events;
- self-regulation;
- self-transcendence.

Gaining confidence from success and achievements

It is not only the failures from which the leaders learn. They also mentioned many moments when they felt really good. These moments were mostly times of success or achievement, such as gaining a promotion, or successfully completing a challenging job. A former president of a professional organization described her experience:

> [W]hen I felt at my best was when I was elected as the president [of a professional body]. It was great. In the past, people were appointed to the presidency without any contest. But I went through a contest. People encouraged me to go for the top post. I was the first woman to be fighting for the presidency. My son told me to go for it too. During that weekend, the sermon in the service at church was also on the topic of leadership. I got 60 percent of the votes and I was very happy. I was president for three years.

The president of a major consulting firm considered her best moment to be when she assumed the position:

> When I became the president of this company! Previously, I was an executive vice president in charge of business. When I became the president, I could see that I could make a difference. I then found that I really could make a difference, in many ways.

The managing director of a consulting firm described a similar experience:

> I think the defining moment was when I was made the managing director and we were in the midst of a crisis. I felt that we had two options and they [the board] had said that if this does not work out in a year's time, you have to decide to shut down the office, so I took that as a challenge … things went well; we picked up a lot of big jobs very quickly.

A significant moment in life of the CEO of a developer's company was when he completed a challenging project in China:

> There was one occasion that brought tears to my eyes. It was a project. When no one believed that we could do it … in China … where a lot of developers have gone bankrupt, contractors have gone bankrupt and come back to Singapore crying. [At the

opening of sales of residential units] the moment I opened the door … the crowd rushed in. It was a huge sales gallery … the crowd was so enthusiastic and was so adamant on getting units in our development that the two salespersons we had were overwhelmed, and we couldn't conduct any sales for a while … I was standing there and letting the whole thing happen around me. I was saying, hey, we finally proved to people that we can do it.

A CEO mentioned that completing some major projects was a very fulfilling experience. He felt that he "was fortunate to be given the opportunity to take on major projects. It was very fulfilling and satisfying to complete those projects." A few leaders also shared about talents which they only realized they had when they were attempting to accomplish some things and succeeded. An interviewee shared: "When you succeed in something, I think sometimes you realize that you have the capability that you didn't see in yourself."

Leading through intense organizational and environmental changes

The interviewees related many events to organizational and environmental changes they had faced during their careers. Some shared about the fallout from recessions which Singapore went through in 1976, 1986, and 1998. Some shared about the corporatization of their companies. Others mentioned growth and geographical diversification as challenges and learning opportunities. Discussing the corporatization of his company, a senior executive of a large consulting firm, formerly a government agency, shared:

> When we became corporatized … we really had to change. And I was given more responsibility also at that time … from being a civil service organization, we had to become a private company. And so, the whole mentality changed, the business completely changed. I mean, previously we had projects whether we wanted them or not. We didn't worry about having no work … And, in fact, if nobody gave us any work, we would be very happy, because we were salaried. So many of the staff would try to work slow … and they'll be perfectionists … they try to do things over and over again to make things better … But then once you become a private company, time is of the essence, and money is of the essence.

A senior director in a quantity surveying firm described the Asian financial crisis of 1998 as a major challenge:

> The [other] major event would be the Asian financial crisis where I was given the task of choosing two officers in the exercise of retrenching staff regionally. That was really a tough one for me. It was very emotional as well, as you know. Business-wise, it was a very tough decision to make; and especially when I was responsible for building up the office in the first place. You have set it up and you have invested so much time and effort but also in terms of relationships with the staff over there. And then you have to tell one of them one day that his time was up. So, it was very tough for me.

The group chairman of a large consulting firm also mentioned the period of the Asian financial crisis as a time when he faced numerous challenges:

The 1998 financial crisis [was a big challenge]! … within six months, every country was flattened. And we could just see the problem coming our way but there was no way of preventing it or doing something to stop it, or even effectively react to it. So, every one of our offices was hit. We had to downsize in Vietnam, downsize in Indonesia, even the Singapore office was affected to some extent. So, that event in 1998 taught me that we have to have a structure so that we can respond to change quickly …

Handling professional challenges

The interviewees also described the professional challenges they faced during their careers. Tackling these challenges equipped them with the skills and knowledge required for effective leadership. The CEO of a developer's organization described one such professional challenge:

> There was a period in 1998 which is just in the midst of the financial crisis, and there was a lot of turmoil in the marketplace … There were a lot of market manipulators trying to destabilize the market with their trading as well as rumors. Our company is a very important part of the business ecosystem, even though we are not a listed company. A false report said our company was facing financial problems. If we had collapsed in debt, the market would have been completely unhinged. It would have created the impression that property prices would go down even further. The banking system would also be facing a serious problem. There was a lot of malice there. We had to come out very positively and affirmatively; to meet up with the media, banks, explain our position. If these rumors had gone on, unchecked, we would have lost the trust of many people, our business partners, and many other organizations. I was never prepared for this at all. But I had to deal with it. And dispel the false stories … It was an interesting point in my life as a businessman.

The executive director in an architectural practice found keeping up with developments in the industry quite hard:

> I think the biggest professional challenge is keeping up to date. I think the building industry has changed quite a lot and if we don't update ourselves, we are left behind. With the type of new buildings coming up, the new technologies they involve and the style with which the buildings are designed, everything has changed, and they continue to change. So, it is important; we have to maintain interest in renewing our knowledge, constantly.

Some interviewees had difficulties in dealing with people. One example was a senior executive in a developer's firm. He maintained:

> The greatest professional challenge is to manage professionals and other people. The biggest challenge is not how to design the best buildings, not how to design the most efficient and best engineering systems. The biggest challenge is to manage the expectations of the building professional and the people around him. You should not forget that you can motivate one leader, but that leader must also carry your ideas through and motivate his subordinates. You simply do not have the stamina to go down and motivate people three or four layers down. If you do that, you may demoralize their leaders. So, you must respect the leaders below you, and do the motivating at the appropriate level.

Processes during the performing phase

Positive reframing of past events

"Positive reframing" is a process in which authentic leaders positively reframe their past experiences as learning events. Leaders structure their life stories according to their self-belief, self-identity, and self-perception. Leaders continuously frame and reframe their life stories depending on the nature of those experiences. A leader's life story is shaped by the past leadership antecedents and it then determines how the leader experiences the future antecedents that subsequently arise. These experiences give meaning to how the leader develops particular behaviors. Simply put, much of leadership development is influenced by the way different individuals interpret certain events. Thus, leadership development is not fixed or stable. If individuals positively respond to these events (regardless of the nature of the events), the outcomes are likely to be positive in the form of self-development and the behavioral responses. On the other hand, if the leader responds to these events in a negative way, the consequences are likely to be negative in the form of negative leader-self-development, self-distortion, cognitive dissonance, inauthenticity, or negative behavioral responses.

A negative event may also result in positive reframing, depending upon how the event is perceived. It was found in the analysis that authentic leaders constantly reframe their past experiences into positive events and treat them as learning episodes. This positive reframing of past events – both positive and negative occurrences – helps the authentic leaders to remain positive about them and continue to grow as individuals. Positive reframing begins with recognizing any failures which occur. A vice president of a developer's firm noted:

> Lead by example in how you accept failure ... and if you can do something well today, you can do it better tomorrow. Success is not the end of the story. It is important to use it to gain enough confidence to pick yourself up in the next failure ... I believe there is no such thing as a failure, it's always just a setback to overcome before the next success.

Another senior executive noted that failures are part of leadership. However, they should not constrain the leader:

> For every failure, he [the leader] should take the failures and move on. I think failures are part and parcel of practicing leadership. So, move on ... To me, there's no point in crying over failures and no point in celebrating and being too comfortable over success. So, ... therefore, this may sound stressful for the people that work around you, but you just have to realize that you must run a never-ending race.

Due to positive reframing of past events, leaders are able to draw lessons for their future actions. A director in a consulting firm shared about an instance when he failed on a project:

> It taught me a lot of lessons. I recall that the reason was that I had distanced myself from the staff who were working on the project. Throughout the project, I had no feel where it was heading. The lesson I learnt was that I should not dilute my attention by taking on too many projects but balance it in spending the necessary time with each of my project teams.

Self-regulation

Authentic self-regulation takes place as a result of authentic self-awareness. Leaders, when fully aware of their purpose, conviction, values, desires, and aspirations, make a conscious effort to regulate their evaluation of self- and others-related information, behaviors, and relationships without any personal prejudice, or denial. Self-regulation is an inner-directed and others-focused process. They accept their strengths and their flaws. The regulation of behaviors engenders authentic behaviors in which authentic leaders act in accordance with their personal values and needs, and are true to themselves. They also achieve authenticity in their relationships by being more open, truthful, and honest. Self-regulation is explained in detail in Chapter 7.

Self-transcendence

Self-transcendence, again, is a function of self-awareness and results from increased self-related knowledge. However, as compared to self-regulation, self-transcendence refers to actions, behaviors, relationships, which are triggered by humility, selflessness, humanness, follower-focused strategy, collective good, customer service, and social responsibility. Therefore, self-transcendence, in a way, is the advanced form of self-regulation in which the focus is completely others; the "self" is not important. Self-transcendence results from the highest level of self-awareness (purpose, vision, goals, and so on) and takes the leader to the next level where he or she serves his people. Self-transcendence is also explained in detail in Chapter 7.

Systemic perspective

"Systemic perspective" emerged as a process in which leaders engage, usually during their "performance" phase of leadership. It refers to various sub-categories such as: all-encompassing success, conducting the orchestra, producing a play, the versatility of skills, seeing the bigger picture, and so on. During the performing phase, leaders look at leadership as a "whole sum" and overarching concept in which they are able to understand the objective, comprehend the context, set the direction, demonstrate strong leadership within and outside the organization, and produce a harmonious play of leadership. A further description of "systemic perspective" is given in Chapter 9, under "sustainable leadership."

Passing leadership

"Passing" refers to the phases when the leaders are nearing the end of their tenure, either in the form of retirement or by way of relinquishing an office. "Passing leadership" emerged as an advanced level of "performing leadership" during which leaders engage in succession planning and developing followers through mentoring and coaching. The leaders actively engage in two principal processes: meta-reflection; and role modeling.

Meta-reflection

"Meta-reflection" is a process through which authentic leaders are able to comprehend the whole picture of leadership by reflecting on their own experiences as well as the experiences of many others around them. They are able to reflect on various socio-economic, cultural, professional, and organizational influences on their leadership. They draw linkages among various happenings in their past and draw conclusions with regard to their leadership development, behavior, purpose, and conviction. Based on these linkages, they are also able to understand what they want to leave behind as their own leadership legacy. They are also able to draw a future vision for their organizations. Meta-reflection refers not only to remembering the past and learning from it; it is also about drawing links among various events and foreseeing the future in the light of their experiences.

Therefore, meta-reflection is essentially the superlative form of reflection which results in the development of a future vision in the light of experiences. Meta-reflection is a lens which leaders use to envision the future by making use of their past experiences. Take an example:

> Talking about my journey, the first day I started work, I looked at what is down the line and realized it could be 45 years of working life. So what is it that keeps me going after all these years? I think the underlying thing is the difference I … make in many situations and in the lives of so many people.

The chairman of a consulting group who was nearing retirement shared his thought about how he looked at leadership and how he was planning to hand over the leadership to the next generation of leaders:

> When you are in stewardship, that means you know you are holding the organization in trust for the owner; you really have to take care of it, preserve it, hand it on to the next steward. So we operate on a stewardship model in terms of ownership … we grow it and we grow our successors, grow the next three layers so that there are people to take over … we are already in the process of handing over … I am still around to sort of mentor and manage all the relationships and take care of problems where necessary. So for the last three years, in relation to my direct followers, I have been operating on the basis that "I will do what you cannot do, I will do what you can do but cannot see the need to do, and I will do what you can do, can see the need to do, but cannot bring yourself to do."

Take another example of how meta-reflection by leaders results in legacy building:

> It took me 30 years to learn that I needed to leave a legacy. As engineers, we need to learn to transfer the knowledge to other generations; help people and improve processes. Build templates based on this experience and knowledge … I don't want to make life difficult for the younger people.

It was noticed in the study that most leaders, particularly those nearing their retirement, are able to perform meta-reflection and are concerned about the future of their organizations. The chairman of a quantity surveying organization stressed the need to build up a sustainable leadership together with making efforts to enhance the capability and competitiveness of the firm to ensure its long-term growth:

> I think I would like to build a sustainable leadership that can last for decades. Building a better business and building better people is what I would like to achieve here. The greatest challenge would be to build up the leadership role of the company in the

region … to maintain what we are, to maintain the talent within the firm. Singapore is a very small country; there is no way we can grow or perhaps even survive by being in Singapore only. We have to go out into the region. I have maintained that others must grow with me. If you only look after yourself, both you and your organization are not going to last very long.

Meta-reflection also provides a view of what a leader can contribute to the organization. A director in a firm of architects was of the view that leaving individual legacies in professional corporate firms was not that easy due to the lean hierarchy and the prevalence of idiosyncrasies in the architectural profession. In such circumstances, if some legacies endured, they could damage the cutting edge of the firm. However, he emphasized the need for the establishment of systems within the firms to ensure their sustainability:

> In a practice where we have six directors, who are all very mature and level-headed, I do not think we can leave our own legacies in the firm. But what we can leave behind is a good structure of the organization, capable staff, and our design capability. After learning from experience and from challenges, we want to put a system in place so that, as a firm, we do not repeat the mistakes. I would like to be remembered that I left the firm in good shape.

Meta-reflection is also an important process which makes the leaders think about succession. Since authentic leaders are concerned about maintaining their organizations and succession of leadership, they build up the people who can succeed them. The president of a consulting group shared this perspective:

> I will be nearing my retirement. I enjoy my work. I am trying to have somebody else in this position. Leadership must know how and when to exit. I have to plan who can take over my role. I am also ready to reverse the role so that other people can have a chance to take on the role.

The chairman of a quantity surveying firm, who was also nearing his retirement, shared his viewpoint about how he wanted to see the leadership in his firm after he had retired. He said:

> I think very important also is that the leadership position should not disappear when you retire; that position must stay with the firm, with the new partners, or new management taking over. That, to me, is really leadership. That is my real philosophy in leadership—to help all around me, and to help myself in the process, but very important is that the leadership must be sustainable … When I retire, those in the management in the firm must retain the leadership, that to me is the true quality of leadership. So, all along, in my various trials, my passion, my working has been very driven by this philosophy. That's why the firm is strong as a whole, not only in Singapore, but all over Asia where we work.

Meta-reflection also helps authentic leaders to comprehend what they want to contribute to society.

Role modeling for followers

During the "passing" phase, authentic leaders are particularly focused on developing their immediate followers and the next generation of leaders who can succeed them. In order to build their followers as authentic followers and future leaders, authentic leaders perform role modeling. They also focus on tasks, such as mentoring, coaching, teaching, and leading by stories in the later phases of their career as they prepare to transfer the leadership to the next generation. "Authentic role modeling" is covered in detail in Chapter 7.

A framework for authentic leadership development

Figures 6.1, 6.2 and 6.3 integrate the above discussion in the form of a framework, referred to as the Grounded Model of Authentic Leadership Development (GMALD). Figure 6.1 presents a summary of the Grounded Model of Authentic Leadership Development (GMALD). It shows that authentic leadership development essentially takes place in various social institutions under the influence of people and experiences. "Social institutions" include family, educational institutions, professional organizations, and so on. "Support group" includes family members, close social educators, and distant social educators. "Experiences" range from early childhood difficulties to professional challenges (see Figure 6.3).

The influence of support groups, institutions, and experiences initiates some socio-cognitive processes at various phases of leadership development shown in Figure 6.1.Figure 6.2 shows the socio-cognitive processes that take place during various phases of authentic leadership development. Figure 6.3 details the experiences, people, institutions, and the resulting attributes that leaders develop at various phases of development. Figure 6.3 also presents various categories of experiences during the phases of leadership development.

The frameworks in Figures 6.1, 6.2 and 6.3 show that authentic leadership development is socio-cognitive in nature and comprises four major phases: preparing, polishing and practicing, performing, and passing. During these four phases, individuals go through various social experiences, and respond to these experiences by engaging in several socio-cognitive processes which lead to authentic leadership development. GMALD is further explained in the following sections.

The leadership support group

As shown in Figure 6.1, and discussed above, a leadership support group is an important category in authentic leadership development. Gardner et al. (2005) assert that one or more individuals who act as positive role models (such as a parent, teacher, sibling, coach, or mentor) are likely to serve as important antecedents of a leader's development. Several other researchers agree with the notion that certain characters in the lives of leaders play significant roles as antecedents of leadership (Bass, 1960; Gibbons, 1986; Popper et al., 2000; Zacharatos et al., 2000; Tesser, 2002; Luthans and Avolio, 2003; Avolio, 2005; Cooper et al., 2005; Gardner et al., 2005; Shamir et al., 2005). The current research has shown that the leadership support group usually involves significant individuals, including family members, close social educators (such as teachers, mentors, bosses, and colleagues), and distant social educators (such as role models, political leaders, social leaders, and famous professionals). As discussed above, the members of the leadership support group influence the leaders in many ways. The leaders gain inspiration, learn values, and seek emotional support from the members of the leadership support group at different phases of

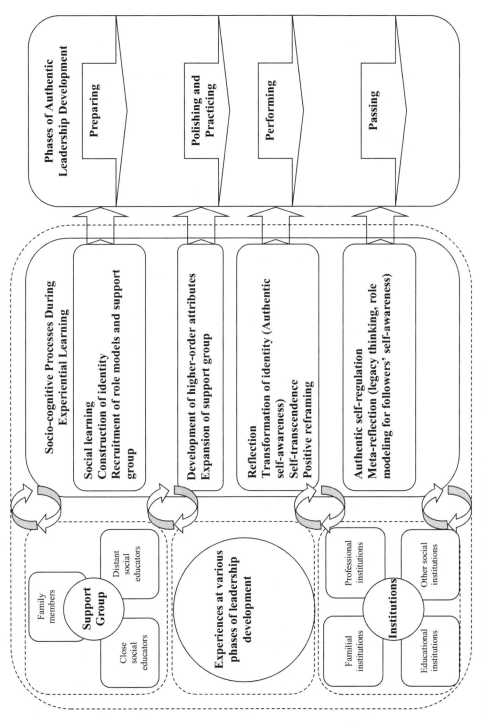

Figure 6.1 Grounded Model of Authentic Leadership Development (GMALD)

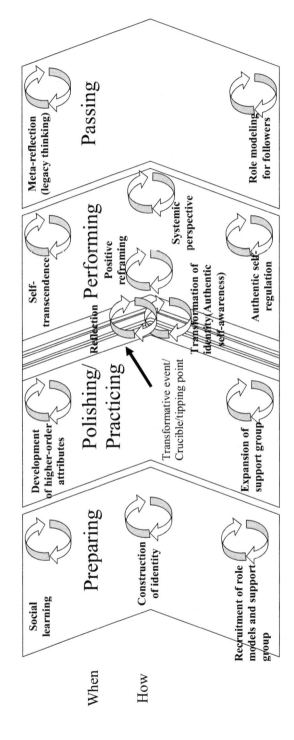

Figure 6.2 Four phases of authentic leadership development

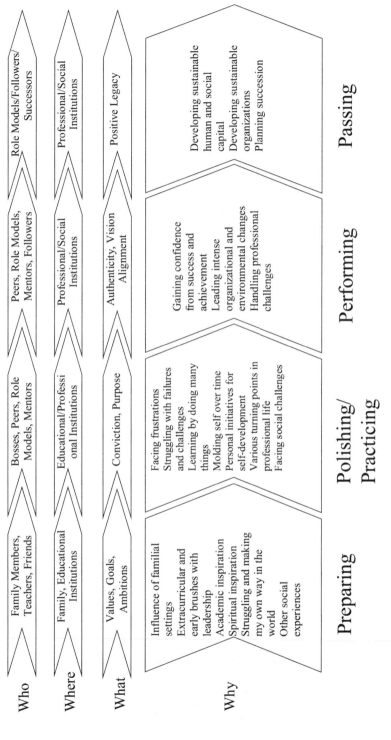

Figure 6.3 Influences in authentic leadership development

their leadership development. From this discussion on the leadership support group, some propositions are drawn.

> **Proposition 1**: Over the course of their authentic leadership development, individuals actively recruit role models who act as positive social forces.
>
> **Proposition 2**: Role models of authentic leaders are not always distant charismatic leaders; instead, they are family members, teachers, colleagues, bosses, coaches, and mentors.
>
> **Proposition 3**: Individuals belonging to the leadership support group are instrumental in developing values and inspiring purpose in authentic leaders, and provide them with emotional support during various phases of their development.

Social institutions

Research in sociology observes that a person's life course involves interlinking events, happenings, transitions, crossroads, crises, and defining moments in relation to a multiplicity of social institutions including family, marriage, education, employment, economy, religion, recreation, government, friendship, and other things with which a person may be involved (see, for example, Lopata and Levy, 2003). The current research suggests that the members of the leadership support group almost always belong to significant social institutions in a person's life. These institutions may include familial, educational, professional, and other social institutions (see Figures 6.1 and 6.2). Research shows that these institutions play a significant role in developing the individuals as social beings (see, for example, Elder and Rockwell, 1979; Lopata and Levy, 2003). Individuals' ideologies are formed over the life span and they are influenced by a complex array of the people and places in a leader's life. Socio-cultural and political perspectives and beliefs are deeply influenced by social institutions, particularly the home, academic institutions, and work organizations. Since leadership has been purported to be a social phenomenon (Parry, 1998; Yukl, 2002; Kan and Parry, 2004), and leadership development a socio-cognitive process (Harter, 1999; Chan, 2005), the role of social institutions is indispensable, as can be seen from the categories that emerged from the data.

> **Proposition 4**: Authentic leadership development is heavily influenced by the social institutions, such as family, educational institutions, and occupational organizations.

Experiential learning

The study of leadership antecedents is pertinent to attempts to explain leadership development and influence. Antecedents are usually perceived as dramatic and high-profile events (Cooper et al., 2005) that immediately precede certain behavioral changes. These stimuli may or may not be discriminative for a specific behavior. These events have conventionally been viewed as negative events though a modern perspective emphasizes positive growth events (see: Avolio and Luthans, 2006; Luthans et al., 2007).

Bennis and Thomas (2002) call such vital episodes of life "crucibles." They consider a crucible "a transformative experience through which an individual comes to a new or an altered sense of identity."[1] The crucible can be a trial and a test, a point of deep self-reflection that forces individuals to question who they were and what mattered to them in life. Hence, crucibles require individuals to examine their values, question their assumptions, and hone their judgment. Tesser (2002) argues that early experiences may be more difficult to undo than later ones; and traumatic events may leave an indelible mark on the individual's personality. Moreover, the consequential self is observed by selecting and molding situations so that they afford the possibility of self-expression. Aspects of the consequential self may be associated with specific genetic predispositions or they may be the result of particular experiences during a developmental period in which there is particular sensitivity; traumatic or prolonged stressful experiences; or, even the way in which the experiences have been processed (ibid.).

In this research, "experiential learning" refers to those life episodes, events, incidents, happenings, transitions, and occurrences which have been directly or indirectly caused by a single or a combination of people and places. There can be a large variety of significant experiences including, but not limited to, reading a book, facing a failure or success, studying in a particular institutions, losing someone special, serving the community, or adversity in childhood. Such experiences can be both positive or negative and lead to complex forms of experiential learning. They cause pleasure and fulfillment, on the one hand, and trauma and stress, on the other. These events may be under full, partial, or no control of leaders. However, their impact is significant, and brings variations or modifications in the behaviors of the individuals. Moreover, this impact can be short or long-lasting, depending on the nature of the experience.

This research has shown that significant experiences occur during various phases of leadership development. However, authentic leaders learn from these experiences and reframe them as learning episodes. The discussion shows that leadership developmental experiences play a significant role in shaping leaders' personality, niches, behaviors, values, motivations, goals, purpose, and convictions in life. The findings in the current research show that authentic leadership development is a complex process in which authentic leaders constantly respond to their social environment and engage in various social and cognitive processes while going through a variety of learning experiences.

Proposition 5: Significant experiences are directly or indirectly caused by a single or a combination of various significant individuals and social institutions.

Proposition 6: Significant experiences help in leadership development through increased self-awareness, such as self-assurance, self-clarity, and self-certainty, purpose, vision, and conviction for leadership.

Four phases of leadership development

Similar to the "seasons of life" (Erikson, 1963; Levinson, 1977; 1981; Levinson et al., 1978; Kotre and Hall, 1997) and the seven ages of a leader (Bennis, 2004), this grounded theory study of authentic leadership development shows that the life-span development of authentic leadership comprises four phases: preparing, polishing and practicing, performing, and passing. No clear boundaries exist between these phases; however, they are differentiated from each other by various experiences leaders go through and different socio-cognitive processes in which they engage during various phases. In line with the view of the "self" being both cognitive and a social construction (Harter, 1999), leadership development is significantly influenced by an individual's environment and the individual's response to the environment and the experiences of the individual. Active experiential learning from one's social environment plays a key role in authentic leadership development. Most theories of personality development break down the developmental path into stages. Most of these stage theories are progressive in nature, although in some, a person can fail to complete the stage while still continuing in the person's development. However, according to the theories, this failure to complete a stage will result in difficulties later in life.

Although there have been attempts to discover leadership development in the form of phases, these studies are either theoretical models (see Brungardt, 1996; Toor and Ofori, 2006b), or are based on populations who are students (see Komives et al., 2005; 2006). Moreover, those studies do not really account for the socio-cognitive processes that take place during various phases of leadership development. For example, a review by Brungardt (1996) clusters leadership development into: early childhood and adolescent development, the role of education, the role of on-job experiences, and specialized leadership education. Examining the leadership development of college students, Komives et al. (2006) also suggest six stages of development: awareness, exploration/engagement, leader identified, leadership differentiated, generativity, and integration/synthesis. However, this current study goes beyond identifying the stages of leadership development. In addition to developmental influences, it highlights the key processes that take place during the phases of leadership development. These processes are primarily an active and reactive response to the experiences which leaders go through while achieving authenticity within the self and operationalizing this authenticity within their social environments (see Figures: 6.1 and 6.2).

The phases of leadership development are more complex, intertwined, and cyclical in nature although they are usually presented in a linear form (Komives et al., 2005). Others argue that each stage of leadership development is influenced by several contextual factors related to both the individual and the surrounding environment (King, 1994). This complexity is due to various processes that take place during each phase of leadership development. The processes during various phases of leadership development are cumulative and cyclical in nature. It is difficult to identify where exactly a process begins or ends. Moreover, the processes are either proactive or reactive to certain events, incidents, and experiences. Hence, it is not the events, incidents, and experiences that result in leadership development but the response of individuals to them.

During the "preparing" phase, various social experiences include: influence of familial settings, extracurricular and early brush with leadership, academic inspiration, spiritual inspiration, early childhood and adolescence struggles, and various other social experiences. These experiences result in social processes such as social learning (Bandura, 1977; Sims and Manz, 1980, Manz and Sims, 1982; Shebilske et al., 1998; Ormrod, 1999), construction of identity (Erikson, 1968; 1980; Baumgardner, 1990; van Knippenberg and Hogg, 2003) and recruitment of support group (Sooklal, 1991; Gardner et al., 2005; Toor and Ofori, 2006b; Toor, 2006), as well as role models (Gardner et al., 2005; Toor and Ofori, 2008a).

During the polishing and practicing phase, leaders often go through a range of experiences that include facing frustration, struggling with failure, learning by doing many things, learning from seniors and bosses, molding self over time, personal initiatives for self-development, experiencing various turning points in professional life, and various social challenges (George and Sims, 2007; George et al., 2007). These experiences engender higher-order social learning that results in the development of higher-order attributes. During this phase, they actively engage in expanding their support group by developing and sustaining relationships. Two related processes that take place during this phase are: action-based reflection (Hixon and Swann, 1993; Lee and Hutchison, 1998; Rock and Schwartz, 2006) which is coupled with transformation; or altered sense of identity (Bennis and Thomas, 2002; Komives et al., 2005, 2006).

Authentic leaders engage in these processes in response to the critical episodes or transformative events that take place at some point during the polishing and practicing phase.

Action-based reflection results in heightened levels of self-awareness – awareness with personal values, identity, emotions, goals, and motives (see Luthans and Avolio, 2003; Goldman and Kernis, 2004; Shamir and Eilam, 2005; Ilies et al., 2005). They invent a higher purpose of leadership through encountering transformative experiences (Avolio and Gibbons, 1988) or crucibles (Bennis, 2007). This is similar to awakening from a trance of dominant practices, social validation, a separate self, and role scripts (Parameshwar, 2006) which stimulates the authentic-self within the leader. During "polishing and practicing," leaders also start to demonstrate "self-leadership" which is explained in Chapter 7. This takes the leaders to the next phase of leadership that is "performing."

Developing from transformative experience or the crucible, high self-awareness and anchoring to personal values, goals, and motives also lead to self-transcendence (Bass and Steidlmeier, 1999; Luthans and Avolio, 2003; Michie and Gooty, 2005) and self-regulation (Carver and Scheier, 1998; Sosik et al., 2002; Kernis and Goldman, 2005c; Kernis and Goldman, 2006; Walumbwa et al., 2008). Keeping to their purpose and underlying convictions, authentic leaders gradually become more inner-directed but others-focused – a combination of both self-leadership and self-transcendent leadership. They become more concerned about their followers, organizations, society, and the people they serve. During the "performing" phase of leadership development, authentic leaders master the action-based reflection, leading to more self-awareness and self-regulation.

Another process in which leaders engage is "positive reframing" of their past. Avolio and Luthans (2006) argue that leaders continuously frame and reframe their life stories depending on the nature of the experiences they go through. A leader's life story is shaped by the past leadership antecedents and it then shapes how the leader experiences the future antecedents that arise in the leader's life (Avolio and Luthans, 2006). These experiences give meaning as to how the leader develops a self-image and particular behaviors. By looking at their past, leaders are able to gain more knowledge about their values, ambitions, goals, vision, and purpose. Gaining knowledge about them helps the leaders to regulate their current behavior. Through this, they align their goals, ambitions, and purpose. Through this alignment, they are able to achieve authenticity in themselves and demonstrate it within their social settings.

> **Proposition 7**: Transformative experiences (defining episodes, crucibles, or tipping points) trigger higher levels of reflection that leads to superior levels of self-awareness in authentic leaders.
> **Proposition 8**: A combined effect of reflection and self-awareness leads to heightened levels of self-regulation in authentic leaders.
> **Proposition 9**: Heightened levels of self-awareness trigger self-transcendence in authentic leaders.
> **Proposition 10**: Authentic leaders continuously engage in positive framing and reframing of their life stories, leading to the development of positive self-concept and hence increased drive to achieve authenticity in their leadership.

While engaging in action-based reflection, positive reframing, and self-awareness, and self-regulation, leaders continuously manage the meaning within the self as well as their social environment through an ongoing interpretive process (Locke, 2007). This implies that leadership development is not fixed or stable; it is always in process, and is heavily influenced by the events in the life of individuals who frame and reframe their stories by building upon the positive occurrences and taking failures as learning experiences (Toor, 2006). During the "performing phase," authentic leaders practice "self-leadership" as well as "self-transcendent" leadership (explained in Chapter 7). Whereas self-leadership helps the leaders to lead their inner self and perform authentic self-regulation, self-transcendent leadership helps them to achieve alignment from their followers. Self-transcendent leadership plays an instrumental role for authentic leaders to sell their vision to their followers and engage them in achieving the common goals for the group or organization. During the performing phase, authentic leaders would already be beginning to engage in tasks that pertain to sustainable leadership. These include vigilant decision-making, creative problem-solving, building high-performance and sustainable teams, and initiating and leading change. However, sustainable leadership becomes fully functional during the final phase of leadership development.

The last, and the most refined, phase of leadership development is "passing." During this phase, leaders have reached a point where they look beyond themselves and are fully concerned about their followers, organizations, and society at large. They engage in three higher-order processes: meta-reflection – a superior form of reflection; systemic outlook (Orton and Weick, 1990; Flood and Jackson, 1991; Thompson, 1992; Edgeman and Scherer, 1999; Collier and Esteban, 2000); and positive role modeling (Avolio et al., 2004; Gardner et al., 2005; George and Sims, 2007). During this phase, they are more concerned about developing the next generation of leaders and building a future vision for their organizations. They also engage in sustaining their leadership by effective succession planning (Skipper and Bell, 2008), or passing on the mantle (Navin, 1971) so that leadership can live on after the leader has departed the scene (Kets de Vries, 2003). Leaders engage in several tasks to sustain their leadership. These include sustainable human and social capital development, sustainable change management, sustainable organizational development, and sustainable succession planning. These are discussed in detail in Chapter 9.

There are no definitive points during the life-long process of authentic leadership development at which leaders engage in certain processes or exhibit certain aspects of leadership. These points can come sooner or later in one's life depending upon environmental conditions and the individual's response to these conditions. Nevertheless, authentic leadership development does involve a continuous evolution and development of the individual at various phases of life. The response of the individual to inner changes and outside environmental conditions is in the form of a "dynamic reconciliation of self and social realities," which is a core category in this research. It transforms the ordinary individual into an authentic leader. This core category is discussed in Chapter 9 which integrates leadership development and leadership influence.

> **Proposition 11**: Incremental and cumulative reflection, positive reframing, and higher-order leadership attributes jointly lead to the development of a systemic perspective in authentic leaders.
> **Proposition 12**: Due to the incremental and cumulative effect of heightened self-awareness and self-regulation, authentic leaders engage in positive role modeling for their followers.
> **Proposition 13**: Due to the incremental and cumulative effect of various cognitive processes during earlier phases of leadership development, authentic leaders are able to undertake meta-reflection during the "passing" phase.

Comparison of IAMALD and GMALD

The IAMALD was presented in Chapter 4. It was primarily developed based on the variables identified in the extant literature. In the IAMALD, the main variables included significant individuals, significant social institutions, and significant experiences. The interplay of these variables resulted in the development of significant characteristics of leadership in the presence of various social forces and mediating variables. The major process that is proposed in the IAMALD is reframing. It is postulated that authentic leaders positively reframe their life stories for positive sense-making, leading to a positive self-concept. A comparison of IAMALD (Figure 4.1) with GMALD (Figure 6.1), which was developed from the qualitative study, shows some significant differences.

There are many socio-cognitive processes (social learning, construction of identity, recruitment of support group, expansion of support group, development of higher-order attributes, reflection, self-transcendence, self-regulation, meta-reflection, and role modeling) that are identified in GMALD that were not included in IAMALD. Moreover, the model which emerged from the qualitative analysis in this research departs from focusing on "significant characteristics" to focus on the socio-cognitive process that gives rise to authentic leadership development. Also, GMALD divides authentic leadership development into four distinct phases. GMALD also narrows down the classification of significant individuals and social institutions that was much broader in the IAMALD.

Summary

This chapter elaborates upon how authentic leaders develop over different phases of life. It outlines various phases of leadership development that have emerged from the qualitative analysis using the grounded theory approach. These are: preparing, polishing and practicing, performing, and passing. Authentic leaders undergo social experiences and respond to these experiences by engaging in socio-cognitive processes resulting in leadership development. The analysis in this chapter shows that authentic leadership development is a socio-cognitive construction and is influenced by various individuals, institutions, circumstances, and the response of the individuals to these social forces. Socio-cognitive processes that take place at different phases of leadership development include social learning, active recruitment of support group, construction of identity, development of higher-order attributes, expansion of support group, action-based reflection, transformation of identity, self-regulation, self-transcendence, meta-reflection, systemic outlook, and positive role modeling.

Grounding of the leadership development framework in the literature also shows that the model developed in this study is explicable and combines complex socio-cognitive processes to elaborate upon how authentic leaders develop. It shows that leaders make mindful choices with regard to their development as they encounter crucibles and defining moments that change their lives. However, they are fully committed to their purpose, conviction, and vision. This tenacity helps them emerge from the difficult circumstances and develop higher-order attributes that are necessary for authentic leadership. Authentic leaders are not perfect "heroic figures"; sometimes they fail in their personal and professional endeavors like everyone else. However, what makes them authentic leaders is their capacity to learn from their mistakes and failures and to reinvent themselves through self-awareness and self-regulation.

Note

1 See https://hbr.org/2002/09/crucibles-of-leadership

7   Self-leadership

Introduction

Leadership begins from the "self." An authentic leader is one who is deeply aware of the leader's values and life story and is able to align these with the purpose of leadership. An authentic leader engages in self-regulation through the leader's authentic behavior and in the relationships the leader forms and maintains. The leader engages in authentic role modeling and makes conscious efforts to grow in professional knowledge and competence. The leader also makes an effort to learn about the people in the organization and uses this knowledge to continuously hone his or her leadership approach. By engaging in inner-directed self-leadership, an authentic leader connects with the leader's inner self and follows a moral and ethical approach to leadership.

This chapter considers the processes that relate to self-leadership. An individual should be able to connect to the person's core self, in order to be able to move to the next step, that is self-transcendent leadership, that is explained in Chapter 8.

Leadership challenges (context)

The authentic leaders interviewed in the study described various challenges they faced in their working and personal lives. They ranged from broad industry-specific challenges to organizational and personal ones. The leaders also shared their views on the dynamism of the functional context in which they were performing the leadership role. Such accounts were useful in enabling one to understand the general, specific, historical, and distal contexts in which these leaders were living their lives and the complexity and dynamism of the jobs they were doing. It also helped to compare different authentic leaders across various contexts and ascertain how they were able to deal with a variety of situations they were facing.

Industry-specific challenges

The leaders interviewed noted a range of challenges they encounter in the context of the construction industry. These challenges can be categorized under the following themes:

- dynamism and change;
- the industry structure and environment;
- the difficult working conditions in construction;
- the industry's social image;
- the talent crisis.

The leaders invariably mentioned specific features of the construction industry which made their leadership tasks challenging. They expressed their views on the following topics:

- the dynamic nature of the construction industry;
- the fast pace of globalization and its implications;
- the prevailing growing demand for construction;
- the multiple stakeholders in construction;
- the bespoke nature of the projects;
- aspects related to technology and business.

One CEO commented on the fast-changing business world and technologies that were also having an influence on the construction industry which provides the physical facilities for business operations:

> I feel that the work is changing. What is good today may not be good tomorrow … previously, we used to have three months to design a building. Now … three weeks is very good and we are happy … we have to move faster to grab the opportunities.

Another CEO viewed globalization as a big challenge that the construction industry is also facing: "The construction industry is in the globalization game. We are already in many different countries. As we do that, you realize that you are facing competition from global players."

The construction industry was described by some interviewees as being complex and multi-faceted and posing particular challenges to leaders. A deputy CEO of a company noted, in the context of operating overseas:

> I think you have to perform at various levels of intellectual capacity. For construction, you deal with technology, law, accounting, finance … you must understand the working mechanisms operating on the ground in order to understand how a construction organization can survive in that operating environment … understand how the related government organizations work in conjunction with the construction businesses.

Some leaders spoke about the industry's structure, the operating environment, and the challenges. Mention was made of the following problems:

- lack of innovation in the industry;
- lack of professionalism;
- poor management practices;
- adherence by clients to traditional procurement approaches (which has many weaknesses; for example, it does not enable the industry to make the best use of the accumulated experience of some of the key players on projects);
- the adverse impact of economic cycles on construction businesses.

Issues pertinent to the social image of the industry were also brought up. Some leaders noted that the industry "does not pay well" while the others expressed concern about "increasing litigation in construction," and construction "not [being] a respectable profession." The executives also showed deep concerns about the construction's inability to attract and retain talent. One CEO noted: "One big issue I am finding now is that the construction industry is not really attracting the top people ... top graduates are opting for other occupations ... engineering is not their top priority now."

Emphasizing the need for a better social image of the industry, a director in a large consulting firm noted:

> As an industry, we need to improve our own profile. In other countries, engineers are well respected. The companies don't go through the vicious cycle of cutting costs to make money. That practice needs to be brought into our industry. Doctors and lawyers don't undercut each other. We do! We do not work professionally.

On lack of professionalism in the industry, another interviewee observed:

> In the construction industry, we should emphasize professionalism. We all have to play our roles well in our own different capacities. The client must pay well, the designer must design well, and the contractor must build well. This way, companies would not need to cut corners to complete projects they have won in fair competitions.

Discussing the construction industry's working conditions and environment, interviewees mentioned the "long working hours," "poor safety conditions," "uneducated and unskilled workforce," and construction being a "tough, dirty and dusty" industry. An executive outlined issues which she thought were "major put-offs" for women:

> In Singapore's construction industry, you deal with workers who tend to have less education and who tend to be rougher. We deal with men most of the time. The working environment is tough; the site is dusty, dirty, and so on. Therefore, we also tend to dress accordingly, which might drive some women away.

Contextual challenges

The leaders mentioned many contextual challenges that they faced as they performed their leadership role. These mostly pertained to cross-cultural issues, dealing with different situations, and dealing with several stakeholders. "Understanding the context" emerged as a higher-order attribute; the leaders noted that contextual challenges were ever present as they performed their functions. One interviewee shared: "I am dealing with different situations all the time; with different clients; with different consultants from around the world; and with different subcontractors and suppliers."

Another leader mentioned the many levels at which he had to deal with people and systems. This multifaceted context was a major challenge:

> You have to engage people constantly. If you are at a higher level, you will be playing a strategic role and you have people who you can give tasks to undertake, together with some ideas they should apply ... At different stages of the projects you play various roles ... In some cases you may have to be hands-on and join in discussions of construction problems, and draw out sketches of possible solutions. You need to revisit some ideas constantly. In such a varied context, you also have to believe that things can go wrong at any time ... and that you may have to step in ... at short notice.

People challenges

Understanding people was mentioned by the interviewees as an important factor. Whereas the leaders outlined their strengths of leading people and garnering their support, many noted that they found people management a big and constant challenge. Communicating with people, motivating them to serve clients, growing them for leadership roles in future, dealing with the issues of lack of competent followers, and non-performing employees were among the issues the leaders highlighted. Many considered "understanding humans and their behavior" as a constant challenge for their leadership. The CEO of a consulting firm expressed these views: "Every member of staff is different and each has different strengths and weaknesses. I now make an effort to understand my staff's strengths and weaknesses so that I know what to assign to them and when to help them."

Another executive shared that he had difficulty in growing his followers as leaders:

> Bringing up people to be leaders is the most difficult part ... And then asking them to teach the people under them. That is my responsibility. However, I myself have diffi-culty in teaching the persons below me. That is one of the biggest problems.

One CEO shared that managing teams was a big challenge and there was no fixed or for-mulaic way by which this challenge can be addressed and overcome:

> Every project is different in its nature and complexity; and even the team you have to work with also changes from project to project. A team is not a hardware; managing the team involves the whole complex process of getting different groups of people to deliver different aspects throughout all the stages, from design to completion.

A senior executive shared his perspective on the challenges of leadership in construction:

> There are always challenges ... about communicating with your staff and understanding their needs. Sometimes you do have to balance different styles of leadership ... One would be the carrot and stick approach which is more the management way of working. And if you don't have to use the carrot and stick approach, if your staff understand you ... the vision that you have for achieving your goals and they are ready to fulfill that vision, then you probably are a successful leader.

Describing the challenge of understanding human beings, a vice president observed:

> Getting people to do more than they think they can [is a challenge]. If a person can do more, you can feel it, and you can perceive a situation in which the person will do better. There are different people. Some people need to be cajoled, others need to be encouraged, [some] others just need to be told. You have to find different things that move people and motivate them.

Organizational challenges

The leaders discussed organizational issues they frequently faced and had to resolve as part of their leadership role. One CEO noted:

There are times when you have team members who have problems and you cannot help them. You find that this team member performs very well and is capable but you can't reward him because it is not within your control, or would be against the company's current norms. You feel that you are pressed so much, but certain things are beyond your control. I lost one of the team members because of such a situation. It was simply because of my inability to influence his remuneration.

A CEO running a family business raised the issue of decision-making. He mentioned that his subordinates were not brave enough to take big decisions as they did not feel they had any ownership in the business. Although he tried to give them the freedom to decide and to act on all appropriate issues relevant to their particular levels of seniority, the approach was still not working. He explained:

If the leader is very decisive, the subordinates may keep waiting for the decisions to come from the top and not want to take the decisions themselves. In understanding the context of my own company … I have to accept that the people who come here as professionals and managers feel that they are not owners and they are afraid to take decisions, as some decisions may cost the client or the firm millions of dollars. They are afraid to make decisions as they don't want to do the wrong thing. And we have to understand this as owners, that it is not that they are shirking the responsibility. Often, it is a big thing, and they are not comfortable with taking the decisions.

Another CEO expressed his concern about losing personal, individual contact with the employees in his firm because the firm had grown considerably in size over time. He shared: "As the firm grows big, I lose my personal touch with my senior people … at the moment, we have 120 staff. There is no way I can know everybody outside my immediate office and team." This makes clear the need to develop appropriate leadership approaches and corporate organizational structures as the company grows.

Personal challenges (weaknesses and uncertainties)

The leaders interviewed in this study were elected as leaders in their professions because of the reputation they had among their fellow professionals, or were nominated by those who knew them in the industry as authentic leaders. Internal triangulation of interviews also supported the conclusion that the interviewees were being authentic in their responses to various questions asked. They discussed their weaknesses and vulnerabilities with frankness, and did not show signs of social desirability which researchers might encounter in such situations. It was also evident that they were not indulging in self-promotion or self-praise. The leaders were not only cognizant of their weaknesses, but, from their narratives, they were constantly working toward overcoming them. A CEO related her perspective:

I started as an engineer. My weakness is that I may be a little bit too hands-on. If I find that things are not properly done, if the team does not do certain parts of the job properly, I end up doing some of their work. You must have heard of the monkey bag. I sometimes carry it.

Another CEO stressed the delegation of work:

> Sometimes I do take on too many tasks. I used to start work at 7.30 a.m. and did not leave the office until after 7.30 p.m. However, work would continue to pile up. I realized that I had to do proper delegation. Also, I now have what I call "block off days" or "block off periods" where I am not to be disturbed and I use these blocks of time to catch up on my work and also take up my more strategic tasks and decision-making.

A senior executive shared that his weakness was that, sometimes, he would lose his temper, but he was working to overcome it:

> I consider myself as a calm leader. But there are instances where during the process of passing down instructions and guidance, I noticed that there were some team members who are a bit slower than others, and I was a bit harsh on them. I realize that it is not good. I am dissatisfied with that part of me. I needed to tackle my temper and treat all the members of my team better.

A vice president shared that she was not really good at emotional quotient. However, she had been working hard on it, and was improving gradually:

> My EQ is not good. Although it has improved quite a lot, I think it is still a weakness. I do not feel at ease talking to a stranger. I don't feel at ease talking to my bosses even. By nature, I am not good at talking.

The CEO of an engineering consultancy also noted the following:

> All of us have weaknesses. I think in the early years, I was too hands-on and I wanted to do too much and I did not have trust in the abilities of the members of my teams. I could not let go of things enough. That prevented the people from being given the exposure and the opportunity to learn. That may have alienated some people and it is possible that we lost some good people because of that. That is often a weakness of certain CEOs and their senior people, particularly those in places like mine; a difficulty with delegating tasks and decision-making.

A CEO who ran a family company accepted that he had no idea of how it feels to be an employee. He expressed this:

> I am working on that. I think one of my weaknesses is that I have not worked for any other organizations. I don't have this real understanding of the aspirations and anxieties of my people, the lower executives. I don't have a strong enough appreciation as I have not been there, where they stand today. In trying to frame policies and strategies to meet the corporate objectives, I regard this as a weakness.

The above challenges provide a variety of contexts in which the authentic leaders interviewed in this study performed their leadership. These challenges demonstrate that leadership is complex and leaders are part of the broader systems within which they lead. Leaders face a range of challenges which together shape a wide context for their leadership.

Social challenges

Social challenges were typically mentioned by female leaders who participated in the study. These pertained to social discrimination that the female leaders faced in the construction industry: challenges related to the tough working conditions and environment of the construction industry, gender-based discrimination, and the social image of the industry.

Open and axial coding of the interviewees' responses generated three major categories that illustrate how authentic leaders influence their followers, organizations, business partners, and external stakeholders. These categories are referred to here as leadership processes under authentic leadership. The processes noted in Table 7.1 are: self-leadership, self-transcendent leadership, and sustainable leadership. Table 7.1 presents these major categories and the sub-categories that emerged from the data.

Self-leadership: leaders' features

Self-leadership emerged as a major category when the leaders discussed how they gained knowledge about themselves by understanding their own stories and engaging in critical reflection after they had faced transformative events, crucibles which formed their leadership, or significant experiences in their lives. Owing to their increased self-knowledge (self-concept, self-identity, self-worth, self-presentation, self-efficacy, and so on), the leaders consciously employ self-regulation to achieve their possible authentic self. "Self-leadership" primarily emphasizes "leading oneself" through increased self-awareness, self-regulation, authentic role modeling, and self-growth. Table 7.2 presents the sub-categories under self-leadership which emerged in the analysis.

Authentic self-awareness

Self-awareness was the most frequently mentioned issue by the leaders interviewed in this study; many of them mentioned that knowing oneself was critical for leading others. Therefore, a major finding is that leadership begins from self-awareness. The following sub-categories emerged under the broader category of self-awareness:

- underlying conviction for leadership;
- understanding the leadership purpose through the life story;

Table 7.1 Major categories under authentic leadership influence

Self-leadership	Self-transcendent leadership	Sustainable leadership
Authentic self-awareness	Soulful leadership	Vigilant outlook
Authentic self-regulation	Servant leadership	Creative problem-solving
Authentic role modeling	Spiritual leadership	Building high-performance and
Self-growth and competence	Shared leadership	sustainable teams
Higher-order positive attributes	Service-oriented leadership	Sustainable human capital
	Socially-responsible	development
	leadership	Sustainable social capital
		development
		Leading and sustaining change
		Building sustainable
		organizations
		Sustained succession planning

Table 7.2 Sub-categories under "self-leadership"

Category	Self-leadership				
	Authentic self-awareness	Authentic self-regulation	Authentic role modeling	Self-growth and competence	Higher-order attributes
Middle-level categories					
Lower-level categories	Underlying conviction for leadership Understanding leadership purpose through the life story Aligning ambitions and goals with the leadership purpose Aligning personal legacy with the leadership purpose Emphasis on personal values	Impartial self-evaluation and evaluation of others Authentic behavior Authentic relationships	Walking the talk Demonstrating self-discipline Mentoring, teaching, and guiding Leading by stories	Continuous personal growth Building positive rapport Professional commitment Professional knowledge and competence	Social intelligence Understanding people Understanding context Self-confidence Positive self-image Self-drive Optimism Hope Courage Resilience Benevolence Creativity Wisdom

- aligning ambitions and goals with the leadership purpose;
- aligning personal legacy with the leadership purpose;
- putting emphasis on personal values.

Underlying conviction for leadership

"Underlying conviction for leadership" emerged in response to the question: "Do leaders develop by chance or by choice (are they born or made)?" The leaders shared in depth their dreams and underlying inspirations that were instrumental in stimulating leadership in them. The leaders took up leadership roles because they chose to lead. Although in a few cases leadership came to them by chance (such as through family ownership of the business), these leaders had a choice to either accept it or do otherwise. In most of the stories the leaders told about their leadership history and inspiration, they indicated that they had convictions which drove them to step up and lead. The life stories of the leaders and their reflections on their past had given them dreams which they wanted to achieve through their leadership.

The majority of the interviewees noted that leadership, to them, was a matter of "choice" and that it could be "developed." They viewed leadership as a quality that everyone possesses to some extent. For example, one executive noted that:

> There is an opportunity in every field. And everyone can be a leader in their own right in their own way. But they need to be there to take up the opportunity, and be bold enough to pursue their dreams. I would say, "Don't worry about failure, don't be disappointed by it if it does happen, and don't worry about people laughing at you because you have failed." Then I would add that, "You should learn from your failure; you will become a better leader if you do."

A senior executive in an architectural practice opined along the same lines:

> It's a matter of willingness. Everyone has a certain endowment of leadership in them. Then it depends upon whether you are willing to go the extra mile. It very much depends on personal initiative and the willingness of the person … it is fundamental: to be a leader, you have to be willing to go beyond your comfort level. If you are not willing to try … you cannot go far from there if you are in your comfort corner.

Some interviewees noted that one should be sufficiently capable to anticipate, spot, and seize emerging opportunities. The CEO of a consulting firm noted:

> Good leaders are able to seize the opportunity. If you put someone with no leadership qualities in a position even at a good and opportune time, he won't be able to take that opportunity. So, I think that's the difference really. Leaders are able to see and seize the opportunity. And they must have those basic ingredients that are necessary for leadership … They say that opportunities come every one's way, but only smart people can smell them, seize them, and take advantage of them.

When asked whether leaders are born or made, a leader shared that in her view: "everyone is born equal. It's a matter of exposing yourself to opportunities." A CEO noted that leaders are developed by choice and they can create opportunities. She went on: "By choice … you can create a chance … I think it is totally by choice." A CEO noted that, in his case, taking up leadership was a choice:

> I say that success is never an accident; you have to work for it. You expect that success in a lottery can be an accident and anyone can be lucky enough to strike it big in a lottery but not in a profession, not in a business like this, not in construction. [Here] you have to work for it.

To another CEO, leadership is always by choice and one's potential can be measured by one's attitude and aptitude:

> Even when I was a student, I always aspired one day to be a CEO or MD. I also believe that if you think you can, you will. In my view, leaders are developed by choice. Through the performance, attitude, aptitude of the individual, you can tell the potential of the individual [to be a leader].

Similarly, the general manager of a contracting firm shared that he had always wanted to be a leader and had taken part in many preparatory extracurricular activities in his youth:

> I was involved in many societies in engineering. And I was doing many things … From my school days, I had the yearning to become a leader. My classmates would say that I was involved in everything under the sun. I think that prepared me well.

These responses illustrate that "underlying conviction" played a key role in developing these individuals as leaders. For the interviewees, leadership is a matter of choice and not merely chance, although the importance of chance and opportunity cannot be entirely denied.

Understanding the leadership purpose through the life story

Getting to know one's authentic self is not an easy process. Self-awareness may come in hard ways. When an individual faces a critical event and takes time to reflect upon their past, the person will get to know what they really stand for and what they want to achieve in their lives.

The authentic leaders interviewed consistently mentioned that they struggled to understand their life stories in order to gain higher levels of self-awareness. They constantly tried to understand the events that had shaped them and why they had come to be the leaders they were. For example, a CEO shared that it was her childhood story that had subsequently shaped her entire life:

> It is the hardship that I went through during my childhood. I always believe that we have to work out our own path in life. The guiding statement of my life that I remember all the time is that "Buddha merely shows the way, we have to work on our own path" … we have to help ourselves.

Similarly, a CEO shared that his early life had a great influence on him. Understanding his life story enabled him to determine how he was going to lead his company:

> my early life is something on which I am often very reflective … there were struggles, difficulties … but I remind myself that I was lucky to be here [in this firm] and that I

had to prove myself to my brother and father who were my personal mentors. That is what has kept me going.

Another senior executive stated that his early life and past events formed a lens through which he considered matters. It had a significant influence on his perspectives. He observed:

> That is how I view things. And therefore ... I have an origin that is pretty hard and low. Through my years, whatever I encounter and have to deal with, and however my life has changed ... the early years ... that is always at the back of my mind ... that's where I came from. Even today, I must say that the impact on me has been deep. Although I drive an air-conditioned car, and spend my days in an air-conditioned office, I still cannot sleep in an air-conditioned room; I can never get used to it.

Another leader explained how his leadership was influenced by events in his past:

> Actually, this relates to leadership. It depends on how you grow up, whether you always aspired to be what you are not ... I did not come from a rich family, I went to study in UK on a scholarship ... I'm a very determined person. I always want to be the best in everything that I do. This personal drive has paid off. For example, I was the top student in the whole university. It has got me to where I am.

These stories show that different leaders anchor their life stories to different events of their past. Understanding his or her life story helps each of them to understand what they want to do in life and what they want to achieve in their leadership role. Past events and critical episodes also shaped the values, goals, purpose, and desired legacies of the leaders.

Aligning ambitions and goals with the leadership purpose

The analysis showed that the authentic leaders in this study were able to align their ambitions and goals with their leadership purpose. This alignment helped them to perform better and avoid any internal conflict. Sharing his desire to become a better professional in his own field, a senior executive noted that: "My strongest desire ... is to be a really, really good designer. And I think I am a good designer but I don't think I am as good as I could be."

In many of the interviews, the authentic leaders expressed their desire to bring improvements into their organizations and the construction industry at large. They wanted to grow sustainable firms which would be well recognized in the industry as doing good work. They also wanted to influence the professional and industry practices in a positive way. Describing his purpose for leadership, the CEO of an organization shared that: "If I have to say what I want to achieve out of my leadership, I always wanted to make this a global company, and I think we have done that, to a very large extent."

Another leader shared his purpose of leadership:

> I would like my current organization to achieve success at a certain international level to be able to be regarded as somebody on the world arena. That is something I strive for, or, to me, it is worth striving for. I don't want this practice to remain a local firm.

A director in a consulting practice noted his purpose as:

> My motivation is to see that my team or my organization is forever hungry to improve themselves and to attain the next level of achievement … Every day, we want to be one step further and farther than where we were yesterday.

Many of the leaders stated that they wanted to grow their followers and develop them into future leaders by mentoring, teaching, and guiding them. The CEO of a quantity surveying firm shared what he considered to be his greatest desire:

> [I want] to have a few more like me. I am not saying that I am the best, I am not saying that I am good … I want people to have this caring attitude for the clients and for the projects they are in charge of. Clients always tell me, "We give you projects not because you are the best QS, or because you are the cheapest QS. But because we know, be it rain or shine, you will still be around to finish the job." I have, since then, often told our employees this story of why people give business to us … I think it's a matter of the trust that we have built up over the years. We have to continue to build on it.

A CEO remarked that he wanted to develop his followers so that they can take up leadership roles in the future. He noted:

> I have reached a stage where I hope to mentor my second liners and third liners, and hope that, as an organization, I would have succession plans and allow second-liner, third-liner professionals or project managers to take over to continue to let this organization run and grow. So, I think my emphasis on mentoring, my emphasis on looking at skill development, is very critical in our leadership succession.

The authentic leaders mostly related their ambitions and leadership goals to their organizations and followers. They wanted to contribute to others' lives, their organizations, and to society. They were driven by a desire to improve the lives of their team members and make a difference in their organizations, the industry, and wider society. Their goals and purposes were others-centered and hence were self-transcendent in nature. The leaders were also able to align their personal ambitions and goals with their leadership purpose. This harmony gave them a consistency and constancy in their actions and behaviors.

Aligning personal legacy with leadership purpose

Thinking about, and working toward, leaving a positive legacy also emerged as a sub-category under self-awareness when the leaders shared about the legacies they wanted to leave behind. Many noted that it had taken them several years to really know what they wanted to leave as their legacy. The legacies the leaders were aspiring to leave behind were deeply influenced by their past. They were able to align the legacies they aspired to leave behind with the leadership purpose they had found through their life stories. For example, describing the difficulties, challenges, and problems he had encountered in his life, a senior executive revealed that he had been through a hard time in the past and he did not want it to happen to others, especially to his followers. He wanted to create a system which could help to make life easier for others. He observed that:

It took me thirty years to learn that I needed to leave a legacy. As engineers, we need to learn to transfer the knowledge to other generations … I don't want to make life difficult for the younger people.

The legacies the authentic leaders want to leave were mostly related to their followers, organizations, and society. It was not a case of seeking glory for themselves as individuals. For example, the managing director of a contracting firm noted: "I want to feel that I have prepared my staff to be able pick up knowledge of the field and to see the big picture."

An executive from a consulting organization expressed her desire for the well-being of her followers to be enhanced. She noted:

I want to see more happy workers in my organization, really … people who are happy to work as hard as they can. And who do not just look at the pay scale and the dollars but those who like their work. You have to reward them appropriately … Also, give them recognition and freedom at work so that they can stay with you in the times when the firm can only offer them less than the expected reward.

Some of the interviewees had not previously thought about the kind of legacies they wanted to leave behind. However, most of the leaders did have a clear idea of the legacy they wanted to leave behind. The subjects of these legacies varied. Some of them mentioned how they wanted to be remembered after they left the company while others were keen concerning the impact of their leadership. Most of the responses affirmed the loyalty of the leaders to their professions. A CEO of an engineering design firm noted:

I would like to be remembered as a practical engineer. Someone who can work well with all parties involved in the project. I would like to be remembered as someone who has a heart for people and someone who can make a difference in their lives.

Similar sentiments were expressed by a senior executive from an architectural design firm, who noted:

I just want to be known as a good and fair architect. There are very few of us from this minority group holding key positions in architects' firms in Singapore. And I wish to leave a good impression on others. That can correct some unfair prototypes. Also, I want to be a good mentor, a good listener.

An executive related a story:

I would like people to say that I was a great boss. I always remember my favorite boss's farewell. All the staff in the firm cried and that was an incredible scene. I saw all the men crying; as you know, that is not common here [in this country]. Personally, I was also very upset because he was the one who gave me the break to be a senior engineer and then a partner. I thought the open show of emotion was the best compliment one could give to a boss. If you achieve that, you have done a great job. I would like to see the same thing happening in my case when I retire.

These statements on legacy show that the authentic leaders in this study wanted to relate their legacies to their professions and leadership purpose. Although they had reached the levels in positional terms in their firms where they were playing leadership roles, they still considered themselves as architects, civil engineers, mechanical engineers, project managers, or quantity surveyors, and wanted to be remembered as being good in their particular professional firms at that. It can also be seen that there is a close congruence between the goals, ambitions, purpose, and desired legacies of the leaders. It can be observed that authentic leaders are others-centered and self-transcendent. They are not motivated by extrinsic motivations. Instead, they take pleasure in developing others and growing their organizations. They are committed to contributing to society and making a difference wherever they can.

Emphasis on personal values

The next level of self-awareness is to know and understand one's values. Knowing one's values not only helps in anchoring the leadership principles, but also provides a clear direction in which the leader wants to go with his or her followers. Authentic leaders know, understand, and follow their values in their professional work and private life. Some of the essential values for authentic leadership that the leaders mentioned are now discussed (also see Table 7.1).

Emphasizing personal values, one interviewee noted that: "The first and foremost thing is integrity and honesty. You have to have a sense of integrity for people to believe you … then, tenacity and perseverance." The CEO of a large consulting firm emphasized: "I think humility is very important … maybe, another way of putting it is human understanding … Humility is a fundamental part of a leader." Another senior executive underscored the importance of empathy:

> You must be able to have a certain amount of empathy as well … you must be a person who respects people … not one who is known as being nasty. If you get the work done but you make people miserable, it is not right.

Several interviewees noted that values were important for a leader. However, even more important was consistency in the values. The CEO of a developer's firm noted:

> Consistency … people should know your value system. We have to overcome the internal conflict which can occur in the company … thoughts among some of the staff that if others are not working, why should I be working? As a leader, you have to mobilize people. In a crisis situation, you need to be responsible. Learning to be calm in a chaotic environment is important. Consistency in living your value system is the key in this.

Although the interviewees emphasized the personal values of the leader, they also high-lighted the vitality of organizational or group values for leadership. They also stressed the alignment between individual values and organizational values for authentic leadership. For example, the CEO of a designer firm noted:

> The values we have are important: the sense of integrity, sense of the market and sense of value. People must believe you in order to work for you. If they don't believe you, and appreciate your values and your firm's values, they won't come in to work for you. We would not be able to attract the best architects.

Another CEO shared that he wanted to build a durable organization that stood for something worthwhile. He shared that values such as diligence, excellence in the business, integrity, loyalty, and unity were fundamental to his organization, noting that:

> We think that these values reflect our vision. We want our people to embrace these values. Clearly, in this age of social consciousness, and deeper corporate conscience, we have to be an ethical business. To me, it is not only because it's good for business, but also, it is because it is the right thing to do.

Similarly, describing his company's values, the chairman of a consulting group noted that, "We have these core values in our system: trust and integrity, perseverance, teamwork and relationships."

The discussion shows that authentic leaders have positive and solid values in life and they endeavor to influence the development of appropriate organizational values. They endeavor to develop adherence to these values among the employees. They align their personal values with those of their organizations. This alignment helps them to practice their authenticity in life and in their organizations.

Authentic self-regulation

Authentic self-regulation is a process in which authentic leaders watch their behavior and relationships and evaluate information relating to themselves as people. They do not deny, distort, or exaggerate information about themselves. Instead, they endeavor to understand it so that they can take appropriate measures to improve their own behavior as well as their relationships with others. Therefore, self-regulation can be further divided into three categories: impartial self-evaluation and evaluation of others, authentic behavior, and authentic relationships.

Impartial self-evaluation and evaluation of others

The first requisite of authentic self-regulation is impartial self-evaluation and evaluation of others. This refers to one's understanding of one's strengths and weaknesses and unbiased interpretation of information about the organization, oneself, and others. The leaders interviewed in this study referred to impartial self-evaluation and evaluation of others. For example, describing the importance of being impartial, a CEO noted:

> I have learnt that I am not necessarily right all the time and that others are not wrong every time. I listen to people. Yes, that is important! When I do the annual performance review of my company, I will let the managers do the performance evaluation of their units themselves. I will then speak to my employees openly [at a general staff meeting] and I talk to them about the problems as well as the achievements. I will also highlight the areas where we can make changes along the way.

The CEO also emphasized that:

> You must feel that what you say is important. In the way you deal with your colleagues and do their performance assessment … If you want the whole group to grow, you must really feel that it is important and not just because you say it. When you talk to the person in his or her performance assessment, you must really want to help him or her.

Another senior executive noted that the leader has to put the leader's personal ego aside in order to be fair and just to others. He noted that:

> [S]ometimes our ego gets the better of ourselves. Nothing is wrong with having an ego, but one has to learn how to temper the ego for the right cause, and how to do this in order to achieve the desired results.

Appreciating personal strengths, weaknesses, and limitations was also raised by the leaders in the context of impartial self-evaluation and evaluation of others. One young leader interviewed revealed:

> I am just 32 years old. There is still a lot I have to go through … So, I am always humble because I realize that there are people out there in the industry that I learn from. Sometimes, there are clients or consultants who are really *nasty*, but you have to be calm in such situations and support and encourage your team. You must then learn the appropriate lessons from what happened and what worked in that situation.

Authentic behavior

Authentic behavior refers to actions which are in accordance with one's values and motivations. Authentic leaders do not act merely to please others. Their behavior reflects their vision, motivations, aspirations, and values. Authentic leaders are true to themselves and others, believe in what they do, behave in accordance with their values, and are consistent in their behavior. They do not have a say-do-gap and they do what they preach to others. They are not afraid of saying things as they are. They have no "losing face" issues. They openly communicate with others. Authentic leaders are not necessarily charismatic. Instead, they are selfless, and they do not copy others' behavior to gain social approval. A CEO interviewed in this study described his authentic behavior:

> To me, it is really honesty … Be honest with my feelings when I talk to my staff … I would tell them exactly how I feel. If I'm not happy, I'll tell them. Because, actually, in such a situation, it's not them that I'm resenting, it's their actions. I don't blame them but try to see how we can avoid similar situations [if things have not gone according to plan]; I try to improve them.

Another leader said that he is not afraid of letting others know that he is wrong. He explained:

> I never had a face issue. I don't have a problem with telling someone I'm sorry, I was wrong, or with taking an idea from someone very junior and then developing it, or improvising with it, and giving the credit openly to that employee. I don't have to worry about taking all or the main credit in every case; it does not always have to be that "Oh! It's my idea."

A vice president outlined the attributes of an authentic leader in a succinct way:

> A leader should be one who has integrity and honesty. That means the leader should be one who does not tell people one thing but does the other. One who tells the truth all the time … Someone who practices what he preaches. These are important

attributes. The authentic leader is honest and says things as they are. He says what he means and he does what he says.

Similar views were shared by another leader: "I am who I am. I give 100 percent of what I need to give. There is no point preaching and not doing what you preach … it's not leadership."

To another leader, authentic leadership was about projecting reality. To him, an authentic leader is "one who tells the truth and one who tells the same story all the time." A CEO also shared this perspective and noted that being truthful and being blunt were two different things. She explained that one must be truthful but one must also be careful that the truth does not hurt others. Therefore, one must learn how to put the message across even if it happens to be uncomfortable. She noted: "All leaders must be true. It is more than walking the talk … One cannot be trained to be truthful. Being truthful is not just being blunt." She also noted that one does not necessarily have to be charismatic to influence others. In her view, sincerity was more important. She added:

> To be a good leader, you don't need to be the best speaker. You just need to be sincere. Treat those on the other side as human beings. They would like to know you a little bit more, so be open. If you do that, you have less to worry about yourself in business relationships.

Authentic relationships

The interviewees underscored the importance of authentic relationships for leaders. Themes relating to development and sustenance of relationships also emerged in the analysis. The interviewees referred to authentic relationships which were established on the principles of pure purpose, honesty, and trust. One leader noted:

> I would like to see in my team that we are good friends. Sometimes, team leaders push their followers too hard and stress them up. No matter where I am, work is work. But I want to have peers and followers as friends and would like to be remembered as a "friend leader."

To one CEO, building relationships was important for her leadership. She observed:

> I always think my strength is my ability to connect to people. Since my early working days as a site engineer, I always managed to get things done faster as compared to my colleagues … To lead is not about giving commands and instructions to the people but it is about understanding them and making them understand you. I have worked on site. I learned to speak the languages of the workers when I talk to them. I speak to them in their [Chinese] dialects. I tell them: "I need your help" … I always say that I need their help. To me, making the connection with people is important.

A senior executive noted that her basic philosophy of a good job was the relationships that she developed on the job:

> Being really proud of a job means, at the end of the day, your whole team is proud of being associated with it … Also, it is the relationships you develop on the job. You are happy to be associated with the job and with the people you worked with to get it done.

Analysis of the interviews also showed that relationship building is closely related to the establishment of mutual trust among the parties. One CEO commented on this:

> The most important thing in being a leader is that you must be liked and respected by your people. That is a challenge. How does one get into a position where your people trust you, and believe that you will take the organization to success, and at the same time be liked by the people as someone who has their interest at heart?

Another senior director stressed that having direct personal contact with employees was important for building the trust required for connected relationships. She observed:

> Personal communication is very important. The personal touch, a personal e-mail or a short conversation [matter]. Informal contact with staff is also helpful. In projects, we have meetings. However, when it comes to leadership, I always try to speak to every-one individually and have the personal touch and make sure that people are able to freely share the problems they have. We often share some jokes in the midst of the rush to finish the project ... Sometimes, you need to break the stress, have a word or two in a light atmosphere, or talk about things over a meal. I think that helps.

The discussions indicate that, in order to have better and healthier relationships, the leader should try and establish trust and earn the respect of the people he or she leads.

Translation of purpose into impact

Figure 7.1 shows how self-awareness translates into self-regulation and hence the desired impact of authentic leaders. Figure 7.1 illustrates the association of leadership purpose with the desired leadership legacy and its eventual impact on others. It shows that understanding of "purpose" translates into "desired leadership legacy" through personal values, motivations, and goals. Practicing solid values, and developing intrinsic positive motivations and goals further translate into long-term legacy thinking. This means that leaders with a sense of purpose want to make a difference in their organizations and in the lives of their followers by creating, living, and leaving powerful and positive legacies. This translation of leadership purpose into desired leadership legacy mediates the leader's attributes,

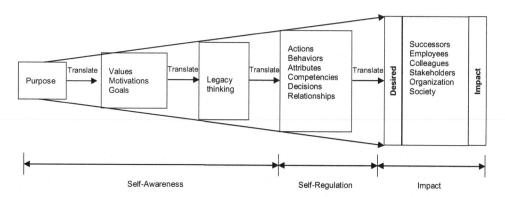

Figure 7.1 Translation of authentic leader's purpose into desired impact

competencies, decisions, actions, behaviors, and relationships. These features of the leader translate into the desired impact the leader wants to have on others.

Figure 7.1 shows why understanding leadership purpose and the route to its achievement through the leader's self-awareness and self-regulation is the key to success and desired impact of authentic leaders. In order to achieve their leadership purpose, authentic leaders continuously grow in self-awareness and hence self-regulation by aligning their values, motivations, and goals with their desired impact on others. They regulate their outer-selves – dispositions (actions, behavior, attributes, and relationships) and skills (competencies, decisions) – to achieve congruency with their inner-selves (values, motivations, and goals). Congruency between inner-self and outer-self translates into authentic leadership; that is the key to attaining the desired impact on others. However, achievement of authenticity within the inner- and outer-self is not sufficient for the desired impact of leadership. Authentic leaders also achieve congruency between self and social realities through dynamic and constructive reconciliation that is explained in Chapter 10.

Authentic role modeling was a key category that emerged from the analysis of the information gathered. Leaders interviewed in this study were convinced that leadership was not about rhetoric, speechmaking, or the exercising of authority. In their stories, while describing their leadership philosophy and purpose, talking about their approach and strategy, and discussing their daily lives, the leaders appeared to be persuaded that doing what one has promised, leading by example, and "doing it before preaching it" were essential for leadership to be effective.

Authentic role modeling

The leaders were continuously presenting themselves as role models for others to emulate. They persuaded their followers by their actions and not by talking only. They believed in what they were doing, and knew they were influencing their followers. Under the category of "authentic role modeling," four sub-categories emerged of how to become a role model:

- walking the talk;
- demonstrating self-discipline;
- mentoring, teaching and guiding;
- leading by stories.

Walking the talk

Under authentic role modeling, the dominant category that emerged was leadership by example or "walking the talk." "You cannot instill in people what you do not have, you have to be the person and walk the talk," noted one CEO during the interview. One senior manager stated: "You must be a person whom they can respect and trust. When you tell your staff don't do this, make sure that you don't do it yourself otherwise whatever you say to them would be of no use." The CEO of a consulting firm made the point:

> I believe in leading by example. You actually do get involved in the projects and guide the younger ones along by what you do. If the people close to you see that it's a good way of getting the things done, they will internalize it, and try to moderate their behavior, if that is necessary. They will also learn from this philosophy.

A senior manager in a contractor's firm also noted:

> In leadership, you have to set an example for your followers or subordinates ... when you show your enthusiasm for the work and demonstrate that you have care or concern about the work, your followers will also follow. Therefore, you must set an example. You need to be hard-working, make fast and right decisions at the right time, and show the subordinates that you are the leader that they are looking for.

The general manager in a contractor's firm shared that he would frequently visit construction sites to encourage the firm's workers at all levels (rather than spending all his time in the comfort of his office). He observed:

> I try to lead by example. If necessary, I go to the top level of buildings under construction and prove to my staff that I can do the stuff. To convince them that things can be done, you sometimes have to step forward to show an example, and I do it for them.

Many other interviewees, while sharing about self-discipline, mentioned leadership by example and positive modeling for their followers. A director in a consulting firm said that leadership, to him, was leading by example. He observed:

> I take part in relevant aspects of the work to show my subordinates that it is only hard work and diligence which take you to the next level of your career. That hard work is necessary in order to apply the valuable experience and skill that they have.

Demonstrating self-discipline

For authentic role modeling to be effective, the interviewees emphasized the need for self-discipline of the leader. They referred to self-discipline in the leader's personal, professional, and social life. They emphasized that they did not compromise on their principles. Discussing the issue of self-discipline, the chairman of a large construction conglomerate gave the example of a principle, which the philosopher Confucius had said:

> Great leaders of all times first set up good government in their states. Wanting to have good government in their states, they first established order in their own families. Wanting order in the home, they first disciplined themselves. Desiring self-discipline, they rectified their own hearts; and wanting to rectify their hearts, they sought precise verbal definitions of their inarticulate thoughts. Wishing to attain precise verbal definitions, they set out to extend their knowledge to the utmost.

Many leaders interviewed in this study shared that they like to have a balanced life in order to perform better at work. For example, one CEO shared:

> There are lot of things that you have to balance as a leader; academics (in some cases), practice, and also family. I have two kids and a wonderful wife. I make sure I spend sufficient quality time with them so everything in our lives should be balanced. I am taking them for a trip, on a cruise in November.

Other interviewees shared about the importance of discipline in life and the profession in general terms. In this context, the interviewees mentioned several personal choices they had made in their lives. For example, a few mentioned that they do not provide entertainment to the staff of clients to win jobs, not only to avoid any impression of corrupt practice, but also to ensure a balance between work and family life. An executive shared that she had certain principles that would not allow her to entertain clients in night clubs. The CEO of an engineering consultancy firm related:

> I will never go to night clubs to entertain clients … it's not my cup of tea … that's not my style, I can't help it. Also, I cannot let a partner go and do this sort of thing either. I say, "No, don't do it." If you lose 50 percent of the jobs because you will not take clients to restaurants and night clubs after work, it's okay; I want you to go back to your family.

The discussion here shows that self-discipline is critical for positive and authentic role modeling of leaders.

Mentoring, teaching, and guiding

The leaders viewed mentoring, teaching, and guiding as essential parts of authentic role modeling. They were developing their followers, industry peers, and even their competitors, and were engaged in some sort of mentoring with each of these groups. Many noted that mentoring, teaching, or guiding others was essential to earn the respect of followers and to establish trust and confidence among team members. The CEO of a consulting firm revealed that he loved teaching others and had the patience to do that:

> I have the patience to sit down with the young ones, work through a problem, and teach them how to do it. I take pains to go through their work without ever losing patience or my temper … People give me their reports and I go through them carefully, sometimes painstakingly word-after-word … correcting the English, and then run through these reports with them, and teach them how to do it better by themselves next time.

Discussing her philosophy of leadership, an executive noted that:

> It is very much mentoring and coaching the fresh graduates from the polytechnic and university and getting them into the culture of the firm. We have our own ways of doing things. I must … guide and coach the new staff to become accustomed to the local culture.

Another CEO of a consulting firm, who indicated that he was close to retiring, shared that he spent most of his time mentoring his followers so that they could take up leadership roles in the future. He noted:

> During the last three years, and especially in the last year, I have actually been more of a mentor, rather than a manager. So you can lead by managing, and you can lead by mentoring, I guess, and by challenging others [to enhance their performance and capability].

Leading by telling stories

As a part of authentic role modeling, the leaders interviewed shared that they led their people by telling them stories, talking about the history of the organizations, and particularly, times of achievement and crises; sharing metaphors and analogies; and giving the followers examples from their own professional development as well as their personal lives. Leading by stories helped the leaders to get closer to their followers and build mutual trust and respect with them, and among them. For example, an executive director in a consulting firm noted that he usually used stories to guide and motivate his followers:

> I always tell my people to watch this particular show [name omitted] as it is about leadership. The sergeant is a person who has a heart for his men. There is one show in which he told his soldiers to have a rest on that day and not go out … Nobody went out; they had that discipline. Later in that episode, he was leading the charge against the enemy. He planned everything himself and he was in the front. The men took comfort in the fact that he was there. They were willing to fight and die for him. That is leadership. On our projects, I am always there with my people. Sometimes, I stay back, clear my work and be with them … This way, the people feel that "my boss cares."

The CEO of an engineering consulting firm revealed that she had found a way to connect with her followers:

> My engineers tell me that I can connect with them much better [than other partners and associates]. That is because I shared with them the story of my childhood and upbringing. I share with them stories about other aspects of my professional and personal life. They tell me that they are inspired by my life story and how my career has developed. They identify with many aspects of my life story.

A senior executive of an architectural practice, also stated that he usually asked his junior staff members to tell him stories about the challenges and issues they had faced. He said he always asked them:

> What is your story? Where are you coming from? What are you trying to achieve? What do you have to offer? It sets people to think about having a sense of purpose in their careers. It puts the young architects very much at ease and lets them feel like they are having to deal with a challenge. Architects like that; the young ones will feel they are back at university, working on a Year 3 design or Year 5, Final Year project … I get great results this way because it unleashes their creativity.

The kind and subtle acts of the leaders usually helped them to make a connection with their followers and do authentic role modeling for their followers. They helped the leaders to develop an environment where their followers felt at ease and were willing to learn and perform better. Authentic role modeling also helped the leaders to develop a positive organizational culture where knowledge sharing and continuous self-development were the norms. Finally, authentic role modeling of the leaders also resulted in loyalty and support from their subordinates.

Self-growth and competence

"Self-growth and competence" also emerged as a middle-level category under the broader category of self-leadership. Authentic leaders interviewed mentioned that self-growth and competence were important for a leader to lead a knowledge-intensive organization and inspire knowledge workers. It begins with a commitment to continuous personal growth. "Self-growth and competence" also include building positive personal relationships within the organization and demonstrating professional commitment and competence by understanding the domain as well as client-specific needs. Some of the lower-level categories under self-growth and competence are now discussed.

Authentic leaders recognize their weaknesses and take steps to address them. The CEO of a developer's firm noted:

> We can't do things perfectly, but one has to have the desire to perfect it at the end of the day and one has to also accept that in the interim, there may be flaws and they should not put one off. They [the flaws] should not slow us down.

With appreciation of the weaknesses, one can work to build upon the strengths and skills one has. The managing partner of a quantity surveying practice noted: "You have to constantly upgrade your skills and be knowledgeable of what is happening in the industry to keep pace in front of the competition." Another senior executive opined about continuous personal growth: "To move up the ladder, it is important to gradually hone your skills."

Some leaders also explained how one can upgrade one's skills and enhance one's professional and personal competence. Most of such responses pertained to participation in professional bodies. A director in an architects firm suggested:

> I think for us who have been practicing for many years, the best way to deal with the information is to participate in the committees [that are formed by the government agencies] and [professional] organizations. And that is when you get first-hand most current knowledge of what is happening in the industry.

Building positive professional relationships

In addition to taking the personal initiative to hone their skills and competencies, authentic leaders interviewed noted that they also actively build a positive rapport with their colleagues, followers, and peers in the industry. To establish this set of positive relationships, one has to show consistency in values, behavior, and leadership capabilities. One senior executive noted:

> The important thing is that you have to show people that you are fair, capable, reasonable, and sincere. Just being fair is not good enough. You should also show that you are capable and reasonable. All these qualities must be expressed and shown. Saying is very easy but convincing others and doing it are very difficult.

In building positive relationships in the organization as well as with the industry, the leader should also hone the leader's persuasion skills. One CEO noted: "It is also leadership in trying to convince shareholders, trying to convince the people to be prepared to support you." He was of the opinion that the leader must be able to convince others of what he or she believes in and what he or she does. Another executive outlined a similar perspective:

One must develop professional skills and have the confidence from having those skills. This is how people get to see what you know and what you believe in. Sooner or later, people will find out whether what you have been saying is the truth [or otherwise] … Professionals should develop that credibility, and exhibit a level of confidence and then they will be able to get people to follow them, and have confidence in them in their business relationships.

Another aspect of building positive relationships is earning respect. Leadership, in the opinion of many of the leaders interviewed, is earning the respect of people and getting their support. The CEO of an engineering consulting group noted that: "the most important thing to being a leader is that you must be liked and respected by your people." Another leader noted that earning the followers' respect was not easy and one cannot learn how to set about to get it. In his view, leaders earn respect through the demonstration of their values, acts, and behavior in their daily actions. He noted: "People can be effective if they have got the authority and they are being respected … If you don't have respect, authority, and experience, you cannot go and learn it from books or by attending lectures."

In order to build and sustain positive relationships, the leader should be able to deliver what he or she has promised to deliver. The CEO of an engineering consulting firm noted that she was strict on herself with respect to the promises she makes regarding the delivery of jobs her company works on: "When I promise somebody that I will do it, I tend to keep my promises … I get it done even if it puts more pressure on us."

Professional commitment

It was found that authentic leaders are dedicated to their professions and invest a lot of effort in their professional commitments. They are keen on attaining professional excellence, and do not compromise on their professional ethics. The chairman of a consulting group related a story on design safety and occupants' health which he had shared with his colleagues to inspire them about professional commitment:

> This is the story I tell my staff to advise them to be humble. I tell them that all of us are graduates … And to get the degree or diploma … you score 50 percent and above … But, even if many of you graduated with a first class honors degree, with 90 percent marks, we cannot afford to have 10–15 percent of our buildings fall over or catch fire, or have air handling systems that do not work … So the only way that we can avoid that is to strictly follow all the design codes and criteria and the safety factors, and so on. And then we have to apply the quality management system, and do the design checks. We must be good engineers.

Another leader noted: "First and foremost, in construction, you must be very committed to your profession. The vision of becoming a good engineer helped me to be what I have become today."

A senior manager noted:

> [Y]ou need a leader everywhere. You need the captain in the aircraft and in the ship. In a construction site, if your leader is more committed and is more results-oriented, you will have a more successful project compared to those who are not so results-oriented or cost conscious and do not do things right the first time.

Commenting on commitment to the profession, the deputy general manager of a contracting firm said that a good leader should be committed to learn about various trades to take better decisions on his own work. He noted: "[Y]ou have to make an effort to understand the specialist construction trades. By understanding the work of others, you can appreciate their problems, and make better decisions."

Professional knowledge and competence

The authentic leaders interviewed highlighted professional competence when discussing their strengths. Most of them had reached their leadership positions by rising through the ranks, and they valued their professional competence. One of the interviewees noted:

> The key thing is you have to be extremely good at what you are doing … From there everything will flow. Because it is natural that if you are so good, you will become a leader in your own right in your field. If you don't have that [competence] or you have not got professional excellence, then I think it will be difficult for you to become a leader.

The deputy general manager of a contractors' firm underscored the significance of knowledge:

> Knowledge is important for leaders. I go to great lengths to learn more about my field. You must have a whole understanding of the relevant matters if you want to lead a team. You must keep up with new ideas, and with changes in regulations and design codes … You cannot say that you don't know how to get any particular thing done. In Singapore, your followers will expect you to have the answers. You must have sufficient knowledge so that you can at least lead in the effort to find the solutions, even if you do not have it offhand yourself.

A director in an engineering consulting firm stressed the importance of his own technical knowledge to his leadership:

> I show my subordinates the hard work and diligence which take you to the next level of your career. From that hard work, you amass valuable experience and skill; that is my philosophy in this technical field. Your ability to lead depends on the technical competency you have acquired during your working life as you took on design and management challenges.

This experience comes from dealing with people and gaining knowledge over time, according to a senior executive in a consulting firm. He noted: "Dealing with people who have greater expertise than you have and more experience and knowledge than you is helpful … The more you interact with such people, the broader your own knowledge base becomes." He also noted that to gain the required experience and competence, it was important to go through the whole process, not only in a project, but also in the profession:

> In my opinion, to take up a leadership position in a consultant's firm, you need to go through the whole process. You need to be a design engineer, then be a senior

engineer, and before then becoming an associate and a director. There are no short cuts if you want to be a good engineer, respected in the industry.

Next to having domain knowledge, several leaders mentioned that their main strength was that they were able to understand the needs, expectations, and demands of their clients. A director in an architectural practice gave the example of his senior boss whom he thought was also an authentic leader and was able to understand the needs of the clients:

> He really understands what the client wants. We do a lot of things but if they [clients] don't understand the end product, the whole endeavor will not be successful. He [the leader in the example] has spent a lot of time in the industry and he understands a client's psychology. So he is able to guide us in our work to meet the desires of the particular client on each project.

Higher-order attributes

The interviewees were asked to mention their own strengths and some essential attributes of authentic leaders. The responses of the leaders resulted in a range of points that are shown in Table 7.3.

Table 7.3 Lower-order attributes of authentic leaders

Ability to make changes and adapt to change	Compassion	Making unpopular decisions
Ability to engage people	Confidence	Not being scared of failure
Ability to foresee	Conscience	Objectivity in approach
Ability to guide	Consideration	Openness
Ability to inspire	Consistency	Passion
Ability to lead in crisis	Courage	Patience
Ability to see the whole picture	Credibility	Perseverance
Ability to stand up to the world	Decisiveness	Persuasion
Ability to take risk	Dedication	Problem-solving
Accepting responsibility for personal mistakes	Diligence	Professionalism
Ability to adjust to the situation	Diplomatic ability	Reasonableness
Being multi-skilled	Empathy	Selflessness
Being positive	Fairness	Sincerity
Ability to build trust	Hard work	Sympathy
Ability to be calm in situations of chaos	Honesty	Tenacity
Caring attitude	Humility	Transparency
Character	Innovativeness	Understanding others
Commitment	Integrity	Wisdom
Common sense	Listening	Working ahead of time
Communication		

Many of the attributes in Table 7.1 represent values, competencies, and traits of authentic leaders. These lower-order attributes were combined on the basis of their similarities. This resulted in several higher-order attributes of authentic leaders which were found to be common among the interviewees. These include:

- social intelligence (understanding the people and the context);
- self-confidence;
- positive self-image;
- self-drive;
- optimism;
- hope;
- courage;
- resilience;
- benevolence;
- creativity;
- wisdom.

Social intelligence

Social or emotional intelligence emerged as a prominent sub-category under the broader category of higher-order attributes of authentic leaders. Leaders were familiar with the term "emotional intelligence" or "emotional quotient" (EQ) and were generally convinced that EQ was imperative for effective leadership. They mostly related EQ to the issues relevant to social relationships. One interviewee noted: "You need to have EQ and you need to know what motivates people. You need to know how to cascade your vision down to many teams."

Discussing EQ, the CEO of a consulting group noted:

> I like to empower my staff ... I like to give them the authority to make decisions. So long as the correct process has been followed, I trust my staff to make the right decisions, and I make them accountable for their decisions and actions.

Another senior executive also emphasized that:

> Having a higher level of EQ is essential for getting along with other people. I tell my staff that we work as a team. And for a good team to perform well, there has to be an emphasis on EQ.

The executive director in a quantity surveying firm noted the significance of emotional intelligence:

> Understanding the concept of EQ has shown me a path that I should take to avoid making mistakes. I cover my inexperience by the way I present myself with confidence, and speak on topics on which I have knowledge and experience, and make sure that I do not dwell on those things in which I don't have strength. I leave those topics to others who have the necessary expertise, or experience.

In the interviews, the leaders related emotional or social intelligence to two further major concepts: understanding the people, and understanding the context. These two concepts are now discussed.

Understanding the people

"Understanding the people" as a category emerged from the points shared by the interviewees on "people" whom the leaders considered to be the central aspect in their conceptualizations of "leadership." They mentioned that leadership was all about "people," and "understanding people" was an important part of the leadership process. Many interviewees explained why "understanding the people" was so important and how they performed it. The central idea that the leaders stressed was the uniqueness and individuality of each person.

The concepts that emerged as a part of "understanding the people" included: acknowledging people individually; motivating them; inspiring mutual respect; and showing care and concern. On individual acknowledgment, a director of a large consulting group observed:

> My major strength would be "understanding my staff." You need to remain grounded, understand the staff individually, and treat them as individuals. When you make generalizations about people in the organization as "staff," you would miss the individual needs and requirements. Only when you treat them as individuals would you acknowledge their specific desires and career objectives and aspirations … It is with such an approach that you would be able to achieve greater results.

The CEO of an engineering consulting company explained his viewpoint of understanding people as:

> It is important for us as leaders to understand our people … After all, whatever we want to do, we have to do it through people … I think in any industry, whether you are in services … or in a capital-intensive production organization … in the end, it is the level of understanding you have of human behavior that helps you to realize success. By understanding our people, I think we also need to accept that people are frail … I don't mean they are weak physically, but while they have their strengths, they have egos, emotions, and weaknesses. This is very complex.

A director in an engineering consulting firm noted that mutual respect was also a key to understanding people, motivating them, and getting the best performance out of them. He observed:

> We always attempt to give due respect to the person's expertise. We are not too keen about the personalities of people; rather we are concerned about how they perform. In that sense, you are looking at the person's expertise, about what has to be done … When teams come together, if they are aware of what each person can contribute, and they respect each other, then they can realize the best results.

Similar to "understanding the people," a related category that emerged under "sustainable leadership" was "sustainable human capital development" under which the leaders explained how they attempted to develop their followers as authentic followers and what strategies helped them in doing so.

Understanding the context

In parallel with "understanding the people," the other major emphasis of the leaders interviewed was on "understanding the context." The leaders often noted the challenges that various contexts posed to their leadership. Leadership, for many of the interviewees, was simply understanding the conditions or the context, getting the support of the team, and formulating an appropriate strategy to achieve the goal in the circumstances.

The interviewees were operating in a challenging business world in which competition was not limited to the local construction industry. The CEOs and directors of multinational corporations among the leaders interviewed were concerned about the challenges they were facing in different parts of the world. Therefore, dealing appropriately with the context was critical to them; they had to solve each problem in accordance with specific "expectations and boundary conditions," as one interviewee noted. The leaders generally observed that different situations called for different kinds of leadership. The CEO of a large firm, like many others, noted:

> Different leadership styles are required for different companies at different points in time. A company which is already developed or well-managed and maintained is different from one that has only recently started. So, at different points in time in the life of the company, you require different leadership approaches ... a person who has the knowledge and skills for the particular situation tends to be a naturally accepted leader.

A director of a firm noted that the leader of a company needs to change according to the situation and adapt to new conditions:

> It is a very dynamic situation these days; the market conditions change rapidly. Today, there is a construction boom; next year could be different. So, you must be able to take stock of where you are and what you need to do to suit the new conditions, the changing conditions.

Similarly, another CEO observed: "I do change my style according to the situation. I don't think I am a bulldozer by nature. I am actually a very effective person and I do vary my style according to the situation." Again, one executive noted that the context governs the demand and the kind of leadership required. A good leader in one situation may not be even an ordinary leader in another situation. He observed:

> Everything is linked back to a time and context ... A good leader five years ago may not necessarily be a good leader now. He may become irrelevant in today's context, even in the same industry and in the same profession.

Leaders who operate successfully across the boundaries of their own culture understand and appreciate the importance of context. A director in an architectural firm noted: "I had many problems in the beginning due to cultural issues. I was dealing with all kinds of different people." Discussing what it takes to deal with the context and understand the situation, the interviewees noted that resolving paradoxes, considering the stakes involved, as many noted, was a part of the leadership role. A senior executive in a contracting firm noted a paradox which was a usual condition to his leadership:

> When we tender, our top management wants us to achieve a certain profit margin with lower risk and also to aim for higher profit if possible; the estimating department is under pressure to be able to win the tender in a competitive industry; our project manager wants to deliver a project with a lower cost ... Everyone in the process has a different objective in the effort to attain the company's goal of wanting to secure the same project. That is a difficult part of my job; to satisfy the

seemingly different objectives of units with different functions and roles. When we are facing the client, we have to convince the client that our company provides the client with quality (we interpret that as lowest price, and superior service). On the other hand, we have to satisfy the company and its stakeholders by getting jobs at a good profit. It is my responsibility as a leader to find an appropriate balance among a variety of objectives of our units, our clients and our key players, and other stakeholders.

The leaders also described the paradoxes in which they had to find a balance between their tasks and considerations, such as between their leadership and management roles. Explaining this, a CEO noted:

I struggle with the lack of time, having to juggle between daily management issues and having the time to focus on what a leader should be doing. I think this is my weakness: how to cope with the management responsibility and the leadership role.

Another CEO noted another paradox he faced as a part of his leadership role: "I struggle all the time with how you manage different interests of different people of different age groups and in different positions in the company while keeping them all motivated and supportive of your mission and vision."

Self-confidence

The leaders mentioned that self-awareness or self-regulation alone were not sufficient for leadership. A director in a large architectural practice noted that: "There must be a lot of belief in my own self to be able to jump into challenging situations and to say that I'm going to come out of this well."

The CEO of a large engineering firm described an incident when he had a problem in finishing a project on time but eventually succeeded in doing so. This achievement boosted his self-confidence. He shared:

[T]his job gave me a lot of self-confidence. It taught me that no matter how many problems you face, so long as you work hard, and work closely with your partners, the problems can always be overcome. But, you must make sure that you are able to work with people with a spirit of collaboration, not by threatening and bullying.

Another senior director noted that self-confidence helped him to make decisions:

You have to have self-confidence and also, people should believe in you. Because I am fairly confident in what I do, I can control the situation. People follow my decisions. They don't expect me to find answers to all the details. They know you should have people to help you sort out those details.

The deputy general manager in a contracting firm was of the opinion that having good professional skills was not enough. One needs to complement these skills with self-confidence: "You must develop your professional skills and have confidence in those skills. From an early stage, you should develop that credibility, and build up and exhibit a level of confidence in your followers."

Positive self-image

Another higher-order attribute of authentic leaders that emerged in this study was that such leaders have a positive self-image (i.e., self-concept or self-perception), having acknowledged their strengths and weaknesses. They like being themselves. This is a source of both satisfaction and contentment as well as motivation for authentic leaders. A CEO noted: "I am happy with who I am. I think I do get along with people quite comfortably … In a way, you are most comfortable as yourself and not being somebody else."

It is notable that authentic leaders, although they are satisfied and contented with themselves, also constantly strive to become better. "Being themselves" actually motivates them to make every effort toward self-improvement. One CEO highlighted this point:

> To be myself is not good enough; I have to be better than my current self. We are imperfect and to be ourselves is to be imperfect. It is not just to strive for perfection. In leadership, one has to be better than one really is. I want my organization to be better than what it is today and what people think it can be. Personally, I want to be better than who and what I am. I know I should be better. Not just for myself but to be right for the tasks I will have to face and the followers I have and will have.

Self-drive

Another important attribute that authentic leaders have is an enduring self-drive. Without being self-driven and without having a passion to achieve his or her goals, the leader cannot convince others to follow him or her. Many of the leaders interviewed noted that they had a strong personal drive. Explaining the secret of his successful leadership, one CEO noted: "I am a very driven individual; I remember being that determined even in school. I don't know how to explain it. When you want something, you just want it. For me, that is the push factor." Another senior executive noted:

> I know I can do it … you give me any task and put me anywhere, I can solve it. And that is probably the competitive spirit in me … That's me! Sometimes you need a fire in your belly.

Another executive also noted that she was very driven. She observed: "I am very driven. I will go out to do a good job, and I want to do a good job. I think I am more driven and ambitious than a lot of the men in the department." To be self-driven, many of the leaders said that one should have interest in, and satisfaction from, the work one does. A senior architect noted that "Architecture is a long course; you have got to like what you do. It is important for you to have passion for the profession." Explaining the link of passion to self-drive and satisfaction further, the CEO of a consulting firm noted:

> I joined this profession because of my passion for engineering. As the eldest in my family, I had to do the repair work when things broke down at home. Given another opportunity [in life], I would still become a civil engineer … I cannot sit like a doctor and keep checking the health conditions of patients from morning till evening. We can get out and meet more people daily. It's much more interesting and fun. There is a special joy working in this industry. You see a piece of land and develop it into a landmark … it is such a great thing to be satisfied with what I do.

The CEO of a developer's firm also noted: "There is always a sense of satisfaction to see things built, in our business, to see things work well and to see a completed project. We feel good about our successes, of course." A senior director in a contracting firm also expressed a similar sentiment: "There is a certain satisfaction … the difference with construction is that you do see the end product that will last for more than 50 years."

The discussion shows that having self-drive is an important attribute of authentic leaders. Leaders with self-drive are self-motivated to make a difference; they are satisfied with the professions they are in, and are willing to go the extra mile to achieve their goals.

Optimism

The interviewees in this study not only demonstrate optimism in their own lives, they also inspired optimism in others, and motivated their followers with their positive attitudes and actions. They remain optimistic about the outcomes. A senior executive said that leadership is full of challenges. However, the leader should be able to motivate his followers to deal with difficult situations. He noted:

> I think the challenges in the leadership role are as to how I can convince my staff that: "Yes! You can do it." I have to change their mindset … [that] it is not to focus on the end date or the big milestone but rather to divide the larger problem into smaller goals and tackle each of them to attain the big goal. I actually practice this myself; I make sure that we are achieving smaller goals [every day].

In facing challenges, the leaders adopt new ways. A director in an architectural firm stated that he was always exploring ideas and trying to improve the processes in his firm by adopting new ways: "I like doing new things. I am an optimist. I like to go for new things and try new ways of doing them." The general manager of a developer's firm also shared about himself, and the optimism he has: "I am a very positive person and I take the view of 'the cup is half-full.'" Another executive noted:

> I don't think I let them [failures] constrain me. I always try to look at how I could have done better … I suppose I am quite a positive person. I don't normally let failures hold me down for long.

Hope

Being hopeful and inspiring hope; both factors were perceived to be instrumental by the authentic leaders interviewed. Particularly, during times of change, hope plays a crucial role, as noted by several interviewees. Authentic leaders are hopeful and are able to give hope to others during a crisis; this makes it much easier for them to garner support from their followers. The CEO of a consulting firm, which had an extensive network in South-East Asia and had gone through a major organizational change, shared the challenge he had to assure others that the company was going to survive. He observed:

> I think at that point in time, with so much uncertainty, people were questioning whether we can survive, questioning whether we can make it … even in Singapore, not to mention our overseas activities. You had many doubts, fears, and concerns among the staff … The biggest, the best way to convince them to continue to work in the organization is to show

them the organization's tackling the problems; that it is progressing, it is improving despite the challenges; and it is on an upward trajectory instead of a downward one. In the situation, the competency I had shown in previous challenging situations must have helped … to convince the followers to stay with the firm and to follow me.

A senior executive of another company that went through a major change also shared her experience. She noted:

A major task of the leader in such a situation is to give the staff the assurance that the company would survive … we informed them about the financial position of the company. Some staff did leave. And we had a manpower restructuring exercise, but the firm went on to do well.

Another CEO observed that the leader must not look "desperate." The leader should always look hopeful so that he or she can draw support from people and positively motivate them. A senior vice president in a developer's firm echoed that, noting that the leader should always show perseverance in order for people to keep believing in him. He noted:

When things are going crazy … you have to … find a way out so that your people can move forward. I remember someone saying, the best way to improve your error is to state your error clearly. You have to put the questions to your followers in the simplest form. The ability to keep people believing in your ideas. The tenacity. The perseverance.

Courage

Courage emerged as an important sub-category while the leaders discussed their decision-making, implementing a strategy, and leading the change. Interviewees consistently noted that courage was an important factor that helped them perform their leadership roles. A CEO noted that "leaders are people who are not afraid to make very hard decisions." This was echoed by another CEO of an engineering consultancy company:

A leader must not be afraid to make difficult decisions. And if he has to make a decision that hurts somebody, then he still has to make it if it is for the greater good. And if he is afraid to make such a decision, that means he is shying away from his leadership …

A leader must demonstrate courage and confidence in his or her actions and initiatives to convince the followers to follow the leader. Followers have this comfort in that they are following someone who has the courage to back them if things go wrong. Given the courage they have, authentic leaders are prepared to take risks and explore new dimensions in business. One CEO offered his perspective:

I am prepared to take the risk. Because it's just so easy for people to sit down and say that I will only do things that are safe, but interestingly, that means that you will never change; you will never make progress; your company will be starved of fresh ideas and will fail to innovate.

Risk taking is important for any business; however, it needs courage. A senior executive in a contracting firm noted: "Leaders are not scared of failure." Courage also plays an important role in the practice of integrity. A vice president in a developer's firm gave her example to explain this: "I learn something new from everybody every day. That is my philosophy that you don't act as a 'Mr. or Ms Know It All' or 'Mr. or Ms Smart.'"

The leader should have the courage to say the right thing as well. A leader should have the courage to take the responsibility when things go wrong. An executive observed: "I facilitate the decision-making such as those which are market-driven. I act as a coordinator in the journey towards the decision. If I make a mistake, I take the responsibility. I don't push the blame onto others."

Resilience

Resilience is the ability to recover or bounce back from the failures. A director in a firm of architects noted his perspective about how he considered failures: "Of course, from every failure, there are always some lessons to be learnt ... we don't have to be discouraged by ... failures ... We need to ask how we could have done it differently, to do it better."

The CEO of a large group shared that he had faced numerous failures in his career; however, he was able to overcome them and that has been the key to his success. He observed:

> I have been fortunate that all failures and disappointments I have had, I have been able to pick myself up. I do admit that I have seen failures and one of the skills I have learnt is to pick myself out of it and not let it affect my next task. If a leader at my level can't pick himself up after a failure, then it is too difficult for him to do the management task, let alone doing what leadership requires.

Similarly, another CEO spoke about his ability to overcome failures:

> I can have failures and disappointments but I get over them very quickly. It is not because I am very strong; my outlook in life is far beyond what I am doing. I think that is the difference. If this is my world, then with any disappointment I will be shattered. But this is not my world, this is only my job. I am here to do my best and that's all I am offering.

The CEO of a consulting group shared that he does not let failures overwhelm him, but engages in reflection and explores the reasons that resulted in the failure:

> I always do a post-mortem on many of the deals I handle. Along the way, you make some mistakes, but you also take many right decisions. I guess once you are experienced enough, you will have this set of "instinctive red flags" that will alert you to think carefully before you act. I believe such instinct comes from exposure, experience and, importantly, reading people and situations.

Benevolence

Under the sub-category of benevolence, the concepts which emerged include: a caring attitude, compassion, consideration, empathy, generosity, humility, selflessness, and sympathy. Expressing his desire to be known as a compassionate leader, a director in an architectural firm noted: "I [aspire to be known as someone who] was inspiring and compassionate. I think, as leaders, sometimes we can be ruthless. But ... being compassionate can help you in many areas of your life, career, and so on."

A senior vice president in a developer's firm underscored the importance of empathy for authentic leaders. She noted that the leader "must be able to respect people." One CEO also emphasized that the leader should be "generous with praise to the staff if they have done well." A leader pointed out the importance of humility for leadership. She noted that humility had helped her to win over her followers. She related an example:

> I remember I would go to the site early and I would pick up some of the workers from their homes on my way. I must say this was not popular with my fellow professionals. I used to be asked, "Are you an engineer or a foreman?" due to the schedule I was keeping and the way I was treating the workers ... My workers saw me differently ... It became very touching at times. They were connected to me. They gave me presents for my family. They would ... bring me presents any time they visited their countries.

Another leader stated that he had risen through the hierarchy from the entry level for an engineer to be the firm's CEO, had experienced what it is like at every level in the organization. Therefore, he could relate to the problems and challenges of his employees, and had a great deal of empathy for them.

Creativity

The sub-category of creativity strongly surfaced as authentic leaders drew attention to issues such as inventiveness, thinking outside the box, and creativity. Since interviewees belonged to the construction industry, most of them emphasized the importance of creative designs, technology use, and creative solution of problems. A vice president in a developer's firm noted that:

> I always tell my staff that I do not look at processes. When I give them an assignment, I would like them to use their own creativity to do the work. I don't care what time you come to office and what time you go home, so long as you are able to accomplish your tasks in accordance with our firm's performance requirements.

A director in an architectural practice also emphasized creativity and innovation:

> In the architectural profession, we have to be innovative and creative in what we produce. So, I see this as a challenge because if you don't renew yourself ... The peers will look at you as a rather outdated and conservative company which is not able do innovative design.

Another senior executive noted that the vision of a focus on innovation should be shared by all stakeholders on a project:

> So you can see that if we have strong leadership in design and engineering to deliver the vision of the developer, the developer must also be willing to accept new ideas and tread into areas that nobody has gone to yet.

A CEO stated that leaders should be creative and able to "do things that others do not do." Moreover, "there are always no-man's areas and you have got to go there and occupy them ... That is being a leader," said the general manager of a contracting firm.

Wisdom

Wisdom is usually described to encompass concepts such as perceptibility, adaptability, predictability, ability to plan, and knowledge. In this sense, wisdom is closely related to having a "vigilant outlook." However, wisdom is more complex from a conceptual point of view. In the current analysis, the concepts that emerged under the sub-category of wisdom include: foresight, ability to see the whole picture, adjusting to the situation, being multi-skilled, and having knowledge. The managing partner of a quantity surveying firm noted:

> I think an important attribute [of authentic leaders] is vision. The ability to see beyond the immediate vision of other people so you have that little extra in terms of foresight. The second would be wisdom … To be intellectually good and very clever that is one thing but to be wise is something else. So it is vision and wisdom. These are two important attributes. I think with these two you can go a long way as a leader.

Wisdom carries two important aspects: learning from the experiences in the past, and being able to foresee what is likely to happen in the future. Authentic leaders are able to draw important lessons from their past experiences. For example, one leader noted: "We have learnt from the mistakes and the reasons why we failed so that we can address the issues and mitigate against the impacts in the future."

Based on their past experiences as well as the knowledge they gain from their social learning, authentic leaders develop the skill of foresight. "My major strength has been that I have a gift of foresight," noted the chairman of a consulting group. The executive director in a contracting firm noted that his major strength was really simple. He observed that "a lot of it is common sense," which he thought he was "quite good at." He also noted that "not many people will have good common sense." Sharing his ability to look at the whole picture from a long-term perspective, another leader noted: "I feel that in many cases, when I negotiate a contract or take a management decision, I am able to grasp the implications of my decisions. This is important, as it helps me to make the right decision." Another CEO noted that:

> There are a few of these people around who can create something out of nothing. I won't mention any names. Some of them are actually my clients, and even here, there are only very few of them. These are people who have got the vision to see beyond what most people can see.

Discussion

In the existing conceptualizations of authenticity and authentic leadership, scholars have emphasized self-awareness and self-regulation (see Goldman and Kernis, 2002; Kernis, 2003; Gardner et al., 2005; Ilies et al., 2005). However, the analysis in the current study shows that authentic leadership is more than just self-awareness and self-regulation. Categories of authentic leadership derived from the study and explained in the earlier sections show that authentic leadership has three major dimensions: self-leadership, self-transcendent leadership, and sustainable leadership. Self-leadership essentially includes what earlier authors have attributed to authentic leadership. However, it can be noted here that self-leadership comprises several dimensions, including authentic self-awareness, authentic self-regulation, authentic role modeling, self-growth, and competence. Here, authentic role modeling and self-growth and competence are new dimensions which are not underscored in the existing literature on authentic leadership. Therefore, these findings not only confirm the earlier conceptualizations of authentic leadership, they further them by adding new dimensions to the concept.

Scholars have also noted that authentic leaders possess higher levels of positive psychological capacities (self-efficacy, hope, optimism, and resilience) (see Luthans and Avolio, 2003). Scholars of authentic leadership have noted that authentic leaders possess greater levels of PsyCap (Jensen and Luthans, 2006; Luthans et al., 2007; Toor and Ofori, 2008g). The current study echoes these earlier assertions. It can be noted under the category of "higher-order attributes" that authentic leaders do possess self-efficacy (or self-confidence), hope, optimism, and resilience. However, in addition to these attributes, it was found that authentic leaders also possess social intelligence (understanding the people and context), positive self-image, self-drive, benevolence, creativity, and wisdom. Luthans et al. (2007) discuss some of these attributes as potential candidates for PsyCap. The findings in the current study affirm that authentic leaders possess higher-order positive attributes which help them perform their leadership roles with authenticity.

The findings further the existing conceptualization of authentic leadership in other ways. It was found in the analysis that authentic leaders are typically "self-transcendent" leaders. Their self-transcendence may take various forms including, but not limited to, soulful leadership, servant leadership, spiritual leadership, shared leadership, service-oriented leadership, and socially responsible leadership. Authentic leaders are inner-focused and others-directed (self-transcendent). This further strengthens the view that authentic leadership is a "root construct" and may appear in any form of positive leadership. Therefore, authentic leadership is "self-transcendent" in nature and endeavors to attain the well-being of others (group members, the organization, stakeholders, and society).

The central idea of authentic leadership is more concerned with self-awareness, relational transparency, internalized moral perspective, and balanced processing (see Gardner et al., 2005; Walumbwa et al., 2008). Given that authentic leadership is advocated as the "root construct" (Avolio and Gardner, 2005), researchers have attempted to draw parallels between authentic leadership and other forms of positive leadership (such as servant leadership, spiritual leadership, and ethical leadership) (see Luthans and Avolio, 2003; Avolio and Gardner, 2005; Brown and Treviño, 2006; Walumbwa et al., 2008).

George (2003) believes that authentic leadership is about the "heart" and "soul," and he maintains that authentic leaders listen to their inner voice and are deeply connected with their inner values. Servant leadership draws on the leader's self-awareness, authentic behavior, positive modeling, conceptual skills, empowering, behaving ethically, creating value for the community, helping subordinates to grow and succeed, putting subordinates' needs first, and providing emotional healing (Avolio and Gardner, 2005; Liden et al., 2008). Spiritual leadership is built around a concern for others, integrity, role modeling, altruism, and hope/faith (see Fry, 2003). Transformational leadership comprises idealized influence, intellectual stimulation, inspirational motivation, and individualized consideration (Avolio and Bass, 2004). Other dimensions of transformational leadership include leader's self-awareness, and internalized moral perspective (Walumbwa et al., 2008).

Similarly, shared leadership has been noted to involve empowering the team members so that they can self-organize and distribute the leadership tasks among themselves (see Pearce and Conger, 2003; Pearce et al., 2008). Consistent with the findings in the current study, socially responsible leadership has been mostly described in the form of environmental leadership (Flannery and May, 1994; Portugal and Yukl, 1994; Egri and Herman, 2000).

Such leadership involves a dynamic process of influencing, consensus-building, and coalition formation with many internal and external stakeholders (Portugal and Yukl, 1994). Such leaders have strong ecocentric, self-transcendent, and openness to change values (Egri and Herman, 2000). In the opinion of Waldman and Siegel (2008), socially responsible leaders balance the needs and perspectives of multiple stakeholders as well as the interplay between morality and management. The behaviors of socially responsible leaders are not limited to only large initiatives or decisions. They demonstrate positive authentic behaviors in their day-to-day actions and decisions. Similarly, transcendental leadership is characterized by its "contribution-based" nature, meaning that transcendental leaders are driven to develop the motivation of people and they work toward aligning the aspirations of the people with the team's objectives so that they can achieve their common goals (Cardona, 2000).

Various dimensions of "self-transcendent" leadership have been discussed and advocated as effective forms of leadership by scholars. However, findings in this study show that authentic leaders are innately self-transcendent and that wins them the trust of people. These findings reveal that authentic leadership cannot be compartmentalized into one or two positive forms of leadership. Self-transcendence inculcates values, such as altruism, self-sacrifice, unity with nature, and social justice (Sosik, 2005). Having these values, one's attention is focused "away from the self and toward helping others, recognizing one's connectedness to natural or spiritual systems" (ibid., p. 227). In this sense, authentic leaders are similar to Sosik's socialized charismatic leaders who manifest self-sacrificial behaviors and aim at building trust, earning acceptance as role models, and gaining loyalty and organizational commitment from followers. Authentic leaders also reflect self-transcendent-centered values that motivate the leader to promote the welfare of others (also see Michie and Gooty, 2005).

In other words, self-transcendence can take many forms in which authentic leaders seek to do things that are focused on benefiting others, providing a greater good to society, while ensuring the observance of social justice, ethics, morality, and service in everyday actions and decisions. In this sense, authentic leaders are self-leaders as well as self-transcendent leaders who are able to first lead themselves and then lead others – not by authority or control but by service, self-sacrifice, empowerment, and social justice. They are able to connect their inner self and values with their purpose and motivations. This connection enables them to go beyond themselves and engage in self-transcendent leadership.

Summary

The categories of authentic leadership discussed in this chapter and the subsequent ones offer an understanding of how authentic leaders function. This chapter discussed two major categories: self-leadership and self-transcendent leadership. These categories do not reflect different forms of leadership. Neither do they represent any leadership style. The categories illustrate how authentic leaders engage in the leadership process. They first master self-leadership, followed by self-transcendent leadership, and then they endeavor to sustain their leadership. These aspects of the leadership development process do not take place in a linear fashion. They are cyclical and cumulative in nature and can take place in parallel and interrelated manner to each other. The discussion in this chapter shows that authentic leaders are inner-focused and others-directed, which means that they are highly self-aware yet self-transcendent leaders. Chapter 9 focuses on the processes under "sustainable leadership."

8 Self-transcendent leadership

Introduction

The second category from the analysis, as discussed in Chapter 7, is self-transcendent leadership. The authentic leader not only focuses on the self, but also seeks to contribute to the well-being and development of the followers. The concern of the leader extends further to include all in the organization, as well as society as a whole. This chapter considers the processes that relate to self-transcendent leadership. An authentic leader goes beyond the self and makes a genuine contribution in the lives of many other people.

Self-transcendence and its subcategories

During the interviews with the authentic leaders in this study, another major category that emerged strongly was "self-transcendence," which can be described as going beyond the self, rising above the self, or thinking or acting beyond one's own interests.

The leaders interviewed showed self-transcendence in their motives, goals, leadership philosophy, decisions, and actions. They were not keen on following their own motivations, goals, ambitions and achievements. They were more concerned about the welfare and well-being of their followers, colleagues, organizations, and society. These leaders were epitomes of what can be considered self-transcendent leaders. They were keen to help their followers and organizations grow and be successful. They shared how they were actively taking steps to transfer their knowledge and experience to their followers through mentoring, teaching, and coaching. Several leaders enjoyed helping others to achieve their goals. They had already achieved so much in their careers and were ready to give back, to play a positive role in developing the next generation of leaders, and to leave a legacy.

Through analysis of the interview data, six sub-categories emerged under the category of self-transcendent leadership, discussed below:

- soulful leadership;
- servant leadership;
- spiritual leadership;
- shared leadership;
- service-oriented leadership;
- socially responsible leadership.

The study revealed that authentic leaders are inner-focused (self-aware) but their ability to align their values, purpose, goals, and ambitions translates into self-transcendence or the desire to contribute to the lives of others. Self-transcendent leaders give priority to the needs of their followers, organizations, and other stakeholders.

Soulful leadership

In the sub-category of soulful leadership, various themes emerged such as the leader's ability to feel for the followers, show concern for others, lead with the heart and soul, make sacrifices, show fairness and reasonableness, show concern for the organization, and grow others. The CEO of a quantity surveying firm noted:

> I am very sympathetic to my employees, because I know how they feel, because I was there. I am concerned about their working environment, and about their pay as well, because I was like that not many years ago, long before I became a manager and now the holder of this office.

An executive highlighted the important attributes of authentic leaders, and suggested that "heart and soul" are important for authentic leadership. Another executive noted that "selflessness" was indispensable for the authentic leader. The general manager of a developer's firm also underscored the need for leadership that takes care of the human side of work:

> There have to be courses on leadership in the education of civil engineering professionals. We need to teach the civil engineer the human aspects as well. We have to have a heart. We can win over people, and get them to perform, in a better way … and not only from a position of power.

Several interviewees, while talking about themselves and other authentic leaders they had known, noted that leadership involved a "huge sacrifice." Authentic leaders are prepared to make sacrifices for others. One noted: "[Definitely] … when you're in a leadership position, you do have to spend a lot more time, you have to sacrifice a lot of your personal time to foster the interests of your followers and others."

Servant leadership

Several leaders interviewed were serving in top positions in their organizations. Therefore, for many of them, the next promotion or further career growth was no longer an issue they needed to consider. They were more driven to give back, and considered themselves as servants of their followers and their organizations. For example, the deputy general manager in a contracting company also shared a similar perspective:

> I don't consider myself to be a very strong leader to be looked up to and admired by others. I don't have such a desire to be seen as a leader. On the contrary, I prefer to be someone who is less visible, more humble, and less looked up to by other people … Now that I have reached this stage of my career, I wish to fulfill the role I am given to the best of my ability.

The chairman of a consulting group indicated that his leadership philosophy was deeply influenced by his personal faith and he also saw himself as a servant leader:

> I follow what is essentially the path of a "servant leader." And why so, because I guess it fits in with my own personal belief that in the end, if I want to lead, I must be prepared to serve. And that translates into the way we run our business.

In the spirit of servant leadership, authentic leaders do not see themselves as authoritarian leaders. They are prepared to serve others and contribute to the lives of others. Another CEO shared:

> That is the fundamental part [servant leadership]. I think it drives me [to think] that we are here to serve, and we are only stewards ... So you need to remember that servant part of leadership. And I need to grow it, make it a much better thing, and then I can pass it on. And part of the growing it is actually growing the people who can then take up the role.

The general manager of another developer's firm shared that he was trying to instill the philosophy of servant leadership in his followers as well: "We have to be servant leaders, and you must know your people. I am trying to inculcate in my successors the desire to be good servant leaders." A CEO noted: "I am only a steward and a custodian of this business for the future generations." A director in an architecture practice underscored the idea of servant leadership for authentic leaders:

> A leader is one who serves ... who really serves. If you are such a leader, you will get people who will work tirelessly for you and be dedicated to you. The human part of it is very important ... The ability to feel, the ability to give, the ability to serve.

Spiritual leadership

The authentic leaders interviewed often discussed their spiritual or religious beliefs while sharing their leadership philosophy and approach. They referred to God-consciousness, religion, and spirituality as an important feature driving their leadership approach which had significantly influenced their working styles and their lives. For example, an interviewee noted that: "I draw a lot of inspiration from my religion." Another executive shared: "Being God-fearing is ... very important to me."

The CEO of a developer's organization noted that:

> Being a Christian is an important part of my life. We have to make our peace with God. I won't call Him my role model, He is more than that. He [God] sets out for me what is right and what is wrong.

The CEO of a consulting firm noted that:

> For me, it [being spiritual] has taught me the moral values I have. I am scared to do things that are wrong. And I know He [God] helps me when I am stuck. Sometimes, when I am in a hard situation, my thoughts just come to me with the solutions I need.

Another CEO noted that he believed that he had to be righteous in his work; and he often spoke about this to his followers. He noted:

We must portray a sense of righteousness. There is no point saying: "I am a good person, we are a good company." You have to go out and extend this goodness to others ... we do everything fairly, and portray righteousness in our work. That's how I approach these things.

Another CEO observed:

I am a very spiritual person. I spend most of my spare time doing church work. If you love God, you represent God in every situation in your life, especially when you are head of an organization and in charge of all the staff, the representation is an even bigger issue. I am very conscious of this because whatever I do wrong, He [God] gets the blame too ...

A senior executive noted:

I have one very simple motto that I put in all my speeches and addresses: "Pray hard and work hard." For me, there is no other formula for success. The Almighty [God] is very important to me. I think He [God] is the one who gives me all the good thoughts, because I ask for them.

A director in a quantity surveying firm also shared:

Most important, we pray hard as well. I am not here because I am [name omitted]; I am here because God is helping me. There is a divine power. I ... cannot fully explain it. In this job, there are so many problems. Literally, I stop my work, go back home, pray about some of these issues and when I come back to work, sometimes, answers to the problems just come up.

Thus, many interviewees thought their spiritual and religious beliefs provided them with the chance to anchor their actions and deeds in their daily routine. They leave the outcomes of their efforts to God and seek contentment in the results they obtain.

Shared leadership

Under the shared leadership sub-category of self-transcendent leadership, concepts that emerged include: empowerment of the people, generosity in leadership, joint leadership, shared decision-making, and shared leadership manifesting itself in shared opportunity and responsibility, shared benefits, and shared knowledge. The CEO of a developer's firm stated that he had empowered the local executives in the firm's overseas subsidiaries to function independently. In his opinion, this was necessary to enable the business to run successfully abroad. He observed:

[T]he trust and the freedom given to the people to operate through local experience is more important than having strong centralized control. That was a very conscious decision we made. Whether you can call it "leadership," I don't know, but that was what comes naturally to me.

Another CEO observed that he believed in a shared leadership with his organization's employees. He stated:

> I have noticed that some leaders in the industry are driving themselves and their companies but not helping others along. Those people, I observed, had limited success. So, very early on, I took the position that I must help those around with me, those working with me and those working for me; to help them and in so doing to help myself, because together we become a stronger unit and a stronger force.

The president of a consulting group also shared his views about joint leadership and shared decision-making:

> I would like to leave behind a company or group where there is joint leadership. We have a committee of three leaders which is managing the whole company and there is actually a lot of discussion on what is best for the company. Our staff feel that they are an important part of the company and they are proud to be associated with the organization. Other companies have partners. Here, we have joint or shared leadership. It does not mean we have a "no leaders" situation. You have a leader but you can rotate [leadership] … you need one person to make decisions. You can rotate the responsibility for that …

Another dimension that the CEO of a consulting firm mentioned was sharing his share of the firm's profits with the employees:

> I have made sure that our employees are financially rewarded. So I used to share some of my part of the firm's profit with all of them so they feel they are part of the company … I want them to feel that they are part of this big team. [Even] the best football captain cannot play the game alone. You must have a very good team behind you.

The leaders are ready to share their experiences with others, and to provide their followers with guidance and advice, and nurture them as future leaders. One CEO noted:

> In my current position, my major strength is my working experience of over 20 years that I can share with my staff. I can tell them of my experience, and what I did in situations they have not encountered, or even heard about before.

Service-oriented leadership

Another type of leadership that was emphasized by the leaders interviewed was "service-oriented leadership" or, in this case, with respect to the construction industry, "client-oriented leadership." The interviewees expressed their desire and commitment to provide a high quality of service for their clients. They noted the importance of customer satisfaction, relationships, and service. The CEO of a quantity surveying firm observed: "I want people to have this caring attitude for the clients, and for the projects we work on." A deputy CEO of a consulting firm noted: "We are an engineering organization. We want to be recognized as an engineering organization which is customer-oriented."

Many of the leaders interviewed mentioned the importance of "knowing the customers' needs." A director of an architectural firm noted that:

> You need to know what the client's expectations are. A lot of the time, the words they use have a different meaning to me from what they might be expecting, and

similarly, they might not understand the terms I use owing to our different educational and business backgrounds. So, you are not sure if they are really in synchronization with you.

A senior executive in a consulting firm explained the importance of "knowing what the client really wants" by giving the following example:

It [healthcare] is different from commercial projects because usually for commercial projects, the client wants things to happen fast; whereas for hospitals … there are two types of clients. There are big companies which have got extra money … and they are doing the project like a gift to the community. They have no idea what they want … They have no idea who the doctors are going be, and who's going to manage the hospital. And after you have worked with them and asked them enough questions, taken them through the design, outlined the subsequent stages of the project and their own roles, then they realize how much effort and investment it takes to build a hospital. They appreciate your patience and your service.

The CEO of a developer's firm also highlighted the importance of knowing the needs of customers, from the client's perspective:

We have to know what the market and the customers want. If we don't start from that point of origin, whatever we do, it is never really going to be spot on. And you need a good turn of the randomness of luck to turn a potential failure into success.

A director of an engineering firm also underscored the importance of client-focused service, sharing: "I want us to … improve and elevate ourselves to provide more value-added services to our clients." A senior executive in a developer's firm also shared her viewpoint about client service: "the philosophy should be to give the best value for money. We should make sure that … we will give people the best investment in the home they are investing in."

Socially responsible leadership

Under the broader category of self-transcendent leadership, "socially responsible leadership" also emerged as a sub-category. The leaders interviewed exhibited social responsibility as an important element of their leadership. They considered themselves to be responsible members of society. They also viewed their organizations as part of the nation's economic, social, and environmental system and were conscious of how decisions made by them and their teams impacted this ecosystem. Given that their organizations were undertaking major construction and development projects with significant social and environmental implications, the leaders wanted to be sure that their decisions and actions were well informed and carefully deliberated. They were well aware of the importance of sustainability in their projects and were adopting new technologies and developing their organizational capacity to deliver sustainable projects.

The concepts which emerged under this sub-category include environmental protection, social justice, individual responsibility, and aspirations for sustainable development.

Environmental protection

Some of the interviewees highlighted the importance of the contribution of the construction industry to the attainment of elements under the environmental pillar of sustainable development. For example, a senior executive called for greater environmental awareness and consciousness among construction practitioners in Singapore in the work they do at home and abroad, noting that: "We need to adopt environmental practices in our method statements, and offer 'green' alternatives."

Tracing the transformation of the objectives of the firm in its projects, a senior executive in a leading developer's firm put forward the historical development of the main features of the properties his firm builds, as well as a vision of the desirable features of those it will complete in future:

> We started with laying bricks to build small houses in Singapore. For almost 20 years, we did the same thing and laid bricks. However, the last eight years have been different. We have reached a stage where people know that our product quality is good. I hope that in the next ten years, people will say that our properties are efficient to maintain and economical to operate, energy-efficient, and sustainable.

A senior manager in a developer's firm highlighted the need to focus on the long term in the design and construction of buildings, while also calling for change in the approach to the design as well as procurement of buildings:

> There are a lot of things being done in the green movement, and there are many ways in which the construction industry can contribute towards sustainable development ... A real estate developer spends three years to go into a barren site to develop a building. The occupants actually stay there, in Singapore, perhaps for at least 30 years before any significant change is made to the building ... So, I think if I have a chance to change anything, it would be to make sure that whatever we are building can benefit the occupant for a good deal longer than we usually plan them to do.

Several of the leaders interviewed suggested that it was necessary to find appropriate means for evaluating projects which did not consider solely financial objectives. A senior architect noted:

> [W]e have to build good buildings, and balance social responsibility, the quality of the building and financial success. This is the unique part of our company's policies and operations. Sometimes, we look at figures and say that we are making too much profit and not paying too much attention to the design. Sometimes we spend part of the profit we would have made to improve the quality of our buildings. The two are interrelated. If you are not careful about the product you are creating, you will not succeed in the long run.

Social justice and individual responsibility

Several leaders in the study expressed their concerns about the social responsibility of their companies and themselves, first, as individual practitioners, and then as leaders in the construction industry of Singapore. The leaders showed their concern for social justice. They expressed the need to pay attention to the well-being of the employees of their firms, and of all stakeholders in the construction industry. A senior manager in an engineering firm was of the view that there should be greater awareness of safety issues, and efforts to provide better conditions for workers on sites. He noted:

> Safety is important! In construction, what we do can lead to loss of life. And it does happen in the industry. Therefore, a certain amount of understanding of how the industry should operate and how to make this industry safer is required. It is important for all of us to give it more thought ... there should be certain laws and regulations in place which compel us to pay more attention to safety, so that market conditions alone do not dictate the industry's ways and means.

The leaders also highlighted the attention they were paying to their own individual social responsibility. The CEO of a large group of design consultants, while sharing about his leadership purpose, noted:

> I ask myself, "What do we want in this lifetime, what do we really want?" At a certain point, we don't just look at money alone. But giving back to society, helping others, eradicating poverty, all these are worth doing. It would be great if I could do something really big and memorable for my country.

Another CEO, while relating her own difficult childhood, emphasized how she lives out her belief that children should be provided with education in order to equip them to make more significant contributions to society. She noted:

> This [poverty] is what I went through during my childhood and early adulthood. But where I stand today, I firmly believe that education is the most important thing I have got. I now place quite a lot of emphasis on education whenever I am invited to give a talk on any broad societal topic. I stress the need for parents to be mindful about the future of their children. I emphasize that they should keep their children in school to keep them out of the trap of poverty.

The director of an architectural practice shared about his current involvement in charities, and expressed his desire to make even greater contributions to social development in the neighboring developing countries. He observed:

> What is your calling? ... You have to know it. I utilize all the knowledge I have to contribute to the improvement of people's lives, and to give to society. I would like to contribute to water technology. Water is the most important among these amenities. The global population is increasing and we have to manage the urban population and solve water-related problems.

The general manager of a large developer's firm showed his passion for doing his part to develop his profession and the construction industry as a whole, and by so doing, to contribute to the well-being of society at large:

> I believe as a senior professional, you have to contribute to society. Therefore, I participate in various industry development initiatives organized by the government agencies ... to develop programs or software or procedures to standardize professional

practices, in order to improve quality, productivity and safety in the construction industry, whether it's in the design, or engineering, or in procurement.

Summary

The second category which emerged from the analysis of the interviews is self-transcendent leadership. In this category can be placed manifestations of the incidences where the authentic leader does not focus on the self but rather pays attention to people and issues beyond the self. They go beyond their own motives, goals, and individual achievements to help their followers and their organizations to perform better and to grow. They were not driven by a need to add to their own achievements and make progress in their careers. They are actively pursuing this motivation. They are engaged in sharing their knowledge and experience with their followers, and mentoring, teaching, and coaching them. The authentic leaders are concerned about making a difference in society. There were examples of interviewees who give this focus on others priority over their own needs.

The analysis of the interview data yielded six sub-categories under the category of self-transcendent leadership: soulful leadership; servant leadership; spiritual leadership; shared leadership; service-oriented leadership; and socially responsible leadership. These sub-categories are examples of positive forms of leadership. They emerged clearly from analysis of the experiences shared by the leaders interviewed. This confirmed the notion that authentic leadership embraces all the positive forms of leadership, which is discussed in Chapter 4.

9 Sustainable leadership

Introduction

This chapter continues the discussion on authentic leadership influence. As noted in Chapter 5, the third category that emerged under "authentic leadership influence" was "sustainable leadership." Integration of several middle-level categories showed that authentic leaders strive to sustain their leadership in order to ensure that the good aspects of it do not wane over time. They do not consider leadership to be their individual act; instead, they take steps to establish systems and a culture that encourage the people they work with to grow as authentic followers, perform beyond their own expectations, and eventually take over as leaders.

Being vigilant in their outlook, authentic leaders understand that they can sustain their organizations by nurturing their people, forging lasting relationships, and developing a positive culture. This neither happens overnight nor is this a one-off activity. Authentic leaders focus on developing a culture that brings people together, and they do not look for short-term gains; they focus on the long-term sustained success of their organizations.

In this sense, sustainable leadership refers to strategies, actions, and decisions of authentic leaders to develop, maintain, and grow authentic leadership values and culture in the organization. In this chapter, focus will be on how leaders engage in various strategic social processes to sustain their leadership in organizations. Middle-level categories under "sustainable leadership" that emerged from the abstraction of lower-level concepts in the research include the following (Table 9.1):

- vigilant outlook;
- creative problem solving;
- building high-performance and sustainable teams;
- sustainable human capital development;
- sustainable social capital development;
- leading and sustaining change;
- building sustainable organizations;
- sustained success planning.

Vigilant outlook

Lower-level categories that collectively generated the concept of "vigilant outlook" include:

- inspirational vision;
- providing courageous impetus;

Table 9.1 Sub-categories under sustainable leadership

Category	Sustainable leadership							
Middle level categories	Vigilant outlook	Creative problem-solving	Building high-performance and sustainable teams	Sustainable human capital development	Sustainable social capital development	Leading and sustaining change	Building sustainable organizations	Sustained succession planning
Lower level categories	Inspirational vision Providing courageous impetus Systemic perspective Vigilant decision-making	Turning problems into opportunities Objectivity in approach Being futuristic Creativity in solution System approach	Fostering collective efforts Leading from the front Singularity of purpose and direction Building team values Developing team chemistry Harnessing followership	Attracting, retaining, and growing talent Acknowledging the individual needs Motivating constantly Realizing the career growth opportunities Job satisfaction Empowerment Developing organizational citizenship Showing care and concern Informal way of developing relationships Mentoring and guiding Leading by example	Relationships within organization Relationships with other professions in the industry Relationships with clients	Simplifying the uncertainty Strategic business modeling Creating and focusing on the vision Communicating effectively Developing entrepreneurship Involving the stakeholders Operationalizing the existing strengths Flexibility in leadership Institutionalizing the change	Sustainable organizational culture and values Sustainable business growth and success Sustainable client service Sustainable company leadership in the industry Sustainable product/service differentiation Sustainable geographical diversification Sustainable management systems Sustainable HRM Sustainable knowledge management	Developing a succession plan Identifying and growing leaders at home Mentoring and coaching the future leaders Smooth transition for succession

- systemic perspective;
- vigilant decision-making.

Vigilant outlook refers to authentic leaders' ability to understand the complex and dynamic details of their contexts, envision their future, and realize their holistic perspective through their actions and decisions. Through their vigilant outlook, authentic leaders are not only able to inspire a vision, they refine their goals by staying focused on the bigger picture. They are also able to garner the support of their followers by providing an appropriate impetus. They also take the responsibility for their own actions as well as the actions of their followers. Their approach to decision-making helps them to take the decisions necessary to transform their plans into actions and results.

Inspirational vision

When leaders were asked about their leadership philosophy, they frequently noted that having a vision and setting a common goal for followers are essential aspects. Leaders should be able to inspire their followers for something bigger than day-to-day work, providing a goal toward which individuals, teams, and organizations can work. More importantly, leaders should be able to communicate their vision effectively to their followers.

Referring to the importance of vision, the CEO of a consulting group noted: "I believe good leaders are able to formulate their vision and then motivate others to carry the vision through … you must still have people behind you to carry that vision through." However, having a vision is not sufficient; it is imperative to communicate the vision to permeate the whole organization. "You need to know how to cascade your vision down to all the many teams and different levels in the organization," shared one CEO. Knowing the destination and persuading the followers to work together toward attaining the common goal sets the leader apart. For example, a CEO noted:

> We will have followers when we share the same goals and objectives and they see we are sincerely working toward those objectives. We have to try to be as fair as possible to everybody, but I know it's not always that easy to achieve this in practice.

Putting emphasis on setting the direction and getting the support of followers, another senior executive stated:

> You are supposed to do two things [in leadership]. You set the direction, and how things should move. The second part is to get people to support you … Now, in construction today, the part of setting the direction … it demands that you put the stress on quality … but you also need to ensure safety, progress with respect to time, and cost control, and each of these issues is like a priority in itself. As a leader … you must be able to make offsets and balance the priorities and … make the list very clear to the team working with you.

Providing courageous impetus

Individuals in leadership positions must work in their own way to provide direction, impetus for action, and opportunities to take the initiative. As their failure can result in profound consequences, they might have to make unpopular decisions sometimes. They are not always able to reveal their concerns, in order to keep their followers motivated. Thus, only they understand the severity of the matters they are dealing with. Therefore, it is often said that leadership is a lonely path and leaders must find their own way and solve their problems. Therefore, "providing courageous impetus" is important for the leadership role.

The CEO of a large consulting firm emphasized the importance of long-term thinking by the leader, noting: "I suppose, what is most important is that as a leader, you must be able to see what [the] real long term is." While taking the long-term view, the leaders do not always make popular decisions. The CEO of an engineering design firm observed:

> I really think good leaders make all the difference. And good leaders don't work on the philosophy of being popular because, as a leader, you sometimes need to make some unpopular decisions. But that is the role of the leader: to look ahead, to project what is coming, to be able to see into the future, and then lead and motivate the people.

However, the leader must also take responsibility for the collective decisions and actions, as well as those of the followers. The CEO of a quantity surveying firm gave this example of good leaders: "They do not expose you when you're in trouble, because they take the responsibility ... if something goes wrong, they are the ones responsible. I think leadership is taking responsibility, not exposing your team members."

Under the category of "providing courageous impetus," another concept that emerged was loneliness at the top and the weight of responsibility that leaders felt in their positions. One CEO noted:

> He who wears the crown takes the lonely path. [As a leader], who do you turn to? You are alone at the top. And the sad part associated with leadership is that the leader must worry first of all, because of all the uncertainty in construction. It can be a huge sacrifice. On the other hand, as a leader, you celebrate after you and your followers have done it.

Another CEO also noted that he had to work much harder or be perceived to be doing so, in order to set an example, and to keep his staff motivated. He stated: "Very often, I am the last one to leave the office, even until now, in my old age. I stay behind, although I don't need to do any of the work; I will just stay with them." However, while providing courageous impetus, the leader has to show positive firmness. Giving an example of positive firmness, a CEO noted:

> There are people who do not listen to you and seem to have problems in understanding you. You can't help it much. You have to stand firm and tell them that they need to change. Sometimes this works, but sometimes it does not.

Systemic perspective

Under "systemic perspective," several concepts emerged, including: all-encompassing success, conducting the orchestra, producing a play, versatility of skills, and seeing the bigger picture. The interviewees noted that they saw leadership as a "whole-sum" and overarching concept in which they had to understand the objective, appreciate the context, set the direction, and demonstrate strong leadership within and outside the organization.

The leadership of authentic leaders is not limited to their own organizations; they are also playing leadership roles outside their organizations. A director of an architectural firm noted that leadership is about "all-round success." He noted:

> You have got to be handling the people both inside and outside the firm. So, leadership means that you are able to relate to your own people as well as others outside the organization ... You have to bring success to the organization and respect will come when you achieve this all-round success.

Another CEO viewed leadership as a role in which the leaders were not self-centered; they were rather more concerned about others and were driven by the passion to bring success to their organizations. He observed:

> Leadership is a very all-encompassing role ... it's a lot of teamwork as well ... Even if you can map the way, you can point people in the right direction, you must still have people behind you to realize that vision.

A senior director in a consulting firm echoed the notion of "conducting an orchestra" as a function of leaders:

> In a senior position in a profession, you are like a conductor. You have to know all the instruments to lead the whole orchestra, the whole team. If you have not been through all that, you may not know who is making the mistake in the orchestra. If you have not been through it, you don't know whether the violinist is playing at his best or not. I am lucky to have gone through all of this in my earlier career. My experience was very wide and deep in many aspects.

The general manager of a developer's firm considered the roles of the different participants on a project, rather than within a single organization when he noted that: "the developer is only a facilitator; he is a conductor and must be a good listener." The CEO of a consulting firm related leadership to producing a dance item or a play. He noted:

> Among our senior staff, we actually did agree that this is the right way to go; that we must put in systems that will serve our people, so that they can do their dance. [Leadership] is like producing a play, or a dance item. You have to have all those things; the lighting, the music, the necessary stage, props and all that. Once you do that, they can dance. Otherwise, they can't even do what they need to do.

Another director in a quantity surveying firm noted the importance of considering things from a different perspective. He noted:

> I have learnt how to deal with people, to take a step back and redefine the whole thing. I have learnt how to look at things differently ... then straight away, the whole ball game changes. For example, if someone is putting pressure on you on a certain issue until you feel backed up into a corner, if you look at the issue from another angle and you raise a new question, the whole thing can change because the person then has to react.

The CEO of another consulting firm noted that a leader should be "versatile" in order to perform well in numerous roles under different circumstances. In his view, the leader is one who is able to see the "big picture." An executive director in a contracting firm noted:

> I look at the bigger picture in most of the issues but I do go into the details in some cases when I think it is necessary for me to do so in order to find the most appropriate course of action.

The discussion in this section shows that authentic leaders have a systemic outlook. They are able to appreciate the big picture and understand the context in which they operate. This helps them in their day-to-day decision-making in a contextualized manner.

Vigilant decision-making

Decision-making emerged as an important category during the analysis. Almost every leader interviewed mentioned the importance of decision-making. The leaders explained why decision-making was so crucial for leadership, how they made decisions, what factors they considered while making decisions, and how they dealt with the consequences after they have taken the decision. The decision-making process, especially when one is in a leadership position, is often complex, may be multidimensional, and is always contextual. In some sense, leadership is defined by the ability to make the right decisions at the right time in different situations. A senior executive in a consulting firm said that "it can be very frustrating, when the leader doesn't decide."

A senior vice president in a developer's firm observed that the project manager has to make the eventual decision, although this has to be based on the information various stakeholders provide. She observed:

> The quality of decision-making is the key. The project manager is not an expert, [neither is the manager] a specialist in every aspect of the project. He is a generalist. The project manager is a people manager. You rely on your service providers to provide the service and there, you make a decision or in some cases, a choice. That is why I have to respect what advice they give me. From there, I tell them what my decision is, based on my own knowledge or feedback.

The process of decision-making can be intricate and multifaceted. Articulating the process of decision-making, a leader noted:

> I think the main thing really is that you have to be confident enough, you look at the information you have available, and you have to make a decision. And then you have to convey the reasons for your decision to your followers. And if there is any change in the circumstances, then you may have to change your decision. But if people know the reason why you make the decision, they can also help you to look out, in case circumstances change … And if the decision turns out to be wrong, you have to just see what you can do to rectify it.

The interviewees noted that empowerment of followers along with provision of clear instructions and direction were important to facilitate the decision-making process. The CEO of a consulting group noted that empowerment was fundamental to his decision-making. He noted:

My other approach is empowerment ... give your staff the authority to make decisions BUT and a big BUT ... always ask them to think of implications, that is "see things that others do not see." ... Also, I tell them, "If there are problems, try and think through your options and then come to me to discuss the options. Do not come and see me asking for my decision if you have not in the first place, brainstormed through the options." Such training will help them to think better.

During the decision-making process, the leader must remain objective. He or she must take stock of the situation and then decide. The CEO of a developer's firm commented on this, noting:

For decision-making, one has to set aside the ego. The ego is important in some ways. You need to have it so that you can feel that you can do things. But one has to have a sense of balance and choose to do something that is less glamorous for the firm, but is more appropriate to the market at the time.

In making decisions, the leader must be courageous to take decisions which may not suit the majority. Also, authentic leaders, while making decisions, keep the balance between firmness and flexibility to ensure that the decisions are for the greater good of the organization. Maturity and good sense are also required in the decision-making process. Also, the decisions should be shared and in a timely manner so that everyone can work toward the objectives set. An important characteristic of vigilant decision-making is that the leader takes the responsibility for the outcomes. A senior executive in an architectural firm noted that:

If I make a decision, I hold the responsibility for that decision. I won't push the blame onto anyone else if things do not work out ... [Sometimes] to them, I make a decision that is against their interest ... Maybe I make wrong decisions [sometimes]. But I take the responsibility for them.

Creative problem-solving

"Essentially good leadership is about good problem-solving or meditating between problems and solutions to problems," said the CEO of a quantity surveying firm.

The analysis revealed that authentic leaders engage in creative problem-solving while performing their leadership roles. They do not look at difficult issues as problems. They turn the problems and challenges they face into opportunities and find creative and sustainable solutions. As a part of "creative problem-solving," the following sub-categories emerged:

- turning problems into opportunities;
- objectivity in approach;
- being future-oriented;
- looking at the bigger picture;
- using a participative and a system approach to find solutions.

Turning problems into opportunities

Authentic leaders consistently noted that they did not view problems through the "problems" lens. Instead, they treat problems as challenges and turn those challenges into opportunities. Giving an example, the general manager of a developer's firm noted that "you should look at problems as opportunities. Problems are there to get you to attain greater heights."

The executive director of an architectural firm also noted that he "took the bull by the horns" and treated problems as challenges. He observed:

> I have made it a point not to use negative terms, because I believe using negative terms encourages negative thinking. So, my architects know that there's never a problem, there's always a challenge. During our project meetings, I'll always ask them, "So what's the challenge?"

A similar view was expressed by the CEO of a consulting firm who shared that his company opened an office in New York which was not doing well. They took a risk in closing it and opening another office in another city in the United States:

> More often than not, there is an opportunity in a failure … We saw that the New York office was failing, so we set up an office in Dallas and that is doing extremely well … So, sometimes, you know, there is an opportunity in a failure and you need to be able to see it.

Objectivity in approach

Analysis showed that authentic leaders always kept the eventual purpose in mind while approaching a problem. A senior associate in an architect firm noted:

> There are always challenges. Every day is a problem-solving day and you learn from there. Every day, there is a challenge. That is why I always encourage all the team members and ask them what we should do. Instead of giving them a solution, if they have an alternative approach or other proposals, we make the final decision by going back to reconsider our fundamental objective.

Another senior executive in a consulting firm also explained the importance of being objective:

> Different people will come with different ideas, views, and ways of doing things. I always tell them we must set our objective clearly. There could be several ways to achieve the objective. We list all the ways we could do things. Our objective has to be clear and set. Then we find the most viable way of doing things to attain it, considering our resources. If they do not agree [as sometimes happens], I come in to join the action and I decide upon the things myself and explain the decision to them.

Future orientation

While solving problems, a leader should be able to see beyond the immediate problem. The leader should remain future-oriented and ensure that the solution adopted is sustainable in the long run. A director in a consulting firm noted that the leader should be able to foresee the risks before they become issues:

In this profession, you have to go through that process, and be able to see the problems, 10 steps ahead. You need to foresee things in the future. You have been through the process and you can see problems ahead that the younger ones cannot see.

The CEO of a quantity surveying firm also agreed with the above point of view by comparing the followers with soldiers: "The leader must be able to pre-empt [risks], so that his soldiers don't fall into potholes. Your ability to foresee things two steps ahead will help your soldiers."

Creativity in solutions

The leaders discussed the problems and challenges they faced during their daily leadership activities. They were always creative during problem-solving. A director in a consulting firm, while explaining the importance of creative solutions reflected:

> There was this project on which the staff under me were trying to solve a problem and they were demoralized. The whole project team was also demoralized and they were getting nowhere. I came over and took up some of the tasks and I broke each task into smaller ones. At the end, we came out of it, successful. Incidents like that really inspire me. The staff thought they were facing a dead end. I showed them that it was not as bad as that, and it is always possible to find a way out.

The interviewees discussed the complexity and multi-dimensionality of the problems they faced on construction projects, which called for creative solutions. A senior executive observed:

> I try to understand the situation and see what is possible. I realize that many people think in only one direction. [As a leader] what I have to do or I … really want to do is to analyze all the whats, whys, and hows. After I gather the information, I try to find the solution which I think goes down to real problem-solving.

A participative approach

In order to solve problems, leaders do not rely solely on their own skills, they take a collective and participative problem-solving approach. They let their followers come up with viable solutions so that they can select the best solution from the pool. A CEO shared: "Along the way, if problems are popping up, everybody in the team should help to solve the problems; we should do this together." In the context of participative problem-solving, the general manager of a developer's firm noted:

> One of the major strengths [I have] is that I allow the ideas to flow. In order to allow the ideas to flow, [the leader] must have an open mind. If he has decided what to do, then his mind is closed. Even if you have the world's best leaders, the best ideas would be killed if they don't let the ideas flow.

Commenting on the participative approach to problem-solving, another leader opined that: "It [leadership] is about planning, organization, problem-solving, where you sit down, talk, discuss, and just have to find solutions when problems are causing the project to be stuck somewhere."

The CEO of a consulting firm also shared that she believed in the participative approach to solving problems. She observed:

> So everybody's view is important … I have to be fair to everyone and give them a chance to participate, to be listened to, and to be heard. I don't want to be surrounded by just a few who agree with me, but make sure that all the people are able to approach me and feel that they are well heard.

A systems approach

Leaders are able to see the bigger picture and have the ability to put things in perspective so that they can analyze a given situation from all angles; it can be considered that they do this by using a systems approach. Concepts which emerged under the sub-category of "systems approach" include: clearly stating the problem, analyzing the problem, looking at the bigger picture, putting problems in perspective, learning through the process, and methodical problem-solving.

In order to apply the systems approach to solve a problem, one must clearly state the problem to avoid any ambiguity in the process of finding the solution. A director in a developer's firm noted that: "The best way to reduce your error is to state your question clearly. You have to put the question in the simplest form."

The next step is to analyze the problems and the various viable solutions. "You look at the problem, you analyze it, you divide it into different areas," said a deputy CEO. A director in an architectural practice also shared: "We always analyze what the problem is. Quite often, we have to then sit back and see … how we can build or sharpen up our skills to tackle it."

A director in a quantity surveying firm also noted: "My strength is to put problems in perspective and almost without much difficulty I can identify key issues. Lately, I have been able to better strategize how to deal with problems."

The leaders know that not necessarily will all problems result in the right solutions. They consider the learning that takes place during the problem-solving process important. The managing partner of an engineering consulting firm noted: "At the end of the day, you may not get what you want but the process is perhaps more important than the outcome. Through the process of putting effort in to get what you want, you learn."

Building high-performance and sustainable teams

Analysis of the interviews of the authentic leaders resulted in another major category which was called "building high-performance and sustainable teams." From the responses of interviewees, they were devoting efforts to building and sustaining high-performance teams within their organizations. The importance of team-building in construction project management comes through in this comment by a senior manager in a developer's firm:

> In project management, we have a leadership position … we have a team position … the project manager's role is to deal with the complexities of the project … to be a part of the team but also to lead the team … he is more like the link for all the players …

Another interviewee noted that:

> [I]t is easy to overcome [hurdles] if you have a good team of players. [With] a team who are good leaders in their own fields, you can perform much better … you can

excel … A leader by himself can only do [so much]. So, the team-building effort is important, to encourage your subordinates or your team players to gel with you. That will give greater success than your own personal success.

As a part of "building high-performance and sustainable teams," the following sub-categories emerged from the analysis:

- fostering collective efforts;
- leading from the front;
- singularity of purpose and direction;
- building team values;
- building team chemistry;
- harnessing followers' support.

Fostering collective efforts

Leaders in the study believed in team efforts and were determined to achieve their goals by fostering collective efforts. Effective leaders are able to get the best out of individuals; they are also able to harness individual efforts into a team effort. Maintaining his strong advocacy for team effort, the executive director in an architectural firm noted:

> One of the things in which I have been a very firm believer, even when I was a student when we used to have group charettes or group assignments, is that it has to be a team effort. Yes, within any team there has to be a leader, but then the leader has to learn how to encourage contributions and delegate, so that everybody feels that he or she is contributing their required share toward the final output.

The interviewees also emphasized the importance of the team-player role of the leader. They believed that the leader should participate in the work of the team. The deputy CEO of a large developer's group noted: "You must be very approachable so that they can share their woes and issues with you. Sometimes, the staff get frustrated, but you have to work with them to get the job done."

Leading from the front

Another notion that was seen under the broader category was leadership from the front. The interviewees noted that the leader should be able to lead the team from the front and must not be afraid of taking risks, making decisions, delegating the tasks to the right people, and motivating the team members.

A director in an architectural firm observed that delegation was essential for the success of leadership in teams. He observed:

> I think delegation of authority is the key thing. How much you delegate and for what reasons [are important]! That has something to do with trust [among team members] and it has also to do with the company's structure, and the definitions of roles and responsibilities in the organization.

The leader should be able to recognize the strengths and weaknesses of each of the followers in order to perform an effective leadership role. A vice president of a developer's firm noted this, and pointed out that the leader should be deeply grounded in his or her team in order to lead them effectively:

> In teamwork … we are in a partnership … Each person has his own strengths. We need to recognize that. There are some types of people that you can give general guidelines to, and they can follow them through, then you have other players who are not so independent. Here you have got no choice. But not everyone understands that there are such differences. If there is a conflict, you need to look into the team's [problems] in depth to understand the matter and find a solution that works for them.

Also, the leader should be able to command respect and make the people follow his or her vision without having to impose authority and power. One interviewee noted:

> Would they [the team members] be able to carry out their objectives without my imposing decisions with my authority? If your staff are doing what is required on their own accord because they understand the philosophy behind completing the tasks and the organization's goals, then that would be leadership because somehow, you have inculcated in them a sense of achievement of goals without your having to manage that.

Leadership from the front also demands that the leader should be enthusiastic and inspire enthusiasm among the team members. A senior manager in a contracting firm remarked that: "You have to show your enthusiasm towards the work, and that you care about the work … the followers will also do so if they see you being so committed."

Singularity and clarity of purpose and direction

For any teamwork, singularity and clarity of purpose and direction are of vital importance for positive outcomes of teamwork. If the team does not share any purpose or objective, it is not a team in the first place.

A senior executive in a developer's firm gave the example of national leaders to underscore the importance of singularity of direction: "Great leaders are those who managed to convince the whole nation to think as a team and work towards the same national goal." A deputy general manager in a contracting firm noted: "In this firm, the model of leadership is more about consensus. We are consensus-oriented in what we do and how we go about doing it."

Another senior executive observed the importance of alignment of goals. He observed that: "There is always a struggle when you are dealing with different situations and different people. The biggest struggle is making people realize that you all need to align yourselves towards achieving the common goal."

Another director commented: "Having a direction, for us, is important. It is helpful for you to plan and give a time frame to people to complete their tasks." A senior associate in an architectural firm explained the importance of purpose in this way: "You cannot simply tell people to do [something]. You have to tell them why we should we do it and for what purpose. And then they … will be better at doing it."

Building team values

The leader builds the team on the foundation of team values that includes mutual trust, respect, and discipline. Fundamental to any team is the trust team members have among themselves. "Because everybody has his own stake in the whole team," said one senior manager in a developer's firm. He explained:

> If the decision that I make is a little a bit stubborn ... of course they [team members] will try to oppose my decision ... they will just go into a debate, argument and [be in a] temper ... In the end, I prove to them that if I make a decision, I take responsibility ... In the beginning when everybody starts to know each other, they won't know me so well ... Along the way, they understand my style. I ask them to solve the problem with me. I don't play the game of finding scapegoats or blaming other people when things go wrong. So, as the project goes on, they will start to have respect for me and say that ... "OK, this man is a doer; and he does take the responsibility for our outcomes even when we make mistakes."

A senior director in an architectural practice reflected on the racial diversity among the employees of the firm:

> The good thing about our firm is that the staff members are mixed within the firm and they accept people as they are. I think that is one important value ... we must respect our colleagues. This way, I think, you get the best out of each other.

It was clear that team members would contribute toward the collective goal when they have mutual respect for each other. It is the leader who develops the culture of mutual respect, said one senior vice president. She noted:

> I started off with having a lot of respect for each of the respective professions. Having said that, people do not have the same abilities; there must be a fine balance between respecting them and allowing them to do something and trying to ensure that they do their work well. I start with respect, teamwork and then we try ... to move those who are not very good out of the team and eventually out of the firm.

Self-discipline and team discipline are other attributes that were mentioned by the leaders interviewed. The CEO of a consulting group noted:

> Before we can lead others, we must know what we are doing ourselves. And from there, the rest of the attributes all come out basically which are: the ability to make decisions, being able to motivate others to follow that vision you have set, and being able to empower other people to do what they should do rather than trying to micro manage them, yet being able to monitor what they are doing, and being able to make corrections along the route if necessary.

Developing team chemistry

Leaders build teams, inculcate the spirit of togetherness, and instill the enthusiasm required to achieve the set goals. They develop the team culture and know how to create a conducive working environment for the team. Good leaders make the team members feel the deep connection among themselves. A senior associate in an architectural practice observed:

> My team members enjoy working here. Sometimes, there are people from other teams in the firm who want to come and work with me in my team … I think, it is communication. And on and off, I go out with them for drinks to loosen up. They appreciate that too. You just don't have to think about the work all the time. Relaxing sessions are also important. Those sessions are also good for obtaining feedback. They talk to you more freely.

A vice president of a developer's firm also commented that the leader bridges the gaps between the team members and ensures that they communicate effectively with each other. He observed:

> [In leadership] … we … bridge the gaps. We form a bridge between the users and the consultants and the contractors so that we can realize the common idea. You always have unclear thoughts from all sides. You have to create an environment where you facilitate people and enable the project to move forward. You have to be effective. If they can do things without you, there is no point you being around.

Another important concept that emerged under "developing team chemistry" was "job matching." A director in a quantity surveying company explained in detail how she used a "job matching exercise" to choose people for her team. She explained:

> We also do the job matching. We try to match people with relevant experience with the sort of project they are most suitable for. That sort of planning and human resource part of it is also important. You do come across some clients with a certain character. So, you avoid putting people who do not match that on that sort of project. It is important to get the right people in the appropriate places.

Harnessing followers' support

Authentic leaders understand the need to harness support from their followers and they invest their energies in connecting to people and gaining their support. They demonstrate their deep consideration for their followers. The CEO of a quantity surveying firm noted:

> [I]n life, you must try never to hurt a person's feeling; that's most important! Once you hurt a person's feelings, I am just worried; … losing a job is not a problem, we can always bid for the next one, but you devastate that person's confidence if you then take it out on him. That's the biggest problem. He has his whole life ahead of him, and if you make him feel inferior, it's very bad, so I'm very conscious of this.

Authentic leaders understand that people do make mistakes; therefore, they allow for affordable mistakes so that everyone can have a chance to participate and learn from the experience. A senior executive remarked on this:

> I think people, if you trust them, you should allow them to make mistakes, but not major mistakes … you have to have a tolerance for mistakes. However, I have no

tolerance for those who don't do their work. I give them a chance. If they don't perform, that is it. Not that I give them a second chance, I tell them that I don't think we
can go on working together. I get someone else to do the job.

To get people on their side, authentic leaders practice open communication with, and
understanding of, their followers. The leader should learn about the culture, values, and
history of his or her followers to understand them. The deputy general manager of a contracting firm observed that:

[Y]ou need to understand, talk to them, open up to them and know what they really
enjoy and what they really want, and what is their long-term goal. Then you can plan
your work and use your people according to their abilities.

Sustainable human capital development

Authentic leaders strive to sustain their leadership. They ensure that the good features of
their leadership do not wane over time. Therefore, they share their knowledge with their
followers and develop them into future leaders. This way, they are able to multiply their
leadership influence by inculcating their philosophy in many others.

Sub-categories under "sustainable human capital development" which emerged include:

- attracting, retaining, and growing talent;
- acknowledging the individual's needs;
- realizing career growth and opportunities for people;
- job satisfaction;
- empowering people;
- developing organizational citizenship;
- mentoring, teaching, and guiding;
- constant monitoring;
- leading by example.

Attracting, retaining, and growing talent

Good leaders are excellent talent hunters; and they are able to retain these smart and
capable team members. Several leaders in the study noted that "attracting, retaining, and
growing talent" is a big challenge; they also considered this as key to developing authentic
followers who will become future leaders. One of the interviewees, while underscoring the
importance of talent, noted that:

[We] need to acquire talent so that we can do innovative things, so people don't always
have to go to American or British companies. Partly, it is because Singapore has a very
small talent pool and good people aren't coming into the industry so what we need to
do is, honestly, to go abroad. We are advertising this weekend in seven different
countries … We will try to get the best talent, no matter the cost.

In order to retain and develop the talent, the leaders noted that organizations should invest
in human capital development to prepare the talent for the future. The chairman of a
quantity surveying firm noted:

I look at the people in different tiers, because at certain times you have got to move them up. These are the people who are going to do the work and bring in the business, not me. So for me, the challenge is always to do it, I won't call it HR, I call it Human Capital Development. I make it the top item on the agenda for me now before I retire.

The CEO of a contracting firm also emphasized investment in identifying, attracting, retaining, and developing the talent:

We have a lot of engineering and construction graduates; we also have MBAs and we spend a lot of money to further develop the professional and management capabilities of our own staff. We support their development to ensure that we build up their trust and commitment.

Acknowledging individual needs

The leaders also noted that every individual is different and should be treated in a different way. Leaders face this challenge of dealing with people with widely diverse talent, needs, and immediate professional and personal problems every day. One interviewee observed:

The most difficult problem is the human part. There are no set rules and set standards. It varies from person to person. Each person has different requirements. People have to be clearly differentiated and you have to understand every individual.

A senior executive in a consulting firm commented, when talking about developing the followers:

Not everyone is looking at money. We have to talk to them and understand what is their requirement and motivation in their careers. We, once or twice a year, sit down with them and discuss what their expectations of the company are.

The leader cannot understand his or her followers unless he or she is prepared to listen to them. "A good leader must be a good listener," said one of the interviewees. The CEO of a consulting firm also noted: "To lead, I would like to listen to people, give my engineers a chance to contribute their views and after I have listened to all of them, then I will be the one to make the final decision."

Explaining the individual needs of his followers and how he addressed those needs, a director in an engineering design firm shared:

There are different people who join [my] team … The people in my team come from seven nationalities, from different cultures and religions. I enjoy handling this. You do get people who are more stubborn and set in their ways. I would let them try their preferred method sometimes … it gives them learning power.

Motivating followers constantly

One important job for a leader is to generate and maintain enthusiasm among his followers. The leader should understand what motivates his or her followers and how he or she can align followers with the goal of his or her leadership. A director in a consulting group remarked that:

> Human beings are actually … territorial animals; everybody has territory because if you put that person in that seat, he somehow will define his territory in that space. A territorial animal requires a lot of motivation do his best, to get out of the territory. A good leader would have to listen, and, after that, motivate.

He went on:

> There are many types of motivations. And it has to do with the personal qualities of the recipients of the motivations. Some could be motivated by small compensations, some could be motivated by big amounts of money, and some could be motivated by other forms of rewards or recognition. So, determining the appropriate form of motivation is also a skill, and the leader has to treat the followers as individuals.

In order to motivate people, leaders also provide opportunity and responsibility on the job. An executive director noted:

> I strongly believe in giving people the opportunity to reach their maximum potential through training. I am a great believer in delegating work to people but with certain guidelines and leadership! Try and bring up people and give them opportunities to work together.

Realizing career growth and opportunities for followers

When the followers know that they have good opportunities and further prospects of growth in the organization, they also demonstrate promise and commitment. The CEO of a consulting firm remarked that:

> The biggest, the best way to convince the followers to continue to work in the organization is to show them that the organization is progressing, it is improving; that it is on an upward trajectory … Your own competency must be evident, to convince the followers to follow you.

A director of an architectural practice also shared the above perspective: "We want to create a path for each individual architect. In fact, we have a growth path down to every draftsman. We want people to improve. If you are a draftsman, you can become a construction manager."

Job satisfaction

Understanding people, motivating them, and growing them for the future were also related to job satisfaction by many interviewees. A senior executive in an architectural practice noted:

> Job satisfaction [is important]; we provide room for them [the staff] to explore their expertise … If you constantly do residential jobs, for example, after a while, you get sick of it … so in a sense we do offer the staff a certain variety of projects which, I think, is important for the firm's development and also for the staff who work with us.

To achieve job satisfaction, a CEO considered appropriate compensation and financial reward imperative: "This is what we are trying to achieve here and I have made sure that our people are financially well rewarded. I share my profits with all them so they feel they are part of the company."

A senior manager in a contracting firm also articulated his views about the importance of job satisfaction: "I believe that [you need to] motivate the staff [and] you have to give them appropriate rewards. You have to encourage them ... give them encouragement in terms of reward when they have achieved a milestone."

Empowerment

Developing followers is closely linked to empowering them so that they can make decisions on their own and take responsibility for their decisions. Good leaders empower their followers. The CEO of a developer's firm with overseas subsidiaries noted: "The trust and the freedom given to the people to operate through local experience [are] more important than having to have a strong centralized control. That was a conscious decision we made. Maybe, you can call it leadership."

Emphasizing the importance of empowerment, a director in an architectural firm also noted: "I like to empower my staff. I give them support. To me, each member should contribute to the decision-making by the team. These should be rational, and it does not matter if they are wrong."

Developing organizational citizenship

Many leaders expressed their desire to develop organizational citizenship among their followers by instilling in them their leadership philosophy and company values. "I instill my philosophy into my staff," said the managing partner of a quantity surveying firm. The general manager of a developer's firm said that he always strived to build positive values in his followers:

> We inculcate in our PMs [project managers] the understanding that we have to collaborate with others ... We have to be servant leaders; and you must know your people. I am trying to develop my likely successors into good servant leaders.

The chairman of a consulting group also expressed her desire "to inculcate a culture" where all employees can "share and learn as a group." However, in order to instill the values in others, the leaders stressed character and role modeling. A vice president in a developer's firm noted: "You cannot instill in people what you do not have; you have to be the person you want them to be ... When you lead, your character shows. People will have to see your character."

Explaining the issue of developing the followers, a senior executive noted the significance of inculcating "organizational citizenship" among followers. He does this by making them feel needed: "I tell them 'I need your help' ... I always say that I need their help. Making a connection with people is important." A director in a consulting firm also shared that: "You have to make them feel at home. To keep your team and company efficient, you have to let them know that you are keen to get to understand them and to guide them through."

The CEO of a consulting firm also expressed his desire to develop people who are loyal to the firm and are not primarily motivated by the money. He noted:

Our next step is to ... have a pool of people that would not leave the company, ever; they are part of the company. That's something which I am working on currently. Acquire a pool of talent that will be an inherent part of this company, who think that they own the company and who are not motivated by money [only]!

Showing care and concern

To garner the support of people and motivate them, the leader must demonstrate care and concern for them. A director in a consulting firm made a point regarding "showing care and concern" to develop followers:

> I want to talk to them nicely ... There are some bosses who shout at people and scold them because they believe they need to do that to get things done. They make people miserable. I am not going to get things done that way.

Another senior executive noted: "Treating everybody as individuals, trying to understand if someone's vision or objectives differs from yours, and why it is so. Being able to communicate with that person to see your viewpoint, I consider these to be important."

A director in a quantity surveying firm also made a point about showing care and concern:

> Taking care of staff is very important. I learnt this from Jose Mourinho [then the manager of Chelsea Football Club]. He took care of his staff by shifting all the media attention from his staff and his players onto himself. Alex Ferguson [the then manager of Manchester United] also does the same thing [he takes up the media questions to safeguard his players] ... You also have to be open with the staff and explain your intentions to them.

Informal approach to developing relationships

Interviewees noted the importance of informal ways of building authentic relationships to retain and develop their followers. Explaining the importance of informal relationships, the executive director in an architectural practice shared the following:

> We even have our own social club; we organize fortnightly events; visiting old people's homes, some other voluntary activities, going for bowling, celebrating the birthdays of the children of the staff and even their other immediate family members. Every Friday, there is some light refreshments in the office. Every fortnight, we organize a full breakfast for the staff in the office; a proper breakfast ... The reason is very simple; everybody is working so hard, you want everybody to know that their work is being appreciated and is being valued, and that each of them is not just another number in the firm.

The executive director in a quantity surveying firm shared the informal ways he used to develop relationships with his staff:

> I also play hard with my team. Once a month, I take them out for drinks and this involves about 20 directors, and ... I buy them any drinks they want; we enjoy ourselves. And the thing is, as it is said, work hard, play hard.

Mentoring and guiding

In developing their followers into future leaders, the interviewees highlighted the importance of mentoring, teaching, and guiding. Describing the significance of mentoring, teaching, and guiding for developing the followers, a senior executive expressed these points:

> Now, when you have a lot of young architects, one has to really guide them, really watch over them, so to speak, because 19 years ago, I started as a young architect. If I had not been given the opportunity, I would never have fulfilled myself, both professionally speaking and as a person.

The CEO of a consulting firm considered himself to be good at teaching; to him, this was his strength. He noted:

> I am able to teach people. I have the patience to sit down with young people, work through a problem, teach them how to do it and I take pains to go through their work in character without ever losing my temper.

Another executive director gave an example of training sessions he was conducting to grow his staff in the company. He shared the following:

> You need to express yourself well, you need to write well. So this is the type of training I am giving in my office now. What I learnt from the EQ side of things from age 13 till now, I am putting them into practice now ... and sharing them with my people.

Constant monitoring

To develop followers into effective future leaders, constant monitoring and performance evaluation were also given considerable importance by the leaders interviewed in this study. One senior manager in a contracting firm noted:

> It is through observation and monitoring over time that we select people. They may not be with you from job to job. They are transferred to other jobs on other sites and they sometimes come back to you. From among them, you can further pick out those with the right attitudes. If one of them is a leader, that person will be easily recognizable among the pool.

In the process of monitoring, the leaders also help their followers by pointing out important issues in their performance evaluation. The president of a consulting group shared that:

> If you want the whole group to grow, you must really believe that it is important, and not just because you say it. When you talk to each person in his or her performance assessment, you must really want to help the person.

Leadership by example

"Leadership by example" has been explained in the section on "authentic role modeling." It is a significant concept in developing authentic followers. A director in an architectural practice explained why leadership by example is essential to develop authentic followers:

[Leadership] is very much by example. If people close to you see that the directions you give are a good way of getting things done, they will try to moderate their behavior accordingly. But if [you] are the nasty person, there would be a lot of reasons why your followers would turn into nasty people. That develops a culture of "nasty-ism." Sometimes, it happens due to two reasons: (1) the followers want to be nasty to reduce their frustration; and (2) the followers think this is the only way to get things done.

If you see that some of the followers are doing that … you have to tell them openly what the situation is. You have to be frank with them and get them to take note. We do have people like that but over the years they change. But you can't change them too much, too fast.

Sustainable social capital

It was also found in the analysis that authentic leaders believe in building and sustaining long-term relationships within their own organizations, across the industry with other professionals, and with the clients they serve.

Relationships within the organization

Leaders saw developing relationships within the organization as an important tool for building connections across the fictional boundaries. Discussing internal relationships in the organization, the CEO of a quantity surveying firm observed:

> I am a very friendly CEO; I still treat them [my colleagues] as friends. I am not an authoritative person. They call me by my first name and all that … So we say that, we started this whole practice as friends, we must always maintain this partnership as friends … This is something we enjoy and we take pride in. Actually, the philosophy is very simple; we just want to be very honest in everything we do, and we want to be friends.

A director in an architectural practice, while talking about the legacy he was aspiring to leave behind ,also noted that his primal aim was to develop friendships at work:

> I don't want to leave a legacy as a boss … I would like my team to consider that we are good friends, no matter where I am. I want to leave peers, friends … and I would like to be remembered as a "friend leader."

A director in a consulting group also made the following observation about the importance of relationships:

> We are too busy on normal days and we hardly have any exchanges. But we try to have some organizational activities just to promote internal networking and enable our people to get to know each other. We do this to create a better work environment.

The general manager of a developer's firm shared a story about a project where he had built close relationships with other professionals:

> I try to create team bonding in my job sites. On projects I manage, we reach a stage where, after two-and-half or three years, they are at the finishing stage and they are

now going to say goodbye. But they have reached a stage whereby they can shake hands, they can enjoy, they finished the job together, they still remain friends, and can become even better friends if the job has been successful.

The executive director in a quantity surveying firm noted that he always tried to connect with his followers through informal ways. He noted:

I don't know how many teams I have by now but … every single day I walk to their place, talk to them, crack some jokes, ask about their work, ask if everything is alright. I always tell them if you don't see me at your desk or in the office, you should not be worried. My doors are always open. In fact, I don't have any doors.

Relationships with other professionals in the industry

The other notion that emerged regarding relationships was the building, by the leaders, of industry-wide contacts. The interviewees noted that they endeavored and were able to develop good relationships with their fellow professionals in the industry. The deputy general manager of a contracting firm shared his views on this issue:

I have wide contacts in the industry which allows me to have access to more information which is very helpful for us when we have to choose projects. Mostly, relationships with people in various companies are the key to obtaining contracts … [and] my major strength.

A director in an architectural practice noted that leadership in architecture was more than leading the team within the organization. He noted:

In an architectural firm, you can't just be a leader internally. You have got to be handling people both inside and outside the firm. So, leadership means that you are able to relate to your own people as well as others outside. This is one of the most important criteria.

The CEO of a quantity surveying firm shared his perspective about how he considered other professionals as collaborators and not competitors:

Many would say … "Why do you teach your competitors?" But people do not realize that the more you do that, the better it will be for you. Because people see you as a leader, as a provider of information, they look at you differently from the rest. I think very differently, and I have a very positive response from the industry, from the community, from my fellow professionals. And they have always helped me in return, when I have had difficulties, when I approached them.

A director in a large architectural practice also shared a story on how he had to lead a big project in which the parties were not very friendly to each other in the beginning but he built relationships among the professionals, and led the project to success:

I think I have good skills when dealing with large teams, whether it is in the office, or on a project, where it includes a contractor, other consultants, and so on. When on the [named] project, for instance, we really had a difficult situation to deal with. Because it was a very big public project, we had a steering committee which included the minister and the President of Singapore at that time. And we had a design aesthetics advisory group which included some architects and artists, and so on. We had a users' advisory group which included performing artists and people from the fine arts community, and so on. All these very prominent talented people ... And to manage such people successfully, I think it takes a certain skill which I found through that project. And that has stood me in good stead, working with contractors and trying to get them to do what we want without getting into an antagonistic relationship.

A senior vice president of a developer's firm also shared a story where she had to take over the leadership of a project on which people had developed hostile relationships. However, by building a friendly and collaborative environment and good relationships with, and among, team members, she was able to accomplish the project goals.

Relationships with clients

The other aspect of relationships that emerged in the conversations with the authentic leaders was the relationships with clients. Interviewees, who were mostly very senior executives within their firm, noted that they always kept their relationships with clients high on their agenda. One CEO, commenting on relationship building with clients, noted: "Whenever a client comes to [our company], we don't just talk about business; we do social things and try to see where our common ground is. Then afterwards we see if we can help them with their business request."

The CEO of a quantity surveying firm emphasized the need to build a cordial relationship with clients. He noted:

> We try to make every client a friend. Of course, we don't succeed all the time ... I think this desire to develop cordial relationships with our clients right from the beginning is also held by all my partners, including those overseas.

Leading and sustaining change

Change is inherent in the leadership process. Leadership is about bringing about and sustaining the change. Authentic leaders are not only open to change but are also able to help their people and organizations transition smoothly through the change.

Leaders in the study, while describing the context in which they were operating, often mentioned the process of change and how they were leading the change. Particularly in the context of construction, where firms' operations are based on projects which are, by nature, short in duration and involve almost constant changes, leadership should be able to effectively respond to change. The CEO of a consulting group noted the importance of change:

> I think the ability to make changes and accept changes is a crucial part of leadership. A lot of leaders are happy with being static but I don't mind making fundamental changes and no matter from where the need for it comes. If I see that it's a good idea, I am happy to implement it and include it in our systems and procedures, the way we work.

He noted further: "Some leaders are fixed in their views; I think leaders must be open to answer questions and to deal with change; and they must believe in the need for change." A vice president of a developer's firm noted that change was imperative, and very much part of leadership. She expressed her concern that many people do not want to change and are usually constrained by rules. She noted:

> [For any assignment] take some time to think how you will do it in a more efficient and effective way. If it was done like that before, you don't necessarily have to follow it this time. The rules are made by, and for, human beings, and therefore they can be changed by us. In my previous job, even when I was not so senior, I did change some guidelines saying that they were made by people like us, and so can be changed.

These interviews show that change is imperative in business organizations and leaders should demonstrate the ability to embrace change and make it an opportunity to realize progress.

The following concepts emerged under the broader category of "leading and sustaining change":

- simplifying uncertainty;
- strategic business modeling;
- creating a vision and remaining focused;
- communicating the change effectively;
- developing entrepreneurship;
- involving all stakeholders;
- operationalizing the existing strengths;
- remaining flexible in the process;
- institutionalizing the change.

Simplifying uncertainty

The most important thing a leader does while managing change is to simplify the uncertainty or ambiguity with respect to the critical issues so that the employees can understand. The CEO of a developer's firm, while describing his own experience of leading change in his company related:

> [It's a challenge] when you try to achieve too many things in times of uncertainty, in times of difficulty … At times of great uncertainties, at times of great change, the message from the person at the top must be clear and simple; it should be easy for people to understand.

The president of a consulting firm who also led a major change in her firm noted that:

> We also shared [with the people] the type of business the company will have in the first few years after the change. We made a commitment and assured them of certain things like pay and compensation. We did not do any downsizing. How do you decide who to release? Many of our staff members had been working with us for a long time and there was a fear that many will have to leave. We were convinced that we could not afford to lose a lot of good staff. The decision was to ensure that we bring the whole group over, although it would be tough.

The general manager of a developer's firm noted the importance of calmness in a leadership position. He observed: "In a crisis situation, you need to play a responsible role. Learning to demonstrate calmness in a chaotic environment is an important attribute for a leader."

Strategic business modeling

The leaders emphasized the importance of strategic business planning to deal with change. The CEO of a consulting firm, which had been a public-sector entity, noted:

> In this change process, [we] had three major challenges ... to overcome. I called it the triple jump. The triple jump means we needed to change from the civil service culture to a business culture; we needed to change from a consultancy firm—which provided consultancy services in architectural, engineering services—to a development firm; and more importantly, the third part of the triple jump was to expand our business from Singapore to overseas.

In order to achieve the triple jump, the CEO explained that they had to have a strategic business model: "[We had to] exhibit the kind of entrepreneurship to convince the employees and other shareholders, including our regular business partners, that we had a good case, we had a good business model and there was money to be made."

Discussing strategic business modeling, another CEO noted:

> Leaders must be able to adapt their business model to change in certain circumstances. You cannot just manage whereby you cut yourself down to size. The other is you diversify ... and change the business model to suit the changing circumstances of the market.

Creating a vision and remaining focused

Leaders noted that creating a sense of urgency and remaining focused were key to leading change. The CEO of a consulting firm noted:

> I told them [the leadership team] to be very, very focused; we don't want to stray too much from our core activities, and end up one day and suddenly realize that we are everywhere and we don't know how to manage ... we define very fairly where is our competency, and we use this competency to our advantage.

The president of a consulting group opined on change management: "Even in a situation of change, you need to help draw the boundaries ... Otherwise, there is no focus. You have to focus people's attention and set the boundaries people should keep within."

Communicating the change effectively

In order to lead the change effectively, a leader must underscore the need for open and clear communication to avoid ambiguities and confusion about what is going to happen. The president of a consulting group noted:

> As a leader, you must go over to them. We had a communication committee and a communication plan. We identified key people from the company in the new

organization. We organized workshops to tell the people the direction of the new group. The CEO was a true leader of the whole group. We arranged a lot of meetings for him to present to the staff what we were going to do.

A director in a quantity surveying firm said: "I stress communication. You are totally handicapped without communication. After being in the industry for so many years, a lot of them as a director, I still can't find [anything] more important than communication."

Developing entrepreneurship

The leaders also emphasized the importance of the entrepreneurial approach while leading a change. An executive from a consulting firm, that was corporatized and went through major changes, noted:

> When we became corporatized, you know, we really had to change. And I was given more responsibility also at that time … from being a government department, we had to become a private company. And so, the whole mentality had to change, the business would be completely changed. I mean, previously, we had projects whether we wanted them or not … and we did not worry about having no work … But then once you become a private company, time is of the essence, and it is necessary to make money to keep your job. So the whole mentality changed … we had to learn to tackle the profit and loss and other aspects of business. And then you have to hunt for jobs … We became profit centers, each division became a profit center, and we had to live on our own revenue.

The CEO of another consultant firm also underscored the need for entrepreneurship to sustain the company:

> At that point in time I would imagine, the biggest challenge was to exhibit entrepreneurship; the ability to sniff out the opportunities, analyze the risks, get the people to seek opportunity, get the company to see what the opportunity can mean to the organization, and whether or not it is worthwhile to put in the kind of time, and effort and resources in order to achieve that.

Involving the stakeholders

In order to lead the change effectively, it is essential to garner support from the stakeholders involved in the change process or those who will be affected by the change. The CEO of a consulting firm remarked that:

> It is also [the ability of] leadership in trying to convince shareholders, and [getting the] people to support you. We went through a period, a trying period, eventually for our business in China, also for India … where we are operating as developers. More importantly, we also managed to persuade our shareholders.

The president of a consulting firm which had been a government department also noted that: "When you are in the private sector, you have to be equally professional. We involved the union. We needed to have the buy-in from the union."

Discussing the need to gain the confidence and support of stakeholders, the CEO of a large contracting and development firm noted:

> There was a period in 1998 which was just in the midst of a financial crisis, and there was a lot of turmoil in the marketplace ... There were a lot of market manipulators trying to unhinge the market. [Our company] is a very important part of the business ecosystem, even though we are not a listed company. If we had collapsed in debt, the market would have been unhinged. It would have created the impression that property prices would sink further. The banking system would also have been in a problem. There was a lot of malice ... We had to come out very positively and affirmatively. We took into confidence the media, banks, [and other stakeholders] and explained our financial position. If these rumors were unchecked, we would have lost the trust. I was never prepared for this at all. But I had to deal with it and dispel the stories. That was a very difficult period. It was an interesting point in my life as a businessman. We were a private company and were an easy target for the doomsayers. We were not listed and so our financial position was not clear publicly.

Operationalizing the existing strengths

The interviewees mentioned that change is good for organizations, but one should not forget about the existing strengths which every organization has. Therefore, leaders should build upon the strength of the organization while they implement the change. One CEO, while sharing about the strengths of his organization in the process of implementing a change, noted: "We have a group of very good people; because no matter how good the leadership is, without a team of good people ... the leader cannot do anything at all."

In the context of building upon existing strengths, another leader noted the importance of the organization's human resources. She emphasized that keeping the talent during times of turbulence is imperative.

The CEO of a consulting group also underscored the importance of the track record of his firm while it was going through a major change exercise. He remarked:

> The fact is that ... where we are today, is because first of all we have a very strong team of people ... That is a very important attribute. Plus, the very good track record that we have. What we are doing in Singapore is a good showcase for overseas investors, clients and customers ... We are trying to convince them to engage us to do the design for them. So, I think if you call it strength that would be the biggest strength; a group of good, very competent people working together for many, many years. We know exactly what our roots are. So, this is very much our organizational strength.

The president of another organization that went through change observed the importance of the strengths her organization has. She noted:

> We now go out [of Singapore] to find business. Singapore is small. We go ... to China, India, the Middle East ... as foreign consultants. We have developed quite a few industrial parks here in Singapore. That has helped us in the Middle East. We cannot depend on our previous track record. We have to build our track record quickly in each of the local markets we want to grow abroad. The over 40 years track record in

Singapore is still ours. But we still have to compete like other people. We also have to quickly build [a new] track record in each overseas market.

Remaining flexible in the process

The interviewees observed that the leader should be able to find a balance between flexibility and firmness, and give opportunities to people to adjust to the change. The CEO of a consulting firm observed: "I think, for our leadership … we emphasize a lot on flexibility. The bottom line is important but it is not everything; there has to be some leeway given to people."

An executive also remarked: "I think in my profession, … I would actually say one would be best to have a very flexible leadership style … However, there is no one rule for this." A director in an architectural practice also emphasized flexibility in management in order to cope with changing circumstances:

> We have to have flexible management. It [the business environment] just keeps changing. We never imagined that we will be doing so much work in India. The structure had to adapt to the new situation. Now we have 30 projects in India. Our people travel to India very often. The structure of the practice cannot be the same. It has to adapt to the new situation.

Institutionalizing the change

Another aspect that leaders underscored about leading the change was to institutionalize the change. Some of the interviewees noted that effective change could only be brought about when the organization accepts the change and builds it into its systems. The chairman of a consulting group noted:

> First of all, there is a change, and then there is a changed management, or rather there is an idea for change, and the next thing is you must get agreement for the change. And then you have to manage the change, and then you have to put in the documentation and agreement so that all the stakeholders can work smoothly with each other.

The president of a consulting group expressed his view about institutionalizing change: "Of course, we have introduced new measures which we did not have in the past, which we need because now we have to keep time sheets for our work, and things like that." Another CEO expressed the need for empowering people to institutionalize the change. He said: "I would trust my lieutenants. In fact they are most important to me. Having selected them, I let them have a very high level of autonomy."

Building sustainable organizations

The interviews included some discussion of what the leaders wanted to change in their own organizations and what were their aspirations for their organizations. An important message that emerged from such discussions was that authentic leaders were not all concerned about their own careers or positions. They were thinking about, taking actions, and involved in building sustainable organizations where they belonged.

One of their fundamental motivations was to take their respective organizations to further heights of achievement and success. The CEO of a developer's firm noted:

I have to do the best I can and build an organization better than what it is. We don't run our business for short-term gain. I would like to see my organization becoming one which people would be attracted to, where people share the same values, an organization which is easily understood, an organization with diversity and various levels of skills, and talent. In the way the society is developing today, it [our organization] has to be a self-ecosystem and a microcosm of the society ... a place for everybody.

The chairman of a quantity surveying firm also articulated his thoughts about building a sustainable organization. He reflected that:

The recession of 1976 really taught me about what I should do to have a more sustainable business. It taught me that the quality of service is very important. [Then] to build up the skill base in the various areas of our domain knowledge, and then to diversify geographically, to look at the needs of the client. So, 1976 was, I think, a year that taught me a lot about what I should do in business in order to have a more sustainable business.

He expressed his desire to further sustain his organization:

I would like to build a sustainable firm and provide a sustainable leadership. People make people, people make business. It is not you who gets the business; they are the ones. So you build the people first. Great leaders build companies and build people.

These excerpts from some interviews show that authentic leaders endeavor to achieve sustainability within their own organizations and take steps to build organizations that are sustainable. From further analysis of interviews of authentic leaders, the following subcategories emerged under the broader category of "building sustainable organizations":

- sustainable organizational culture and values;
- sustainable business strategizing;
- sustainable business growth and success;
- sustainable client service;
- sustainable company leadership in the industry;
- sustainable competitive advantage;
- sustainable geographical diversification;
- sustainable management systems;
- sustained knowledge management.
- sustainable human resource management.

Sustainable organizational culture and values

Positive organizational culture and values were underscored by several leaders as being key to the building of sustainable organizations. Without people collectively espousing a positive culture and values, it is not possible to lead the organization toward growth. The executive director of an architectural practice observed:

What is unique, so far as [our organization] is concerned ... is that it is a very warm and cozy environment. It is a very family-oriented firm; it is a staff-oriented firm, in the sense that we even have our own social club.

The executive director of a quantity surveying firm also presented his firm as having a family culture: "Frankly, I hope we can work as one family, one family unbroken. And everyone shares the same culture as I work with them. We maintain cordial relations. We don't scold each other, and we don't shout at one another."

The chairman of a quantity surveying firm emphasized the shared values between the organization and its members. He observed:

> We have come through with the same core values, same vision. It is a very strong family business. It is not that I developed it, if you have a group of people with the same core values, same vision, and then when the leadership thinking is in that mold, I would use the word, it is a very generous leadership ... I am not selfish; I don't keep everything to myself.

The CEO of a developer's firm gave his views about ethical values that are important for the organization's core business:

> We have to be an ethical business. It is not because it's good for business, but it is because it's the right thing to do. We do things right ... by the laws of the country, and the firm beliefs that we have; and we also do things to be right by our customers.

Sustainable business strategizing

This sub-category of sustainable business strategizing refers to strategic business modeling. The leaders noted the importance of strategic management in business development while they expressed their views on sustaining their organizations. Good leaders depend on collective intelligence in the organization to develop and implement business strategies. The CEO of a consulting firm explained this:

> What we call a strategic chalk, we do that every year ... we take all the directors away to Sentosa one day every year ... Every director has to present a paper on what changes we need to make in the company. They must answer: where should we go from here?, what should be our next diversification area?, what is our next move to survive in the face of competition? We do all these things on a yearly basis because, if your [business] model is five years old, it cannot survive! So, this means that every year you must take stock of where you are and make necessary changes to your business model.

The chairman of a quantity surveying firm also shared the importance of strategic management of business:

> We always do a 3–5 years forecast of our business. Not only in financials, but in how the businesses would shape up, what sort of services would be demanded, and how the services would be reshaped by events, by changing demand requirements from developers. For instance, in the mid-1980s, I was developing Design and Build, when the industry was not exactly looking at it. So, in the next 10 to 15 years we took the lead in Design and Build, Gross Maximum Price and other forms of building contract procurement. That's why we were able to work on a range of projects from very small jobs to the very big jobs. And I think I would be proud to say that the firm has done the bulk of all the major developments, not only in Singapore but in the region.

As a part of strategic management, a senior director of a developer's firm noted the importance of finding the right partners in business. He observed:

> In every business, you need to collaborate with somebody. You need to find part-ners who are good in skills your firm lacks which you can put in your team. However, partners can be risky ... You have to ensure that your reputation of ethics is taken care of. People say that Singapore is small. In my opinion, even India is small too. I talk to one person in the morning and then fly to another city 3,000 kilometers away to meet another client. He already knows that I have met the earlier client. I don't know how it works but that is the way it is. We need to maintain our good reputation.

Sustainable business growth and success

The leaders noted the significance of sustainable business growth and success in the development of their organizations. Many of them noted that their organizations were constantly striving to maintain the success they had achieved.

The general manager of a developer's group observed that:

> It is a never-ending race. And if you stop and say that "I am very successful now," it means that you have stopped running. Then you have reached a fatal level of com-placency. Therefore, I don't think I have reached "success," because there are more things to do, more places to go.

The CEO of a developer's group also underscored the importance of sustainable and con-trolled growth. He emphasized focused and controllable growth and sustained success in the business. The CEO of a consulting group underscored the need for excellence in service:

> The best marketing you can do is doing a good job. If you have done a fantastic job for a client, you don't need to market, you don't need to entertain them [the clients]. They will entertain you and you will get repeat business. Repeat business is the best marketing you can do. A lot of people don't understand that, they just think that you must entertain the client and play golf with the client ... but if you do a good job for the client, then he will come to you automatically.

Going further on business growth and success, that CEO noted that consulting companies must achieve a balance between their business success and professional success:

> There are two sides, one is our professional side and the other is our business side. We can't just run it purely as a business and we can't also run it purely as a con-sulting company that makes no money because then you can't invest in technology and you can't invest in people. So, you must have two distinct sides and I have suc-cessfully created that business model to be technically a number one company, innovative, with the best people doing the best projects. Yet, on the business side, very successful in the business manner, continuing to increase its turnover and its profitability. These are the two measures by which you keep your people, and invest in technology and in people.

Sustainable client service

Under the "sustainable client service" sub-category, leaders discussed sustaining relationships and trust with their clients by providing an excellent service. This category closely relates to "service-oriented leadership" that has been discussed in Chapter 8 on self-transcendent leadership. However, sustainable client service also appeared as a key concept under "building sustainable organizations."

Sustainable company leadership in the industry

In order to build a sustainable organization, the leaders noted the significance of company leadership within the industry. In the opinion of many of the leaders, leadership was not about just developing personal leadership. It was also developing the company's leadership in the industry, gaining respect from the peers of the firm' leaders. The CEO of a quantity surveying firm explained:

> My own thoughts about being a leader is not only establishing oneself in the office but also establishing oneself in the profession and in the industry. So, that has always been my target. And in so doing, to drive the firm not only for myself, but drive the firm and all who work with me in the firm to be recognized as leaders in the industry.

He also added:

> The greatest challenge in future would be to maintain the leadership role of the firm [in the industry]—not the role of its individual leaders—and that the firm must continue to prosper and grow … So, the challenge has been to maintain our firm's leadership in Singapore and in the region.

Many leaders expressed their commitment to take their organizations further as leaders in their respective professions. A director in a large architectural practice observed:

> I would like us to get to the point that the best design graduating student in university picks [us] as the first choice … as the firm of choice … if your peers think you're the best, … I think you have achieved some success … if the other architects think you are good, that's very much what we need to achieve.

Many leaders were engineers, architects, or quantity surveys by profession. They highly valued innovation and creativity and believed that their companies should be known as leaders in innovation in the industry. The CEO of a large consulting firm expressed his aspiration for his firm:

> I have tried to make [my organization] a number one company. A company that is highly innovative, and while being a professional company, it still operates like a business. So, I want to make sure that [my organization] will continue to be known as the best engineering company in Singapore … [and that] we have made ourselves at least the best company in Asia, if not the world.

Sustainable competitive advantage

To build a sustainable organization, leaders also stressed the need for developing long-term competitive advantage to differentiate them from other firms in terms of the products and services they offer to stay ahead of the game. The CEO of a consulting firm noted: "We were able to differentiate ourselves from the rest of the firms. I have always thought that we are in the technology business therefore we must be able to do unconventional things, difficult things that others would not do."

A senior director in an architectural firm made observations about the competitive advantage of his firm:

> We are [a] multi-professional firm … There are clients who prefer firms like that. But there are clients who like firms which are specialized in their own disciplines. So, you actually do not always compete with the best firms either in architecture or engineering. We already have an integrated service and we want to sell this ability and derive advantage from it.

The chairman of a consulting group noted the niche areas in which his company offered services:

> Here we have project management, consultancy as well as advisory services. These are three divisions. There is actually no substitute for being very good, for projecting excellence in what one is doing. Because, as a consultancy practice, we have to be extremely good at what we are doing; that is what we are selling.

To sustain the competitive advantage, a senior executive noted the importance of delivering the product. He explained:

> You have to deliver. That is very important at the end of the day … We should be able to say we got this project through our own effort. We have completed it and it is one of the proud moments when we opened the project … Quality is very important for us.

The CEO of a developer's firm maintained that they had differentiated themselves by providing the best quality products which the market needed:

> On a white site, it is difficult for someone to decide what we are going to build … It is difficult for one to say … let's build a residential building. It may look good but an office building may be better in three years' time … But I took the risk and said that we will build a medical centre … We have to be good at product development to know what the market wants and what is sellable … We have to know what that market and the customer want.

The executive director in an architectural practice observed the importance of innovation and creativity as competitive advantages:

> I think the architectural profession is unlike physical construction. We have to be innovative and creative in our service and our products. So, I see this as a challenge because if you don't renew yourself, one of the problems is that the outsider will look at you as a rather outdated and conservative company who doesn't do innovative designs.

Sustainable geographical diversification

As a part of developing competitive advantage, the leaders noted the significance of sustainable geographical diversification. Most of the interviewees belonged to companies which were already venturing out of Singapore and were concerned about adequate diversification. The leaders believed that they should grow their businesses and expand their footprint; opportunities in the local market were limited and therefore they needed to look for projects in the large emerging markets abroad. The CEO of a consulting group observed:

> If we are operating in Singapore and the Singapore market is in decline, what are the options left? Downsize? Or find work somewhere else? Typical guides say that if the total construction business workload comes down from 25 billion dollars per annum to 10 billion dollars per annum you should reduce the [firm] size from 100 people to 40 people. We say no! If we become 40 people we will never be able to pick up a big job again because everybody will say "You are too small." So, we said we want to keep our 100 people. We will look for work outside Singapore; we will extend to China and set up our offices there, in Beijing and Shanghai ... We went to the Middle East, we set up offices in Dubai, Abu Dhabi, Qatar, and other countries and as a result, our size has doubled; and more recently, from 200 people we have become a firm of 350 people.

In growing their global footprints, leaders were of the view that they needed to empower their local offices and not manage them closely from the head office. They recognized that local management teams have better local business intelligence and should have autonomy to localize their business approach. The CEO of a developer's firm also shared the need for the autonomy of foreign offices:

> We adopt the philosophy from our first day ... don't try to have a centralized policy when we come to diversify [geographically] ... I would trust a few lieutenants ... And as they are operating in India, in the Middle East, in China, and elsewhere in Southeast Asia, I cannot have too many things, too many central policies that are limiting them; they must have much autonomy to operate. So the important thing is to select the right people, then trust them, and let them develop.

The CEO of a quantity surveying firm believed in transferring the people from Singapore to the foreign offices when his firm decided to diversify geographically. He noted:

> I identify the people who have been working with us, and I say: "Look, I give you an opportunity. If you want it, please go to India and live there and don't come back [too soon]." I don't like people, who say that "I want to champion the location, but buy the return ticket on Friday night." That's not going to work. That way, to me, you are just a tourist. If you really say, "I want India," then you stay in India.

The same point was also made by the CEO of a consulting firm:

> [W]hen you go to another country, you must be able to send people from here [Singapore]. That brings your culture, your experience, your expertise ... You just can't go to another place and hire a few people and start a new business. Because the work you

are offering is your experience in the home base, your expertise, and your construction technology. You have to physically transfer these to the other, new, place.

The leaders also emphasized the need to understand the local market, culture, and ways of working. A senior executive noted:

> You really need to know what is happening in the market. We can rely on the information on the internet and that from consulting agencies but we need to network and talk to people. We invest a lot of time and resources on market research. "Market intelligence" is actually the expression I should use in our case.

Another interviewee pointed out the importance of understanding the local culture and working style:

> From Day One, you land in a foreign land and you see that everything is alien to you. You have to work there, and perhaps live there for a long time. You have to become adapted to the new environment. It is not easy. Sometimes, our staff face many problems with food and health care. You just have to cope as the clients and the business are there ... In the cocktail parties in India, you would be drinking, then at 10' o'clock they will say "let's go for dinner." You will be eating until 12 midnight and then have to make do with a few hours' sleep. You have to understand the culture and how things have to be done ... You are bound to have surprises. In some places, you have to make sure that you don't shake hands with people of the opposite sex straight away. You can't do such a simple thing like shaking hands everywhere!

The CEO of another organization noted the importance of understanding culture:

> You can grossly underestimate cultural differences; the language you speak could be the same, but the cultural difference is something else ... that you really have got to be there – live there, study there, work there – to understand it ... Understanding of the people, understanding of the market, understanding of the country, the thinking process of the people, what is important to them, what is not important to them – those were the biggest lessons. Don't take the local culture for granted!

Sharing his experience about adapting to the local environment, the CEO of a consulting firm said that his office works on Saturday, which is a holiday in Singapore. They have to adjust to the Middle Eastern schedule to deliver the jobs in that market. He observed:

> They [Middle Eastern clients] were used to using British and American consultants who tell them you can't call them on a Saturday, Sunday, and also these clients don't work on a Friday. When we were there, we told them "You can call us seven days a week, including me. This is my mobile phone number." So, our people [here] we changed their life around. Our people who work here on Middle East projects, work on Saturday and Sunday and take another day off ... The clients say, "You are the first company in the world who has adjusted your work style to suit our working week." We can't tell people that "I am working on your 2 billion dollar job but, sorry, I won't work on Friday, I won't work on Saturday and Sunday and three days in a week you can't contact us." It just doesn't work.

Sustainable management system

Several leaders underscored the importance of sustainable management systems, particularly in the case of their offices overseas. The management system should have sufficient flexibility to adapt to the local environment while having regular reporting to the head office. The CEO of a consulting firm noted:

> You must have a system of accountability … Leaders should not try to micro-manage the people. You also can't let everybody loose either … there must be, somehow, a system of accountability, and KPIs and you need to monitor them.

The chairman of another consulting group explained some of the management systems his company had established to operate across borders. He noted:

> We have set up systems, where basically we have standard drawing templates, standard calculation spreadsheets and templates, rigorously monitored methodologies … There was always a possibility of bringing in some of the knowledge and the institutional framework and systems … we built a stimulus-integrated platform to deliver work anywhere … we can take a job from here [Singapore] … part of the design might be done here, part of it in Auckland … in Wellington or in Christchurch … we call it remote resourcing.

He also explained other benefits of the "remote resourcing" that his company was doing:

> For all our offices, the workload will fluctuate. So, you can have someone sitting in Christchurch, there might be a team of 4–5 engineers in Christchurch for the next 3 weeks loaded up only 60 percent, because they have just finished a major project. But somewhere [else], say in Beijing, there is a loading of 200 percent, and they are totally overloaded … What we can do is that we can skim off that extra loading, put it into a package, translate it through our network, and get our people in Christchurch to do that for the next three weeks. Essentially, as a business concept, you are doing something at zero marginal cost. I am paying all the salaries as well as the rental in Christchurch. It depends on you whether you use 50 percent, 100 percent or 70 percent [of the salaries you pay].

Strategic human resource management

Human resources were noted as the most important source of competitive advantage by the leaders interviewed in this study. Interviewees underscored the importance of strategic human resource management in order to attract, retain, and develop the talent. A director of an architectural firm noted:

> We have to bring in new blood and new directors who are good in design. So we constantly keep track of staff who are good in design work and who have a broad outlook and a good perspective of the future. That's why we have recently recruited some directors who are fairly young, below the age of 40 … we have to get the best people to be in the management team. This is very critical.

The business offering of consulting organizations is expertise and knowledge services for projects. To keep up with the market trends, they need to enhance the capabilities of their professionals and other knowledge workers through training and development opportunities. The CEO of a consulting firm noted:

> There is a lot of training; we send people to other offices and bring people from other offices to work here ... If somebody is leaving, for example, leaving from this Singapore office and he says he intends to go to Australia, we make our best effort to send him to our Australian office, if the person is good. We also have training in every office, anyway ... on projects, we do it regularly in the form of design reviews, workshops, and all that. The other is that when we do the annual directors' conference, we get all the directors from every office, key people, into one place ... once in Bangkok, once in Singapore, next in Australia, where we do a review of our performance in the last year and plan our strategy for the next year.

The CEO of a developer's firm noted some of the human resource management policies of his firm:

> In our way of dealing with HR, we try to be just. According to the viability, being just to the individual ... Our HR policies are different. We are different from the rest of the industry. These things are a reflection of the kind of organization I like to build.

A director of an architectural firm also shared some of his firm's human resource management policies, particularly highlighting growth opportunities depending upon individual needs:

> Most of the staff who work here for two to three years, they stay on. They attain a certain level of comfort and satisfaction in the teams they work with in the organization. It's how we retain the people, it depends on individuals ... The best we can do is to be flexible with timings. We also move people around depending upon their needs. We also move the people within the teams ... We tell people that they will be grown to the next level; we have to inspire and motivate them.

Sustained knowledge management

The leaders expressed their desire for knowledge management. The Deputy General Manager of a contracting firm observed:

> [K]nowledge is important for leaders. I went through thick and thin to gain knowledge about my field. You have got to have a whole understanding of the matter if you want to lead a team. And you have to be thick-skinned too. You cannot say that you don't know how to get something done. If you don't know, you have to ask others ... You have to make an effort to understand the various specialist professions and trades in construction. By understanding others' work, you can understand their problems and then you can make better decisions yourself for the company.

The authentic leaders interviewed were determined to share their knowledge with others so that others can grow and become leaders in their respective fields. Many of the leaders were building knowledge management systems in their firms. A CEO gave an example:

In time to come, with this huge database of knowledge management covering so many areas, educating the staff will come to a point when you will find that even those from the polytechnics are going to come out just as [good], with knowledge as university graduates ... If they can access the knowledge, the playing field is going to be level. So, I am trying to say, if I build a fantastic system here, to support the office here and in the other countries, all my other staff will be accessing it for their work ... Actually, this knowledge management is very important, but people don't realize it ... I see it happening, that's why I spent time building it up.

An executive director also mentioned:

I came up with knowledge management. The knowledge built up in the office should be retained. If a person leaves our firm, his knowledge goes with him. So, we must retain the knowledge and I am conducting classes every Friday now. I find that when I conduct these classes, it helps me too; it reinforces my own learning and understanding.

The general manager of a developer's firm also underlined the significance of knowledge management. He said that scarcity of human resources had necessitated the knowledge management systems and his firm was actively building a system. He observed:

We cannot employ too many people ... we exploit technology to increase the capacity. Less manpower is easier to manage. But it has to be enriched with technology. When we go to another country, we sort out who is who. We establish a system in one country and when it works well, we duplicate the template in other countries, with suitable modifications to suit the context. A young engineer in [our company] learns fast here. I believe in training and knowledge management ... in how to conserve information and how to disseminate and effectively apply it.

Sustained succession planning

Although the leaders interviewed belonged to various age groups, the notion of effective succession was covered in most interviews. Each leader expressed a desire to find someone who could take up their role of leadership. Some leaders mentioned their succession policies and planning. Many spoke about the identification of future leaders, plans for growing the identified leaders, mentoring and coaching programs for future leaders, and the transition necessary for succession. A senior director in an architectural practice shared his views on succession:

At this moment in my career, what I have been telling my colleagues is to be able to pass [the leadership] on to the younger ones to continue with the practice here. People ask me if I am retiring. I guess I would want the younger ones to take on more responsibility. That is something I want to do and [it] should be done as a corporate practice. I see many examples of professional practices that the principal [leader] refuses to pass the mantle on and there is no succession planning. The proprietor takes all the money. When he goes, the firm disappears. That should not happen in this firm. It would be a pity if one does not enable others to learn to lead and take over.

The chairman of a consulting group also shared his perspective about the importance of leadership succession:

Basically, I want to make sure that I leave behind a group of young leaders who will be able to further expand this company and keep the name that we have built alive, and make it a better name because there is always room for improvement. So preparing and putting in position people who follow the same philosophy, that is critical.

The deputy general manager of a contracting firm also noted: "I would like to reciprocate what many good bosses have done for me to certain deserving individuals that may carry on the leadership within this organization."

The group chairman of another consulting firm also underscored the importance of leadership succession, while giving the example of the stewardship model:

It is a process of rejuvenation. And our philosophy was that the young people who are coming up will be starting to drive the business, they should be the ones who will gradually build up their shareholding and grow their ownership of the company ... that kind of underlying philosophy ... actually colors the whole lot of the way you do things. Because you are then taking on a "stewardship model" ... not the "king model" or the "emperor model," if you like ... When you're in stewardship, that means you know you are holding things in trust for the owner and you really have to take care of it, preserve it, and hand it on to the next steward.

Under the broad category of sustained succession planning, the following sub-categories emerged:

- developing a succession plan;
- identifying and growing leaders at home;
- mentoring and coaching the future leaders;
- smooth transition for succession.

Developing a succession plan

The leaders emphasized the need to develop effective succession plans in order to pass the leadership reins on to the next generation of leaders. The CEO of a consulting group outlined his succession plans:

I think ... if something happens to me, there will be no issue. The person who will take over from me is defined and the one who will take over from him is [also] defined. The second and third generation leaders of this firm are all defined, so there should be no issues. I don't believe companies should be based on only one person because of the type of person. So, they must have a very clear definition of leadership.

The managing partner of a quantity surveying firm observed:

Succession planning ... that is a key issue in most organizations. Actually, it is a fundamental policy issue. If it is not attended to, you see in the history of the construction industry, that many firms have wilted in the space of less than ten years. And that has essentially to do with [lack of succession planning]. The leader retains most of the control and authority. So, essentially, the firm dies with him ... This is most relevant to architectural practices where you have a strong leader, and a strong branding with that person.

The general manager of a developer's firm noted:

> I would have succession plans and allow second liner, third liner professionals or project managers to take over to continue to let this organization run. So, my emphasis on mentoring, and on looking at skills development are some of the areas that are critical in any program of leadership succession.

The president of a consulting group also shared her plans of succession: "Leaders must know how to, and when to, exit. I have to plan who can take over my role. I am trying to have somebody."

Identifying and growing leaders at home

When asked how they choose their potential successors or potential leaders in their organizations, the leaders mentioned the identification of talent at various stages of the careers of those who demonstrated some leadership potential. The president of a consulting firm noted:

> One way is to identify those with higher potential to be future leaders and consciously work with them to test their potential. We cannot be developing everyone. We identify potential leaders and give them projects to develop them. This gives us a chance to see how they behave, how they work, and how they perform.

Another leader noted:

> I hope that some of them are willing to step up and take up a role, assume more responsibility, put in more effort to help find solutions to our prevailing problems. I encourage them and push them a bit to go beyond what they think they should. It depends on individuals. Some respond very well ... Some come up to a certain level and stop. For some, it does not work at all.

The leaders preferred to identify potential leaders from within their organizations and develop them to take over the leadership role. This was important as these home-grown leaders shared the organizational philosophy and were aware of the organizational values and working style. A senior director in an architectural practice shared:

> You need to cultivate the right people to succeed and carry on with the structure ... we the first generation of leaders have built the firm to succeed. We don't buy people. We grow our own people. We are conscious that we have to grow successors ... Our people are our most important assets. We have to grow them to be successful.

Mentoring and coaching the future leaders

Next to identifying the individuals with promising leadership potential, interviewees stressed mentoring, guiding, coaching, and teaching. Many of them were actually performing their roles as mentors as they were nearing their retirement. The general manager of a developer's firm noted:

As you grow older, you have something to contribute. You need to mentor and coach your followers. There were people who mentored you, coached you. You should also do the same to others ... I was given the opportunity in Singapore ... a place and opportunity to grow; I, therefore, need to contribute to their [the followers'] development.

Another senior executive noted: "As far as the leadership role is concerned in my organization and to go further, of course, I have reached a stage whereby I hope to mentor my second liners and third liners." Another director, who was nearing his retirement, noted:

At this age and point in time, what I think is that if I can guide some younger people to make them leaders and give them support to improve their leadership, that will be something that I would like to achieve. Give them the opportunity to handle them- selves and guide them. I think that is a good way to train more leaders.

Smooth transition for succession

As they talked about succession and succession planning, the leaders also underscored the need for a smooth transition of leadership. Many had actually started transferring their responsibilities to their successors. The CEO of a consulting group gave an example:

I am still around to sort of mentor and manage all the relationships, and take care of all the problems. So for the last 3 years I have been operating on the basis that !I will do what you guys can't do, I will do what you guys can do but cannot see the need to do, and I will do what you guys can do, can see the need to do, but can't bring yourself to do."

He explained these points:

On the first one, some things, maybe, I can do because it takes relationships, and deep knowledge of certain things which others might not have because I have been pushing the boundaries for the group. So, if there is a problem with a particular client, I can go in and sort it out. The others still can't. They are building up their relationships and confidence in that. That's quite usual, most companies end up like that; most senior professionals, in their last few years, that's what they look for.

The second set of instances is this: they can do it, but they don't see the need to. You know, particularly sometimes in moving into new areas in business, they can do it, but [they think] its "tough," or "Well, really, I don't think it's going to work out. I don't see that this is going to become a viable area of business or a viable location for us to go into" ...

The third case is where they can actually see the need, and see that it's viable, but are of the opinion that it is too difficult. So, in other words, you always have to have people, if you want to break into new areas of work, and sometimes this does not mean new areas of work outside the firm, it's also new areas of work in building up systems internally. So you need to be able to see that it's needed, you need to be able to say "We have people who can do it" and then you have to have the power to say to them, "We are going to make it happen."

A director in a consulting group also shared:

> Of course, you would like to lead the organization with staff who are young and slowly take on the leadership roles. So, there is development that needs to be done. At the end of it, if you have people who have taken on the roles that you were previously fulfilling, I think that is part of the satisfaction you can have, that you leave the organization in a better state of health than you first joined it.

Discussion

Authors such as Hargreaves and Fink (2003) and Ferdig (2007) have used the terms "sustainable leadership" and "sustainability leadership." Using "sustainable leadership," Hargreaves and Fink (2003) add that such leadership responsibly deals with human and financial resources and is concerned with environmental issues. It encourages participation from all stakeholders to solve the problems. It creates and preserves learning; secures success over time; sustains the leadership of others; addresses issues of social justice; develops rather than depletes human and material resources; develops environmental diversity and capacity; and undertakes activist engagement with the environment. Along similar lines, Ferdig (2007) argues that sustainability leaders create opportunities for people to come together and generate answers to key questions. They develop and implement actions in collaboration with others; they make the notion of sustainability relevant in their own lives through their actions. They are realistic, hopeful, and courageous.

Ofori and Toor (2008a; 2008c) also note that such leaders possess the values, attributes, and qualities that help them deal with difficult tasks, and they are well equipped to address the issues of sustainable development. In the drive toward fulfilling the sustainability agenda, authentic leaders are sustainable leaders who capitalize on the opportunities to secure a better future for their people and society as a whole (Ofori and Toor, 2008a; 2008b). They recognize their responsibilities for ethical leadership in sustainable development in relation to their own professions. They recognize that, in order to achieve sustainable development, they – in their capacity as clients, designers, or contractors – must do their part. They are conscientious about their own roles and play their part with full responsibility. By performing authentic leadership at the individual, organizational, and collective levels, these individuals are contributing their part to sustainable development of the built and human environment. Ofori and Toor (2008b) propose that by developing sustainable leadership practices, authentic leaders are not only able to reap the benefits for their own organizations, they also create a better and sustainable future for society.

The analysis presented in this chapter elaborates on various dimensions of sustainable leadership. The findings resonate with what earlier authors have noted about sustainable leadership. However, the findings take the discussion on "sustainable leadership" to a further level. Various dimensions of sustainable leadership are drawn from the qualitative data. Although the existing literature underscores the importance of more or less all the dimensions noted under "sustainable leadership," integration of these dimensions in reference to leadership is achieved in this study. In relation to leadership, there is much discussion on vision (Bass, 1990), the systemic approach (Sternberg, 2007), decision-making (Vroom and Jago, 2007; Toor and Ofori, 2008f), problem-solving (Bedell-Avers et al. 2008), team-building (Hooper and Martin, 2007), human capital (O'Leary et al., 2002), social capital (Adler and Kwon, 2002), change and its management (Kan and Parry, 2004), sustainable organizations (Bradbury, 2003; Manring, 2007; Lozano, 2008; Toor and Ofori,

2008b), and succession planning (Conger and Fulmer, 2003). However, the analysis in this chapter shows how these dimensions come together under the umbrella of "sustainable leadership" and how authentic leaders engage in sustainable leadership and influence their followers and organizations.

These findings differ from the existing literature in at least two ways. First, they show that authentic leaders engage in processes and strategies to sustain their leadership. In previous conceptualizations, authors have put emphasis on self-awareness and self-regulation as two important dimensions of authentic leadership (see Luthans and Avolio, 2003; Avolio and Gardner, 2005). However, the current study expands the conceptualization of authentic leadership to self-leadership, self-transcendent leadership (see Chapter 8), and sustainable leadership. Second, these findings highlight various dimensions of sustainable leadership that were extracted from the qualitative data. These dimensions are more comprehensive than those offered by Hargreaves and Fink (2003).

Summary

In order to sustain their leadership, authentic leaders develop sustainable leadership practices within their own capacity as well as in the organizations they work for. This chapter outlined various sub-categories under "sustainable leadership" that is the third major dimension of authentic leadership influence. Major categories that fall under authentic leadership influence are self-leadership, self-transcendent, and sustainable leadership. However, how these categories are related to leadership development and how both leadership development and influence can be explained together remains a question. This question is addressed in Chapter 10 which integrates all the categories that emerged under leadership development and influence.

10 Integration

Reconciliation of self with social realities

Introduction

This chapter integrates and synthesizes the analysis and findings presented in Chapters 7, 8 and 9. The categories presented in the "paradigm model" are integrated. The conditional/consequential/interactional matrices are developed and integrated. Finally, a story line representing the "dynamic and creative reconciliation of self and social realities," the core category in this study is developed. The focus in this chapter is on selective coding, the final coding phase in the grounded theory approach. Selective coding integrates all the interpretive work of the analysis. Strauss and Corbin (1990) define it as "the process of selecting the core category, systematically relating it to other categories, validating those relationships, and filling in categories that need further refinement and development" (p. 116). The principal objective of selective coding is to explain the story line after integrating the major categories.

The paradigm model

The paradigm model, considered in Chapter 5, was constructed to integrate the processes involved in authentic leadership development and influence. The first step in integration is the selection of a core category that represents the main theme of the research and can be regarded as an overarching concept that can explain all the other concepts emerging within the study. In order to achieve theoretical integration and understand the relationships among the categories, two techniques are offered under the grounded theory approach: the paradigm model and the reflective coding matrix. A paradigm model was constructed to explain the relationships between the causal conditions, the central phenomenon, the contextual factors, the intervening conditions, the actions/interactions (response of leaders in response to the situation, problem, or event), and the consequences (see Figure 10.1). The paradigm model was constructed and used as a tool to understand what was happening when the categories related to leadership development and influences were combined. It should be noted the model was constructed not as a final outcome but as a means to explain what was happening within the data.

The paradigm model in Figure 10.1 shows that authentic leaders function in a dynamic distal (organizational environment, industry context, national culture) and proximal (group and team environment) context. Their leadership development and influence are dependent on their historical context and the transformative events in it, and the experiential learning from it. Authentic leaders engage in various socio-cognitive processes leading to individual and group-based leadership processes, namely self-leadership, self-transcendent leadership,

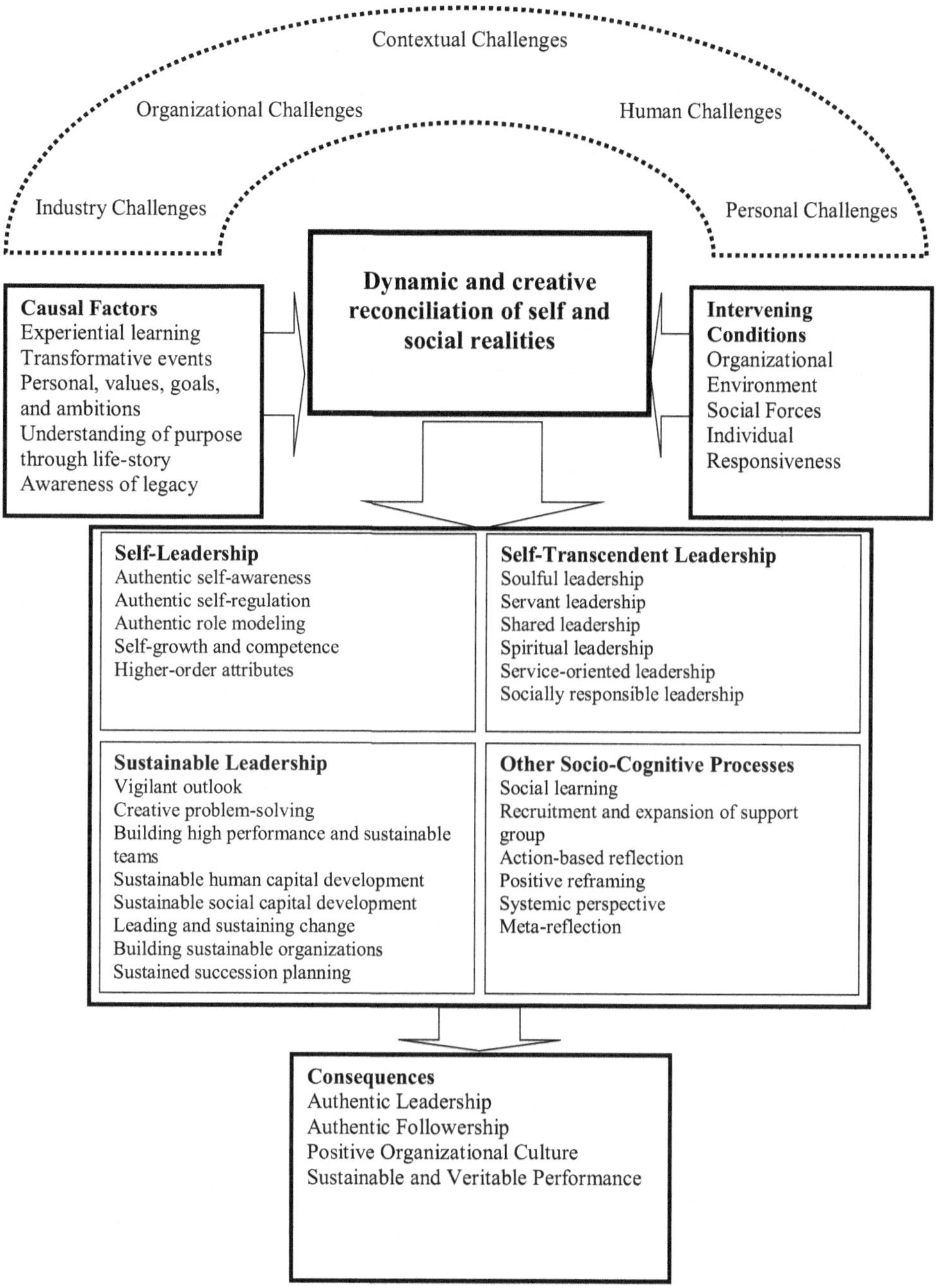

Figure 10.1 The paradigm model for authentic leadership development and influence

and sustainable leadership. In doing so, they constantly interact with their followers, orga-nizational stakeholders, and many other social agents. They also function in a dynamic context while dealing with the challenges related to their organizations and industry. They also deal with their own personal challenges and weaknesses. In doing so, authentic leaders reconcile between the various contexts they find themselves in. This core process of "dynamic and creative reconciliation of self and social realities" is what makes authentic leaders effective and successful.

The reflexive coding matrix

Corbin and Strauss (2008) observe that the paradigm model helps to relate various categories to each other and explain the context. However, the paradigm does not address several theoretical sampling choices; nor does it explain the varied, dynamic, and complex ways in which the conditions, actions/interactions, and consequences coexist. Moreover, the paradigm model does not account for constructions of different actors. These and many other limitations of the paradigm model were overcome by constructing the most advanced level of analysis (McCaslin, 1993) in the grounded theory approach: the reflexive coding matrix. This helped to contextualize the core category, the central phenomenon to which all the other categories were related. The reflexive coding matrix was drawn with the help of simplified forms of conditional/consequential/interactional matrices that are shown in Figures 10.2 (a), (b) and (c). These matrices were developed to capture the higher level of abstraction necessary to move to the final phase of grounded theory analysis, selective coding, and interpretation of the theory in a story line (see Strauss and Corbin, 1990). Figures 10.2 (a), (b), and (c) comprise concentric circles which show that the inner circles are subsets of the outer ones. Similarly, arrows going to, and away from, the center depict the interaction of these circles with each other; the inner and outer circles interactively influence each other.

Although Corbin and Strauss (2008) draw a single matrix representing conditions, actions/interactions, and consequence, these matrices are separately drawn here to explain the dynamics and complexity that are inherent in leadership development and influence. Figure 10.2 (a) shows that authentic leaders exist and operate in various dynamic contexts. They reconcile these contexts while being authentic within the self and within these varied environments they are part of. The "authentic self" of the authentic leader is a subset of the group, organization, society/culture, and cross-cultural environment. On the one hand, these subsets influence the development and functioning of an authentic leader. On the other hand, an authentic leader also influences various contexts surrounding him or her. Similarly, the authentic self of an authentic leader is a function of various dimensions, including values, ambitions, goals, vision, purpose, aspiration for a legacy, and disposition. Interactions of these dimensions in turn influence the development and functioning of the authentic self (see Figure 10.2 (b)). Authentic leaders regularly reconcile their authentic self with their values, ambitions, goals, purpose, vision, aspired legacy, and consequential dispositions. Finally, as shown in Figure 10.2 (c), authentic leadership development cannot be separated from authentic leadership influence (self-leadership, self-transcendent leadership, and sustainable leadership). The continuous interaction of leadership development and influence results in a holistic picture of authentic leadership.

The integration of these matrices is shown in Table 10.1 in the form of the reflexive coding matrix which shows each phenomenon, its properties, dimensions, context, and strategies in order to understand the consequences. Four major phenomena include leadership development, self-leadership, self-transcendent leadership, and sustainable leadership. Their corresponding properties, dimensions, and strategies for understanding the consequences were carefully selected. The reflexive coding matrix was instrumental in the selection and theoretical enrichment of the core category. Guidelines given by Strauss and

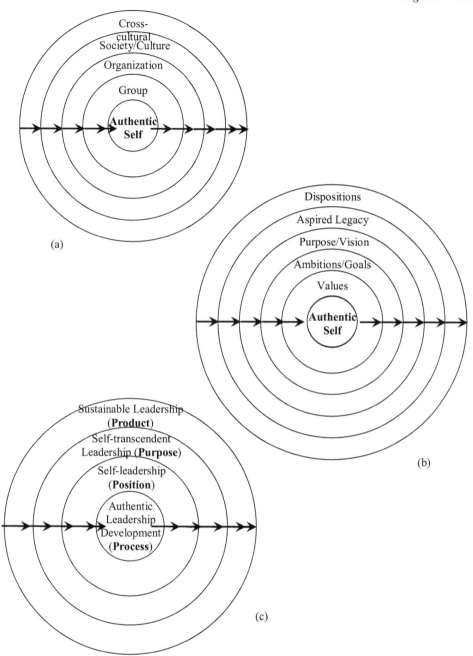

Figure 10.2 (a), (b) and (c) Dynamic contexts in which authentic leaders exist and operate

Corbin (1990) and Corbin and Strauss (2008) were employed to select the core category that could be regarded as an overarching phenomenon representing authentic leadership development and influence. For this purpose, all sub-categories of leadership development, self-leadership, self-transcendent leadership, and sustainable leadership were observed

Table 10.1 The reflexive coding matrix

Core category	Dynamic and creative reconciling self and social realities			
Properties	Process	Position (conviction)	Purpose	Product (achievement/legacy)
Phenomenon	Leadership development (socio-cognitive)	Self-leadership (inner-directed)	Self-transcendent leadership (other-focused)	Sustainable leadership (inner-directed/other focused)
Dimensions	Leadership support group Social institutions Preparation for leadership Polishing and practicing leadership Performing leadership Passing leadership	Authentic self-awareness Authentic self-regulation Authentic role modeling Self-growth and competence Higher-order attributes	Soulful leadership Servant leadership Spiritual leadership Shared leadership Service-oriented leadership Socially-responsible leadership	Vigilant outlook Creative problem-solving Developing high-performance and sustainable teams Sustaining human capital development Sustaining social capital development Leading and sustaining change Building sustainable organizations Sustained succession planning
Context	Experiential learning	Personal goals and ambitions	Understanding of purpose in life story	Awareness of legacy
Strategies for understanding the consequences	Socio-cognitive processes	Aligning personal values with goals and ambitions	Aligning values, ambitions, and goals with purpose and vision	Aligning aspired legacy with purpose and vision

together in the reflexive coding matrix. The most likely core category was seen to be the one that could explain both leadership development and influence.

The reflexive coding matrix showed that authentic leaders continuously engage in the following processes: dynamic, reactive, and proactive social learning; comprehension of inner states, dilemmas, and paradoxes; understanding and aligning personal values, goals, purpose, and aspired legacy; understanding the social context; developing social alliances to pursue their leadership purpose; and seeking authenticity in a given social context. However, an overarching concept, that could represent all categories and concepts that emerged from the data, was yet to be discovered.

A number of possibilities for the core category emerged. Each of them was analyzed to see if it could represent all major categories. It was found that "reconciliation" was a concept with the analytic power to explain what was happening within the data. When related back to the data, "reconciliation" was found in all cases in one form or the other. Initially, the label chosen for the core category was "reconciliation of socio-personal goals." However, this label did not explain anything beyond "goals." Leadership is not only the achievement of goals; it is about seeking social support "within a given context" to achieve a certain collective purpose. The label that was chosen next was "reconciliation of self and social realities," where realities represent every aspect within the personal and social context. When grounded back to the data, this label could not support the dynamic nature of leadership development and influence within a given context. A new label, "dynamic reconciliation of self and social realities" was selected. When grounded back to the data, it was seen that authentic leaders did not merely engage in reconciling the self and social realities; they did it in creative ways. This resulted in the final label for the core category: "dynamic and creative reconciliation of self and social realities."

The grounded theory of dynamic and creative reconciliation of self and social realities

"Dynamic and creative reconciliation of self and social realities" comprises three fundamental aspects: (1) reconciliation of self-realities – reflected in leadership development, self-leadership, and self-transcendent leadership; (2) reconciliation of social realities – reflected in self-transcendent leadership and sustainable leadership; and (3) the dynamic and creative nature of reconciliation in which authentic leaders persistently engage (also see Figure 10.3).

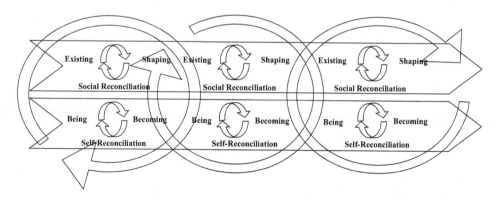

Figure 10.3 Dynamic and creative reconciliation of self and social realities

Authentic leaders strive to achieve authenticity within the self and within the broader context in which they find themselves. Due to transformative events, key turning points, or crucibles that the leaders experienced, they engage in high levels of self-reflection, leading to the transformation of identity or high levels of self-awareness. They recognize their strengths and weaknesses and all about what they are "being." This self-awareness imparts to the authentic leaders knowledge about their personal purpose, values, ambitions, aspirations, and goals in life. Due to their high levels of self-awareness, authentic leaders are able to understand what they really want to achieve in life. They develop a sense of purpose and live with that. This understanding of the self also comes with an understanding of their desired, possible, or future self – what they might become, what they would like to become, or what they are afraid of becoming. In other words, they have a present self – "the being self" – and a possible, future, or desired self. What makes them stand out is that they make a conscious effort to close the gap between their present self and the possible self. This process of constant evolution of self is "the becoming self." They are reconciling their past with their present and their present with their future.

This "being" and this "becoming" are fully grounded in and stem from the leaders' values, purpose, vision, conviction, aspirations, and goals. They are inner-focused, as the principal driver of being and becoming is self-awareness. However, self-regulation is what helps the leaders to achieve their possible, future, or desired self. The inner focus of authentic leaders guides them to perform self-leadership that necessitates constant self-awareness, self-regulation, and self-growth in terms of knowledge and competence.

The other part of reconciliation is "reconciliation of social realities," which represents how authentic leaders engage in "shaping" the existing context. High levels of self-awareness and self-regulation lead the authentic leaders toward self-transcendence. While they are being and becoming, their becoming is to achieve a dual purpose; that is, to achieve their future, possible, or desired self in conjunction with positive or authentic goals for their followers, organizations, and societies. Their actions and achievements are other-directed although they are deeply inner-focused in the first place. Authentic leaders, inner-focused and other-directed, are aware that they can achieve their possible, desired, or future self by the mutual alignment between themselves and their followers and organizations – who are key stakeholders in their leadership. They mold the context by aligning and mobilizing others' opinions and actions. Therefore, authentic leaders strive to shape their existing context through finding reconciliation among social agents, competing forces, social dilemmas, and paradoxes. They reconcile their social existence by translating their purpose into a vision that can inspire their followers. They acquire support from their organizations by aligning their purpose and vision with, as well as influencing and shaping, the organizational vision, mission, and values. They acquire the support of their followers by relating to them and recognizing their individual purpose and desires.

These processes are primarily an active and reactive response to experiences which the leaders go through while achieving authenticity within the self and operationalizing this authenticity by the "dynamic and creative reconciliation of self and social realities." In their functioning as leaders, they develop alliances and coalitions with various social players including their followers, colleagues, other leaders in the organization, and stakeholders outside the organizations. They buy in the support of others by inspiring them through their vision. They reconcile chaos with order, paradoxes with rationality, and management with leadership. They also reconcile themselves with the context which means that they strive to understand the context and capitalize on it. They reconcile themselves with industry challenges, contextual challenges, people challenges, organizational challenges, and personal challenges by turning these challenges into opportunities.

The authentic leaders devise effective strategies to achieve social reconciliation. These strategies help them to perform their leadership within a given context with more effectiveness. Reconciling social realities includes reconciling with: competing demands of stakeholders and various social agents; the paradoxes that exist in social systems; the different contexts in which they function; and the social challenges they face in everyday life.

"Dynamic and creative reconciliation of self and social realities" is also a striving to reduce socio-cognitive dissonance. Authentic leaders, being fully aware of themselves and the context in which they function, are able to sense anomalies that lie in their own character as well as in their social context. In the desire to achieve their possible (future) self and their proximal context, they endeavor to reduce the existing dissonance – the disparity in their current and desired state of existence (in terms of both self and context). This process of reducing the dissonance within the self and the social context can be termed "reducing the socio-cognitive dissonance." The interaction of the reduction in cognitive dissonance and social dissonance is rather complex, cyclical, and cumulative in nature.

Dynamic and creative reconciliation and leadership development

Reconciliation plays a major role in leadership development. Leaders constantly reconcile their past and present with their future selves. They move from their actual self to their possible self by taking appropriate actions, including reconciling their failures, values, ambitions, motivations, actions, and behaviors with their possible or future self. They gradually get closer to their future self by reconciling with their internal conflicts. They are constantly evolving in terms of their self-identity and their changing view of self with others. This cyclical process is the reconciliation of self-realities within the self and with the outer environment within which authentic leaders develop.

Dynamic and creative reconciliation and self-leadership

Authentic leaders harmonize their authentic self through self-awareness and self-regulation. In self-leadership, authentic leaders reconcile with the context they operate in and with the challenges they constantly face and by adopting strategies that can help them achieve reconciliation with self-realities. Reconciliation in self-leadership is essentially about how leaders view themselves in a given context and how they adapt to the context by self-regulation, self-development, and authentic role modeling – without compromising their personal values, belief systems, goals, and desires. Self-reconciliation also refers to how leaders assuage uncertainties related to the self – such as developing competencies, knowledge, skills, and behavior.

Dynamic and creative reconciliation and self-transcendent leadership

Authentic leaders, being self-transcendent in their leadership approach, always seek social reconciliation. Social reconciliation has two aspects. The first pertains to philosophy and beliefs about how authentic leaders want to contribute to the lives of their followers, colleagues, stakeholders in the industry, and society at large. The second aspect concerns the approach or strategies that authentic leaders adopt in order to achieve social reconciliation. In the process of self-transcendent leadership, leaders reconcile their philosophy, beliefs, values, and desires, with those of their followers whom they lead or influence. Reconciliation in self-transcendent leadership is, in essence, how authentic leaders align their spiritual, religious, and ethical belief systems in order to serve those around them. By understanding their purpose through their life story, they are able to align their values, goals, and ambitions with their purpose and vision of leadership. Authentic leaders reconcile their routine organizational responsibilities with their social commitments. They align their professional goals with societal good to practice self-transcendent leadership.

Dynamic and creative reconciliation and sustainable leadership

The foregoing discussion shows that leaders deal with a variety of situations, contexts, challenges, and paradoxes. They solve complex problems and deal with many stakeholders within and outside their organizations who often have conflicting interests and goals. They operate in complex and dynamic environments, and often have to make trade-offs. They resolve conflicts and hostilities among various parties. They should also seek to prevent potential conflicts before they emerge. They ensure the well-being of those they lead and those they serve outside their organizations. In order to sustain their leadership, leaders reconcile their decisions with the goals of their organizations and followers. The ability to reconcile the self and social realities is an important tool of authentic leaders which enables them to function effectively as leaders.

Reconciling with context

Leadership is very dependent on, and mediated by, the context. Therefore, the foremost priority of leaders is to work effectively within the context by fully understanding the situation in order to adopt adequate measures and strategies. They reconcile the situation with the available resources and options. At first, they have to reconcile with context at both the macro-scale as well as at the micro-scale. Authentic leaders realize the importance of the context they operate in and devise strategies and take measures to operate most effectively within the context. For this purpose, they reconcile between resources and various stakeholders and interest groups.

Reconciling with social agents (people)

Next to reconciling with context, authentic leaders reconcile with people at different levels. They reconcile with their followers by acknowledging them, understanding their personal needs, and appreciating their strengths and weaknesses. They also reconcile with people outside their organizations to broaden their support group. Authentic leaders are able to draw support from their followers by aligning their aspirations toward a common goal. This ability to reconcile varied viewpoints is the key to achieving a shared goal.

Reconciling with complexities and paradoxes

Leaders also reconcile with several other paradoxes they come across on a daily basis. These paradoxes range from their performance of daily roles to the exceptional circumstances they come across. A point made by an interviewee indicates an example of reconciliation between the big picture and detail: "I look at the bigger picture in most of the issues but I do go into smaller details when I think it is necessary for me to get into them." Another leader faced difficulties in reconciling between leadership and management: "I struggle from the lack of time having to juggle between daily management issues and time spent focusing on what a leader should be doing."

The interviewees also noted the issue of compartmentalization in which some people in their organizations engage in order to solve various problems. The general manager of a contracting company revealed how he encouraged his followers to overcome compartmentalization and achieve balance in various tasks by having a systemic perspective.

One interviewee discussed leadership and organizational constraints – the kind of reconciliation he often has to try to achieve by helping his followers under given organizational constraints. Another CEO highlighted how he achieved a balance between the business side and the professional side, yet another kind of and means of reconciliation.

One CEO mentioned the importance of simplifying the uncertainty that refers to reconciling with complex circumstances and prioritizing the issues: "At times of great uncertainties, at times of great change, the message from the top man must be clear and simple, and easy for people to understand."

Reconciliation between expansion and sustained success of organization emerged in the following way in the discussion of a leader: "We don't want to stray too much ... We are focused on our competency, we identify and define very fairly What our competency is."

A senior executive in a contractor firm also noted a paradox which he found to be an usual condition to his leadership. He noted:

> When we tender, our top management wants us to achieve a certain profit with lower risk and higher profit; our project manager wants a project with a lower price, the estimation department is under pressure to be able to get the tender in a short time.

Reconciling with social challenges

Female leaders participating in this study noted a number of social challenges they faced in their professional careers. However, in order to succeed, these females were striving to reconcile these challenges in many ways. One senior vice president related her experience of a difficult time with a foreign contractor. She described how she overcame it through open communication and inspiring trust:

> I remember that as a woman it was very very difficult as they [Korean contractors] will not take any instructions from a woman. I remember once going to a meeting and I was very tough. He was very unhappy with me. Eventually, when we got to know each other, we got along well.

Describing the social challenges that female leaders usually face, a senior executive in a consulting firm gave an insightful perspective about how she had reconciled between her professional life and personal life, mentioning the support she receives from her family: "I also have very good support from my mother-in-law and sister-in-law ... So, I don't have to worry that the children are lonely."

Another senior woman executive explained that although she had a difficult time in handling the challenge of being a woman in construction, she was able to tackle them tactfully. She explained:

> The first impression counts ... subsequently, I always find the opportunity to have future contact and communication so that we can work together and allow the person to know me better ... Not most of the time, but I do encounter problems where some senior people say "Who is this young woman you are asking to lead our project?" ...

sometimes, they ask questions they are not likely to ask a man. For example, while they will take the firm's account of a man's experience on paper, they would ask me more questions: "Have you done similar projects before?" Actually, if they ask questions like this, I feel more comfortable. That gives me the opportunity to share my experience with them.

These excerpts show how the female leaders faced social challenges but were able to reconcile their social and professional roles. They had to face the challenge of social stereotyping, professional demands, and family commitments. However, in their individual attempts to reconcile the push and pull of the societal and professional demands, these female leaders were able to practice authentic leadership and excel in their organizations.

Grounding the theory in the literature

The term "reconciliation" refers to the act of "reconciling" and was first used in the fourteenth century. In the online Merriam-Webster Dictionary (2020), reconciling means: "to restore to friendship or harmony," "to make consistent or congruous," or "to cause to submit to or accept something unpleasant" or "to check (a financial account) against another for accuracy." In the Oxford English Dictionary (2008), reconciling means: "to settle into position," "to make (discordant facts, statements, etc.) consistent, accordant, compatible with each other," or "the action of smoothing or planning; removal of roughnesses." Thus, the *Oxford English Dictionary* defines reconciliation as "the action of bringing to agreement, concord, or harmony." Finally, the American Heritage Dictionary (2020) defines reconciling as: "to reestablish a close relationship between"; "to settle or resolve"; "to bring (oneself) to accept"; "to make compatible, harmonious or consistent."

In the study, the label "dynamic and creative reconciliation of self and social realities" was chosen for the major category for many reasons. First, dynamic reconciliation refers to an active and on-going process that takes place as a part of authentic leadership development and influence. Authentic leaders engage in dynamic reconciliation daily, at both the personal and social levels. In this study, reconciliation is referred to as being creative because authentic leaders engage in a self and social reconciliation that does not make them compromise on their personal values, beliefs, and goals. Creative reconciliation does not seek to satisfy self-serving objectives, and does not cause any socially detrimental consequences. Creative reconciliation is neither a coping strategy nor a way of avoidance that leaders adopt to deal with difficult issues or challenging paradoxes. Instead, "Dynamic and creative reconciliation of self and social realities" is a conscious process that has a dominant cognitive dimension when it relates to "self-realities" and a more social dimension when it relates to "social realities." After a search of the literature to ascertain how extant knowledge explains the core category in this study, it was found that dynamic and creative reconciliation of both self and social realities can be explained through various concepts, shown in Table 10.2.

Dynamic and creative reconciliation of self-realities

There is evidence in the literature that the self is never constant. It is constantly evolving, going through various stages of change and development (Kegan, 1982). It is important to regard each phase as a transitional process where dynamic tension pushes one to grow. This developmental process of self is influenced by many factors within and outside the self. In

Table 10.2 Concepts explaining dynamic and creative reconciliation of self and social realities

Reconciliation of self-realities	Through narrowing present self with future (authentic) self	Through comprehending, managing, and sharing inner states	Through inner sense-making	Through resolving inner dilemmas and paradoxes	Framing and reframing the life stories
Reconciliation of social realities	Reconciling social paradoxes and dilemmas through reducing social dissonance	Through constructive and creative politics	Through organizational sense-making and sense-giving	Through management of meaning (for employees and organization) or aesthetics	Through chaordic leadership

the research tradition of Social Theory, the examination of self and identity has a long history. The self is inherently a complex concept and several scholars have discussed it extensively. In Jung's (1959) view, self is a "total personality" that unifies that core of psyche ensuring a balance of conscious and unconscious forces. To Rogers (1959), self is the organized, consistent, conceptual whole composed of perceptions of the characteristics of the "I" or "me" to various aspects of life (p. 200). According to Schlenker (1985), one's situated identity is "a theory of self that is wittingly or unwittingly constructed in a particular social situation or relationship" (p. 68).

Self-concept is defined as an identity that is invoked when one "attempts to answer a personal question about the self posed by oneself" (ibid., p. 67). Giddens (1991) views self-identity as a reflexively organized endeavor designed to sustain coherent biographical narratives in the search for ontological security. It is the continuity of self over the past, present, and future that results from the operation of memory (Allport, 1961). Self-reflections and self-disclosures are the ways and means by which individuals gain knowledge about themselves and demonstrate this knowledge in their activities. Individuals are best understood as social selves (Burkitt, 1991) and their actions are understood within their complex conditions, processes, and consequences (Collinson, 2003). Kondo (1990) views individuals as having multiple selves which are never fixed, coherent, seamless, bounded, or whole. This is because they are members of many social groups and thus possess multiple social identities (Brewer, 1991; Rosenberg, 1997; Freeman, 2006). Different selves are activated under different situations. Individuals develop or subdue different selves as a result of various experiences in life. Therefore, during their life-long development, the process of reconciliation within the self is a natural process.

Kegan (1982) suggests that individuals go through various phases of development of self. These stages include: incorporative self (days of infancy), impulsive self (early childhood), imperial self (early adolescence), interpersonal self (adolescence), institutional self (early adulthood), and inter-individual self (adulthood and beyond). While an individual moves from one stage of life to another, there is a transition in which individuals reconcile between what they have been in the past, what they are at the moment, and what they will become in the future. During various stages of life (Kotre and Hall, 1997), individuals go through various transactions, transitions, and transformations which induce, infuse, or erase different selves in individuals. Cashman (1999) adds that an individual's ability to grow as a leader is inseparable from that individual's ability to grow as a person. Leaders develop through complex socio-cognitive processes which occur at different stages of life, in which they are constantly being and becoming within their environments. This iterative cycle of "being and becoming" is a complex cognitive process and is referred to here as the reconciliation of inner realities (Figure 10.4).

Several authors have attempted to study the process of reconciliation from a socio-psychological perspective (Festinger, 1957; Wyer and Srull, 1989). Reconciliation basically involves a higher level of internal negotiation which leaders engage in to resolve internal paradoxes and dilemmas (Lowy and Hood, 2004; Fisher-Yoshida and Geller, 2008), sense-making (Weick, 1995), comprehending, managing, and sharing inner states (Higgins and Pittman, 2008), creative framing and reframing (Avolio and Luthans, 2006; Toor, 2006) and cognitive dissonance (Festinger, 1957; Hampson, 1998). Key to these processes is self-awareness which leaders gain from their experiences, social learning, and understanding of their past. The consequence of these processes is self-regulation which goes hand-in-hand with self-awareness. Being self-aware and self-regulated, authentic leaders engage in reconciling between their states of "being" and "becoming." The process through which authentic leaders possibly seek dynamic and creative reconciliation of inner realities – in which they are being and becoming – is explained in the following sections.

Reconciling the past, the current, and the possible self

In this discussion, the concept of possible selves which has been examined by several scholars should be highlighted. Fundamental to human development is that it is not what one is now but what one is becoming that matters (Higgins, 2005). Markus and Nurius (1986) are among those who first explored the concept of possible selves in detail. They consider possible selves as referring to ideal selves or desired selves. The possible selves are the individual's cognitive manifestations of enduring goals, aspirations, motives, fears, and threats. Individuals strive to understand their present competencies and their personal reference values as well as imagine their future-self competencies – who they might become, hope to become, dream of becoming, fear becoming – and then use these future-self competencies as reference values for self-regulation in the present (see, for example, Markus and Nurius 1986; Oyserman and Markus 1990). Tulving (2005) considers one's ability to relate the present self to both the past self and the future self as mental time travel.

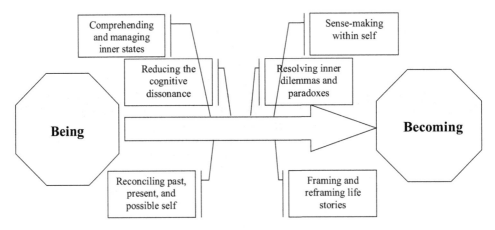

Figure 10.4 Reconciliation in being and becoming

Possible selves furnish the essential link between self-concept and motivation of individuals. They are influenced by past conceptions of the self. Possible selves are simultaneously separable and connected to the current selves. This point of view is also confirmed by Dunkel (2000) who argues that possible selves emerge out of past social experiences; therefore, they also form a link from the present to the future. Collinson (2003) believes that being reflexively monitoring and purposive creatures, humans have the capacity to reflect upon things, which helps in raising awareness of past processes and future possibilities. This flexibility and malleability (Tedeschi and Lindskold, 1976) of self-concept – which is highly dependent on the nature of the social situation – are helpful in attempts to envisage alternative realities and to reconstruct and change the future one can possibly have. This creative potential of self-development enables individuals to reflect upon and exercise some discretion and control over their actions which is formally known as self-regulation in academic terms.

Researchers have noted several points regarding possible selves. Possible selves are significant in the sense that they provide an anchor for personal development and subsequently validate them (Wurf and Markus, 1991). Possible selves mediate long-term motivation and supply direction for the achievement of the desired goal. These representations enable people to achieve some behavioral control and self-directed growth by guiding the development of specific plans and strategies for action (Kuhl and Beckmann, 1985; Wurf and Markus, 1991). Possible selves also serve to focus and direct behavior toward or away from the expected, desired, dreaded end-states they represent (Cross and Markus, 1991). They also provide a vision for future life which individuals tend to perceive from the lens of their future-selves.

Higgins and Pittman (2008) elaborated upon the conceptualization of past-self, current-self, and future-self with an example. They observe that positive responses of parents to what a child is becoming depend on the relationship between the past, current and future self of the child. The notion of "what is becoming," as Higgins and Pittman observe that it is true of parent-teacher, boss-employee, and coach-player relationships, and beyond such cases, is true of business executive-stockholder, politician-voter, and so on. Groups and nations are concerned with what they are becoming and not just with what they currently are. They also note that comprehension of the future self helps the individuals to employ self-regulation techniques that involve making present decisions in relation to future goals and outcomes as desired and undesired end-states. Gardner et al. (2005) assert that the degree to which self-regulation is achieved is assessed by considering the extent to which personal strivings are helpful in realizing possible selves. In this respect, individuals actively imagine how happy or regretful they will feel as a result of a particular decision they have to make at present. They also imagine the competencies of their future selves which give them reference values for self-regulation and self-evaluation.

Avolio and Luthans (2006) explain how possible selves play an important role in developing the individual as an authentic leader. They argue that a dynamic balance between actual self and possible self is essential. Authentic leaders make conscious efforts to reduce the gap between what they are and what they would like to become. They imagine their possible self in their role models and try to bridge their current self with their desired or possible self. Hoyle et al. (1999) postulate that focus on possible selves motivates and guides people's pursuit of goals and typically promotes self-improvement.

To achieve their possible selves, authentic leaders engage in regulating their behavior which is congruent with what they want to become. Since authentic leaders aspire to become better individuals, leave positive legacies, and develop authentic followers and sustainable organizations, they regulate their behavior to match their goals. Individuals continuously wrestle with their past to shape their current self and in order to achieve their possible self. This process, in general terms, is reconciliation between past incidents, current scenarios, and future perceptions that leaders have about themselves. It shows that leaders reconcile between their various cognitive selves such as: failed-self and desired successful-self, average competent-self and competent-self, disappointed-self and hopeful-self, inauthentic-self and authentic-self, unethical-self and ethical-self, feared-self and effective-self, unimportant-self and influential-self, selfish-self and transcendent-self, effective-self and ineffective-self, rigid-self and flexible-self, and so on. They also reconcile between their functional selves such as: decision-maker-self and decision-facilitator-self, transactional-self and transformational-self, team player-self and team leader-self, power-seeking-self and power-sharing-self, hands-on-self and delegating-self, leading-self and managing-self, giving-self and control-self, rewarding-self and egocentric-self, and so on. This yearning to achieve a possible, desired, or ideal future self results in action-based strategies – self-regulation – that leaders engage in. From this discussion on the leadership support group, the following proposition is drawn:

> **Proposition 14**: As opposed to less authentic leaders, more authentic leaders constantly engage in reconciling their past, present, and future (possible) selves in order to achieve authenticity.

Reconciling within self through inner sense-making

The process of sense-making tends to be triggered by a discrepancy between the current and expected state of the world (Weick, 1995; 2001). Schutz (1967) argues that people engage in sense-making by looking backward upon events and engaging in retrospective analysis to arrive at an understanding of their current circumstances. This view is also echoed in the work of Loftus and Loftus (1980), who describe memory as a current view of the past. According to Weick's (1995; 2001) conceptualization, sense-making is grounded in identity construction that means the notion of self is constantly redefined by how individuals act and how they interpret their environment. In this view, sense-making is an iterative process that continually redesigns the image of individuals about themselves. Sense-making is also retrospective in nature and is considered to activate sensible environments. Sense-making takes place in a social setting which means that the person is highly influenced by his or her surroundings and the construction of his or her images are highly contingent on how the social environment is. Sense-making occurs through familiar points of reference.

People have a strong need to enhance their self-concepts by behaving consistently with their actions, statements, commitments, beliefs, and self-ascribed traits (Cialdini and Trost 1998). Therefore, sense-making processes take place in the service of maintaining or restoring a consistent, continuous, and positive self-conception (Weick, 1995, p. 23), which consists of a personal identity (which refers to person-specific attributes) and a social identity (which refers to social group memberships). Allard-Poesi (2005) posits that these identities are constituted out of the process of interactions in which the individuals decide which self is appropriate and which self to present to others. Based on the conduct of others, people enact these selves and define the situation and actions consistently, by projecting them into the environment and observing the consequences.

The sense-making process is essentially a way of reconciling a reality which threatens the smooth existence of a particular self-identity. To overcome the discomfort or anxiety that develops as a result of an extraordinary situation, individuals consciously engage in a reconciliation process to achieve coherence within their inner states. For example, describing the "lived experiences" of an individual suffering from spinal cord injury, Mitchell (2002) observes:

> Some persons begin a journey to reconcile their ingrained belief system with their new reality. This requires serious inquiry and study, and challenging the system. Most, like me, change their belief system as they continue their journey toward one more compatible with their lives. This journey is extremely difficult, even more so than the daily negotiating of body and environment, because it is done alone, without the community's support.
>
> (p. 52)

These observations of Mitchell are convincing in that people who suffer from extraordinary situations or transformative experiences have to reconcile their personal belief system with their existing reality. These events can trigger moral transformation and change a leader's view of life. As a result, instead of single-loop coping mechanisms and defensive routines, authentic leaders engage in double-loop learning that helps them invent their higher purpose (Parameshwar, 2006). This reconciliation helps them to find their place in the society while being able to seek internal satisfaction. For example, Casey and Long (2002) observe that people with mental health problems make sense of their experiences of mental distress through reconciling several voices in life and by making sense of internal as well as external voices in order to achieve coherence and confidence in their own version of life events. Internal voices refer to their beliefs and common sense whereas external voices refer to those of family members, culture, and psychiatry.

The categories emerging in this study also show that authentic leaders generally engage in the sense-making process in their daily life. However, the sense-making process is more conscious and deliberate when they face extraordinary circumstances, such as internal conflicts, ethical dilemmas, and paradoxical situations, transformative events, or tipping points which trigger an ambiguity or uncertainty within the self. Under such situations, authentic leaders engage in sense-making to analyze the alternatives available to reconcile their inner states or realities – thoughts, feelings/attitudes, competencies, and reference values (goals and standards) – with their actions, behaviors, decisions, and actions.

Proposition 15: As opposed to less authentic leaders, more authentic leaders make better use of inner sense-making to understand and manage their inner states.

Reconciling within self through comprehending, managing inner states

Higgins and Pittman (2008) illustrate how human motives of comprehending, managing, and sharing their inner states influence behavior and actions. In their view, these human developments lead to comprehending, managing, and sharing the inner states: (1) social consciousness or awareness that the outcomes or significance of a person's action depend upon how another person reacts to it; (2) recognizing that people's inner states can mediate their outward behaviors; (3) relating the present to both the past and the future (mental time travel); and (4) sharing reality with other people.

Higgins and Pittman explain how this typology of broad categories relates to people's concerns with comprehending, managing, and sharing their inner state – thoughts, feelings/ attitudes, competencies, and reference values (goals and standards). They discuss these issues with reference to imagining future-self inner states, managing how others comprehend one, and sharing knowledge about the world. They assert that people tend to imagine what type of individual they might or would like to become and what competencies they will have after becoming what they imagine or desire to be. Such imaginations about future-self, possible self, or desired-self provide an anchor for self-regulation and self-evaluation for people who then act and behave in accordance with their future, possible, or desired self. Individuals are also motivated to manage the comprehensions about their selves.

> The comprehensions that others develop about an actor ... will influence how those others react to and behave towards the actor, which in turn will have hedonic consequences for the actor. These potential hedonic consequences give actors a strong reason to want to influence how others comprehend them. In addition, because important aspects of the self are influenced by, and dependent upon, the ways in which one is comprehended by others, the wish to create and maintain desired comprehensions of self will also produce motivation to influence the comprehensions of others.
>
> (ibid., p. 373)

This shows that individuals strive to manage others' comprehensions and are driven by how others perceive them.

Authentic leaders are, therefore, motivated to engage in positive self-presentation and self-management actions that are favorable to achieve the kind of self they want to project. In this process, they construct their social realities so that interaction partners are induced to comprehend and respond to them in self-consistent ways. Authentic leaders also engage in managing damaged identities when they encounter a situation in which a damaged or threatened identity must be salvaged or protected, such as when some harm has been done to another that could lead to negative comprehensions of self by that other, then explanations in the form of excuses are an effective response. Actions such as excuses, apologies, and expressions of remorse are useful to create a particular shared comprehension of those circumstances in ways that benefit the hedonic outcomes and sense of self.

Reconciling within self through resolving inner dilemmas and paradoxes

People develop self-views as a means of making sense of the world, organizing their behavior, and predicting the responses of others to attain self-clarity (Gardner et al., 2005). However, they invariably face internal dilemmas and paradoxes and hence strive to resolve such dilemmas and paradoxes to enhance their self-concept and achieve more self-clarity. In the context of leadership, resolving the dilemmas, such as being open and spontaneous and thoughtful and consciously self-controlling, is an important characteristic of leadership (Larsson et al., 2007). Ethical dilemmas are also frequent in the contemporary turbulent world (Gardner, 2007), such as the dilemma of the interplay between morality and management (Waldman and Siegel, 2008).

Other challenging and current ethical dilemmas include: trust, listening, networking, openness to learning, and a willingness to change priorities or practice for the common good (Broussine and Miller, 2005). Sometimes, leaders also face a social dilemma in the form of conflict between serving one's own interest or the interest of the group or collective (De Cremer, 2007). Fisher-Yoshida and Geller (2008) note the self-related paradoxes that leaders and managers face in everyday functioning. These are: the paradox of knowing (knowing the self versus knowing the others); and the paradox of focus (self-centeredness versus community-centeredness). Authentic leaders actively work toward managing these dilemmas and paradoxes. They overcome such dilemmas and paradoxes by actively engaging in developing an ethical mind (Gardner, 2007), incorporating the conscious use of critical reflection, fostering the personal willingness to challenge assumptions, building the recognition of complexity inherent in organizations, challenging assumptions, questioning conformity and embracing difference (Fisher-Yoshida and Geller, 2008).

The toughest dilemma that authentic leaders face is how to manage their authenticity. Goffee and Jones (2005), while explaining the paradox of authenticity within oneself, observe that authenticity is a quality that others attribute to the leader although the leader has full control over his or her disposition of authenticity. They argue that authentic leaders are not only well aware of themselves but also aware of what self they should reveal to others under different circumstances. This cannot be regarded as manipulation of self.

Goffee and Jones (2005) also explain that managing the authenticity of the self lies in two aspects: (1) consistency between what leaders say and do; and (2) finding common ground with their people. Since different people have different world-views, demands, aspirations, and approach to life, the strength of authentic leaders lies in appreciating the uniqueness of individuals yet sticking to their personal values and principles. Authentic leaders recognize others' viewpoints but they find a fine balance between their values and conforming to social and organizational norms. They understand that too much conformity can lead to ineffectiveness while too little conformity can isolate them from others. Authentic leaders, therefore, need to work hard to manage their authenticity, however paradoxical it may sound (ibid.). This perspective of managing authenticity is somewhat similar to the "moral manager" of Treviño et al. (2000). In their study of ethical leaders, Treviño et al. note that being a "moral person" – an individual who models "normatively appropriate" conduct – is not enough for leaders. They argue that ethical leaders are also "moral managers" – individuals who are able to manage others' perceptions of their being ethical by openly and explicitly talking about ethics and also empower their employees to be just and seek justice.

Managing authenticity is similar to what Sosik et al. (2002) describe as self-management through self-monitoring. Leaders who engage in managing their behavior in accordance with a given situation are emotionally intelligent and able to operate effectively in a dynamic environment (Shamir and Howell, 1999). They possess a repertoire of behaviors that can suit a variety of situations and people. This self-monitoring or impression management ability of authentic leaders enables the authentic leaders to manage their authenticity and reconcile their inner states with the external environment without compromising the genuineness of the self.

> **Proposition 16**: As opposed to less authentic leaders, more authentic leaders are better able to reconcile their inner dilemmas and paradoxes by engaging in reflection, sense-making, and management of inner meaning.

Reconciling within self through framing and reframing life stories

Framing and reframing one's life narratives form an important technique of inner self reconciliation in which authentic leaders engage consciously or unconsciously. The "reframing process" is a vital feature of leadership development. Leaders structure their life stories according to their self-belief, self-identity, and self-perception; however, these stories are subject to what happens after. Understanding their life stories helps authentic leaders to recognize "why they have come to be who they are." Life stories are sources of individual motivations (Bluck and Habermas, 2000), and the themes and order that structure one's life events in the creation of the life story schema set the rationale for behavior across time, and guide the formulation of future goals and actions by their links with that life story model (Boal and Schultz, 2007).

Avolio and Luthans (2006) argue that leaders continuously frame and reframe their life stories depending on the nature of the experiences they encounter. A leader's life story is shaped by the past leadership antecedents and it then shapes how the leader experiences the future antecedents that arise in the leader's life (ibid.). These experiences give meaning as to how the leader develops particular behaviors. As Locke (2007) notes: "Meaning is handled in and modified through an ongoing interpretive process" (p. 22); much of leadership development is influenced by the way different individuals interpret certain events. In some instances, leaders invent a higher purpose in life and wake up from the trance of dominant practices by reframing personal suffering in the light of perceived eternal truths (Parameshwar, 2006). In other instances, awakening from the trance of social validation takes place when leaders reflect upon personal suffering by referencing the inspiring standards of others (ibid.).

Thus, leadership development is not fixed or stable; it is a process that is heavily influenced by the events in the life of an individual. If individuals positively respond to these events (regardless of the nature of the events), the outcomes are likely to be positive in the form of self-development and positive behavioral responses (Chan, 2005; Ilies et al., 2005). In this case, it is likely that the leader will reframe and reinforce his or her story positively. On the other hand, if the leader responds to these events in a negative manner, the consequences are likely to be negative in the form of negative leader-self-development, self-distortion, cognitive dissonance, inauthenticity, or negative behavioral responses (Chan, 2005). A negative event may also result in positive reframing depending upon how the event is perceived (Toor, 2006). This positive or negative reframing occurs as the leader struggles to adapt his or her actual self with his or her possible self in the course of life. If positive events occur, the self-image and possible self of the leader are likely to develop, resulting in the development of positive psychological capacities (Luthans et al., 2007). The development of positive psychological capacities or PsyCap (Luthans et al., 2006) helps to reframe the past and perceive the future events more positively.

It is argued here that reframing is a natural cyclic process which occurs over the life span of individuals. However, to derive benefits from this process, it is important to accelerate it and employ it through proactive scientific ways, such as leadership interventions. This is because the natural occurrence of unplanned and unpredictable trigger events involves the risk of exposing the individuals to negative perceptions (Luthans and Avolio, 2003). To avoid such risks, the most effective alternative is intervening through coaching, training, or other types of micro-interventions (Luthans and Youssef, 2004; Luthans et al., 2006) or macro-interventions (Toor, 2006) in which leaders are exposed to planned trigger events where they can learn to positively reframe or reinforce their self-identity and self-knowledge. Therefore, if leaders are aware of their developmental antecedents, it is likely that they will gain extra potency from further positive events and construe positive meanings out of negative events.

Consequently, the reframing process can be highly beneficial, and result in enhanced psychological capacities and improved leadership potential of the person. Considering framing and reframing as a way of inner reconciliation, Toor postulates that leaders continuously frame and reframe their life stories depending on the nature of experiences and develop particular behaviors in accordance with the meaning they elicit from these experiences. As opposed to negative sense-making (or reframing), positive sense-making (or reframing) results in positive leadership development in individuals. Finally, awareness of developmental antecedents helps individuals to positively reframe their significant experiences to reinforce their self-identity and self-knowledge to achieve enhanced psychological capacities and improved leadership potential.

Proposition 17: As opposed to less authentic leaders, more authentic leaders regularly frame and reframe their life stories to reconcile with their failures and disappointments.

Propositions 18: As opposed to less authentic leaders, more authentic leaders are better able to positively reframe their negative experiences as positive learning.

Reconciling within self through cognitive dissonance

Festinger's (1957) theory of cognitive dissonance has been a pillar of social psychology (Shultz and Lepper, 1996) and has led to a large amount of research to understand cognitive behaviors. Festinger developed his conceptualization of cognitions based on three possibilities. In his view, cognitions which contradict each other are said to be "dissonant," while cognitions which agree with each other are said to be "consonant." Cognitions which neither agree nor disagree with each other are said to be "irrelevant" (Festinger, 1957). To Festinger, human beings strive for harmony in their thinking. People make every effort to reconcile their conflicting thoughts and to resolve them in more or less healthy ways (Crigger and Meek, 2007).

According to cognitive dissonance theory, people are motivated to reduce any psychological state of tension – referred to as dissonance – between two cognitions which contradict each other. In other words, if an individual holds two cognitions that are inconsistent with one another, the individual will experience the pressure of an aversive motivational state called cognitive dissonance, a pressure which he or she will seek to remove. Dissonance can be reduced in three ways: (1) by changing one of the dissonance elements – that is by simply changing an attitude, value, opinion, or behavior to another behavior that is typically being the most resistant to change; (2) by adding consonant cognitions that reduce the overall level of inconsistency; or (3) by trivialization – decreasing the importance of the elements involved in the dissonant relations (Simon et al., 1995). However, reduction of dissonance depends on the resistance to change of the various relevant cognitions, with less resistant cognitions being more likely to change (Shultz and Lepper, 1996).

According to Paulhus (1982), individuals attempt to present themselves as favorably as possible and they actively attempt to convey an image of consistency to others. It is assumed that a person can tolerate cognitive inconsistency when alone but through socialization learns to present a consistent self to others. Hampson (1998) argues that people with trait discrepancies are prompted to resolve such dilemmas. This is essentially because people prefer consistency in their beliefs, attitudes, and behaviors and that any inconsistency or dissonance will result in changes intended to restore the preferred state (Meyer and Xu, 2005). Sometimes, individuals seek reduction in dissonance to reconcile two inconsistent traits. This reconciliation is carried out through descriptive overlap and evaluative balance in personality representations (Hampson, 1998).

Other strategies identified by Asch and Zukier (1984) for reconciling inconsistent traits were "inner versus outer" (a person displays one trait outwardly while being quite different inside), "common source" (the conflicting traits are both the product of the same higher-level trait), "interpolation" (the two traits can be reconciled by imagining events that explain why a person with one trait would develop the characteristics described by the conflicting trait), and "segregation" (the two traits describe situationally or temporally independent parts of personality). Cognitive dissonance also gives insight into the reasoning of people who change their ideas, opinions, and behaviors to support situations that appear to be contrary to their behaviors (Crigger and Meek, 2007). Festinger (1957) used dissonance theory to account for a wide array of psychological phenomena, including the transmission of rumors, rationalization of decisions, attitudinal consequences of counter-attitudinal advocacy, selectivity in information search and interpretation, and responses to the disconfirmation of beliefs. It has since been successfully applied to many phenomena in a wide variety of both predictive and postdictive contexts.

How individuals seek reduction in cognitive dissonance can be explained through a grounded theory study of self-reconciliation. Crigger and Meek (2007) attempted to understand the behavior of nurses after they have committed a mistake in practice. Their analysis generated a core category "reconciliation of the self, personally and professionally," which comprises four distinct categories:

> coming to terms with the reality of the mistake (here called "reality hitting"); determining or weighing the need to disclose or report the mistake ("weighing in"); deciding on the best trajectory for responding ("acting"); and, finally, evaluating ("resolving") the event and moving on.
>
> (p. 180)

Their study also found that the participant nurses talked about their struggle to meet personal and social ideals in an attempt to reduce the dissonance regarding the ideal self "and to reconcile the ideal with the actual through either a public response and apology or by internalizing guilt, denial, and rationalization. Cognitive dissonance creates tension and causes mistake makers to wrestle with opposing viewpoints" (ibid., p. 182).

Four stages of self-reconciliation mentioned in Crigger and Meek's study are similar to self-awareness and self-regulation that have been discussed as cornerstones of authentic leadership. This view also strengthens the proposition that authentic leaders engage in self-reconciliation to achieve their desired or future self through reality hitting, weighing in, acting, and resolving. Authentic leaders actively seek reduction in cognitive dissonance to reconcile any inconsistencies between their cognitions. Being highly self-aware, they recognize the inconsistency between what they believe and how they behave. They actively seek reconciliation between what they are and what they would like to be. They reduce the dissonance between their current self and future self through this reconciliation.

> **Proposition 19**: As opposed to less authentic leaders, more authentic leaders are better able to reduce the cognitive dissonance by reconciling their conflicting attributes and thoughts and achieve authenticity within self.

Dynamic and creative reconciliation of social realities

Reconciliation has been used largely in the context of conflict resolution and decision-making. However, the term has enjoyed a rather broader meaning in the social science disciplines. For example, Crigger and Meek (2007) note that "reconciling means to bring to acquiescence or to resolve an issue or situation. In reconciling, one might not be happy with the outcome but still has some degree of acceptance and acquiesces to the situation." (p. 179). Reconciliation ensures the continuation of cooperation among parties with partially conflicting interests (de Waal, 2000). Others observe that reconciliation takes place among two parties and necessitates the resolution of emotional issues (Nadler and Liviatan, 2006). Kelman's (2006), revision of one party's own identity to accommodate the other's is essential to reconciliation. In this view, each party is sufficiently reassured by the other's acknowledgment of its identity so that it in turn becomes free to remove negation of the other as a central element of its own identity. Staub et al. (2005) note that reconciliation must lead to a changed psychological orientation toward the other party. Shnabel and Nadler (2008) also underscore that reconciliation cannot take place "as long as these emotional needs remain unsatisfied." Nadler (2002) terms such an approach to reconciliation the socio-emotional route. Therefore, in human society, reconciliation has been institutionalized, elaborated on, ritualized, and surrounded with a great many societal influences, such as the role of elders, conciliatory feasts, and compensatory payments (Fry, 2000).

Other researchers have referred to reconciliation as the balancing of competing interests – or conflicting demands – of various stakeholders (Kolb, 1995; Waldman and Siegel, 2008). Hung (2004), in his study of innovation within organizations, highlights the reconciliation that has to be achieved between "action-structure" dichotomy. He views innovative action as both creative and cumulative, destroying established order, while simultaneously generating new rules to guide and provide resources to empower future action. In order to do successful innovation, the organization has to reconcile between action and structure. In yet another example of professional reconciliation, Fawcett et al. (2008) make observations about how professional architects reconcile their personal preferences with those of their clients. Their study found a clear difference in preferences of architects and laypersons; however, Fawcett et al. suggest that architects are able to achieve reconciliation between their preferences and those of their clients by producing a design that is successful with respect to both aspects of design, being well-liked by users and also making a high-quality design statement. Figure 10.5 illustrates the discussion, so far, of the complex process of reconciliation of the many factors one must consider in decision-making and the different contexts in which this takes place.

Reconciliation and leadership influence

A person does not become a leader by virtue of the possession of some combination of traits, but the pattern of personal characteristics of the leader must bear some relevant relationship to the characteristics, activities, and goals of the followers. Thus, leadership must be conceived in

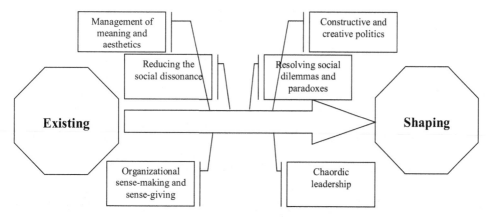

Figure 10.5 Reconciliation in existing and shaping the context

terms of the interaction of variables which are in constant flux and change ... The persistence of individual patters of human behavior in the face of constant situational change appears to be the primary obstacle encountered not only in the practice of leadership, but in the selection and placement of leaders.

(Stogdill, 1948, pp. 63–64)

This assertion made by Stogdill still holds to a large extent. Authentic leaders are able to perform outstandingly as they comprehend the patterns of change, the intricacies attached to the flux of their environment, and the needs of their followers. Understanding these along with a superior sense of self-awareness – understanding the life story, understanding personal values, awareness of personal ambitions, awareness of leadership purpose, awareness of personal legacy, self-confidence and positive self-image – authentic leaders are able to reconcile their social context and their existence in that context.

Leaders operate within complex and dynamic social systems. In this sense, leadership is a process of social influence (Conger, 1998; Parry, 1998; Yukl, 2002; Parry, 2004; Toor and Ofori, 2008c) that engages people, brings people together around a shared purpose, utilizes social resources that are embedded in relationships, and deals with a variety of situations, contexts, challenges and paradoxes. Leaders exercise influence which may travel within a group of people in four different ways: top-down, bottom-up, lateral, and integrated – a combination of top-down, bottom-up, and lateral leadership (Locke, 2003). This demonstrates the complexity of the leadership influence process. In order to generate and sustain influence, leaders face conflicting interests and the goals of people and organizations in highly dynamic and complex environments.

Goethals et al. (2004) opine that individuals are leaders sometimes and followers at other times. Under certain circumstances, an individual may adopt both a leader and a follower role depending on whether he or she is influencing or being influenced, setting directions or following directions, providing support or receiving the support, motivating others or needing to be motivated at a particular point in time. Thus, the role can change in a dynamic, fluid, and ongoing manner. Pearce et al. (2008) observe that leadership is a complex process and involves a dynamic give-and-take. Leaders also face resistance from people and systems who refuse to change or respond appropriately to a given situation. They resolve conflicts and hostilities between various parties and ensure the well-being of those they lead and those they serve outside the organization.

Leaders undertake many activities, such as: creating and communicating a vision; formulating organizational structures, processes, and controls; developing key competencies and capabilities; managing multiple constituencies; selecting and developing the next generation of leaders; and putting in place an effective organizational culture and ensuring the infusion of an ethical value system into the culture (Boal and Hooijberg, 2000). Boal and Schultz (2007) argue that organizations are complex adaptive systems in which leaders are involved in basic processes such as the creation of the context and structure of coordination among members for the purpose of searching out and processing resources vital to organizational survival. Thus, the leader's abilities to reconcile conflicting demands and reduce disorder are key behaviors that a leader should exhibit in order to improve the performance of the team (Kolb, 1995). On the other hand, leader behaviors such as "not being influential in garnering outside support" and "failing to confront and resolve inadequate member performance issues" can be detrimental and prevent the team from functioning effectively. Therefore, the leader must be able to regularly reconcile between the conflicting demands and behaviors of the team members as well as those of outside stakeholders.

Boal and Schultz (2007) suggest that leaders bridge the past, present, and future of organizations. In doing so, they focus on organizational values and identity, ensuring continuity and integrity, developing structural, human, and social capabilities, and simplifying the environmental turbulence and ambiguity by giving a clear vision to the organization to develop. Doing this, leaders typically play two roles in organizations: an architectural role and a charismatic one, in which they redesign their organizations by putting in place appropriate structures and systems in which they envision the future, empower, and energize the elements of leadership (Kets De Vries, 1996). However, to be effective, they also reconcile between their architectural role and charismatic role to avoid ending up having appropriate systems without a vision or dysfunctional systems with an inspiring vision.

In order to perform their complex role in a dynamic social context and to sustain the leadership, authentic leaders regularly reconcile, in everyday decisions, actions, and interactions. They also reconcile between how much they should lead and how much they should manage. They reconcile between how much they should give up and how much they should cling to. They reconcile between how much they should change and how much they should sustain. They reconcile between rational and pro-social decisions, flexibility and rigidity, conflicting interests of various stakeholders, competing demands and goals, needs and desires, go and no-go situations, personal and social dilemmas, personal interests and organizational interests, and numerous leadership and management dilemmas.

Kodish (2006) confirms the above assertion and found that effective leaders are able to reconcile social realities through their seemingly paradoxical behaviors. Such leaders are simultaneously resolute and humble. Collins (2001) terms such leaders Level-5 leaders in his book *Good to Great*. They do their work conscientiously, responsibly, and successfully; they care about the people they work with, they care about the company, and about the community; they shun the limelight and any kind of publicity, and live normal and quiet lives. Unlike narcissistic leaders, they are not driven by image and fame nor by any other accoutrement of a narcissistic personality, such as losing touch with reality or supplanting sound business principles with lofty rhetoric and unfulfilled promises. Kodish (2006) explains the reasons for such paradoxical behavior of effective leaders by noting that leaders do not do it alone; they rather behave while interacting with a dynamic environment around them. They exhibit such behavioral paradoxes; and they adhere to both reality and faith at the same time. They are able to balance skills and dispositions in creative ways (Sternberg, 2007).

Kodish (2006) also refers to Aristotle in explaining how reconciling of universals and particulars is also present in Aristotle's philosophy. He explains Aristotle's viewpoint that the split between the universal and the particular is mostly artificial and unnecessary because both the general and the specific have their respective roles. Furthermore, there is a constant mediation between the universal and the specific, resulting in a dynamic complementarity rather than a schism. In such instances, leaders come to realize that they do not exist in isolation and are part of a larger whole from which they cannot detach themselves. This recognition of the universal self and a comprehension of a whole versus a fragmented reality broadens the leaders' conception of self and connects them with others around them. In such instances, they define themselves in terms of their relationships to others (Lord et al., 1999) develop a relational or a collective self (Uhl-Bien, 2006) which is concerned with social order and reconciliation.

Similarly, Sternberg (2007) observes that

> Wise leaders skillfully balance the interests of all stakeholders, including their own interests, those of their followers, and those of the organization for which they are responsible. They also recognize that they need to align the interests of their group or organization with those of other groups or organizations because no group operates within a vacuum. Wise leaders realize that what may appear to be a prudent course of action over the short term does not necessarily appear so over the long term.
>
> (p. 38)

He gives examples of historical figures – such as Hillary Clinton, Richard Nixon, Sigmund Freud, and Napoleon Bonaparte – who lost sight of one or another set of interests and could not achieve the goals they otherwise would have. These leaders failed as they could not reconcile the self-realities. On the other hand, those such as Mother Teresa, Martin Luther King, Abraham Lincoln, and Winston Churchill were able to successfully reconcile the self as well as social realities and hence made indelible marks on the face of history and were known as authentic epitomes of leadership.

> **Proposition 20**: As opposed to less authentic leaders, more authentic leaders better influence their context through better comprehension of the context they operate in.

Reconciling social paradoxes and dilemmas

According to Barnes (1981), a paradox is any conclusion that at first sounds absurd but that has an argument to sustain it, although these arguments are often buried, ignored, or brushed over quickly. He also observes that the reconciliation of apparent contradictions underlies some of the most truly creative discoveries of science. Conceptualization of a paradox is difficult because of its inherently fractal quality: there are dual paradoxes and multiple tensions within any paradoxical situation (Ofori-Dankwa and Julian, 2004). These paradoxes are part of everyday life and leaders and managers in organizations frequently wrestle with these paradoxes and contradictions (Hock, 1999).

Fisher-Yoshida and Geller (2008) present a continuum of five paradoxes they believe leaders and managers need to consider in order to function effectively. These are: (1) the paradox of knowing (self and other); (2) the paradox of focus (individual and communal); (3) the paradox of communication (direct and indirect); (4) the paradox of action (doing and being); and (5) the paradox of response (time focus: short- and long-term). Sometimes, such dilemmas shape up in the form of much-needed change. Pietersen (2002) argues that the best time for a company to change is when it is successful; although it is the hardest time to do so as the resistance is likely to be at its highest at this time. Such situations again put leaders in dilemmas giving rise to what Pietersen (ibid.) calls the FUD Factor – fear, uncertainty, and doubt. To Lowy (2008), typical dilemmas that leaders face include: head versus heart, inside versus outside, cost versus benefit, product versus market, change versus stability, know versus do not know, content versus process, and competing priorities.

Given the increasingly complex and paradoxical nature of organizational environments (Hitt et al., 1998; Lewis, 2000), leaders need to better understand the nature of paradox, its likely effects, and how to effectively operate when it is present. They need to polarize and synthesize, to see questions in answers, to be both inside and outside of situations, to learn while teaching, and to find unity in opposites as well as opposites in unity (Barnes, 1981).

To resolve the paradoxes and dilemmas, leaders should be able to recognize, acknowledge, and interpret dilemmas in a timely and useful fashion. Where successful, articulation brings meaning, coherence, and alignment to organizational efforts; failure in articulation results in confusion, doubt, and aimlessness of the organization. Effective leaders treat dilemmas as opportunities and are able to draw new directions for their organizations out of the dilemmas. For this purpose, leaders stay focused, remain dynamic, and are prepared to change (Lowy, 2008).

Leaders incorporate the conscious use of critical reflection, foster the willingness to challenge assumptions, build the recognition of complexity inherent in organizations; and finally, challenge assumptions, questioning conformity and embracing difference (Fisher-Yoshida and Geller, 2008). The "dilemma agenda" of leaders addresses the shifting needs, drivers, and opportunities occurring around them and emphasizes two points: direction-setting and culture-setting (Lowy, 2008). At the heart of dilemmas lies a tension between two opposing values because dilemmas are inherently fuzzy. Leaders resolve dilemmas by identifying and mobilizing pockets of support, reminding people about the long-term value of their effort, staying open and responsive to feedback, and communicating actively on plans. Leaders dynamically fix the social gaps – by reconciling between different mindsets, people and politics—and fix technical gaps – competencies, processes, and technology – to resolve the dilemmas they face in leadership positions. Leaders reframe the conditions that dilemmas present and are able to unleash the potentials that dilemmas carry (Lowy and Hood, 2004). By doing so, they gain access to a powerful and timely method for increasing understanding and organizational agility. They are able to make sense of the situation by immersing themselves in the issues to comprehend the competing forces. By developing this understanding, they help shape the context for viable decision-making and action required to overcome the dilemma.

Leaders resolve the dilemmas by persuading themselves that gains would be greater than any perceived loss. Then they move on to persuade people about this by creating a situation for the effective convergence of realities (Kan and Parry, 2004) leading to reduced frequency and fear of dilemmas and paradoxes. This way, they reconcile competing interests and diverging realities to establish order from chaos. Dynamic engagement in reconciling social realities is what distinguishes authentic leaders from ordinary leaders. Authentic leaders successfully engage in reconciling social realities through creative and positive organizational politics (Ammeter et al., 2002); organizational sense-making (Weick, 1995; 2001); sense-giving (Taylor, 1999); management of meaning (Smircich and Morgan, 1982) and aesthetics which refers to felt meaning of leadership which are rooted in feelings and emotions (Hansen et al., 2007); and mastery, congruence and purpose (Ladkin, 2008).

Proposition 21: In order to perform dynamic and creative reconciliation of social realities, authentic leaders employ a variety of strategies that include positive and creative politics, organizational sense-making and sense-giving, and management of organizational meaning and aesthetics.

Reconciling social realities by reducing the social dissonance

Similar to cognitive dissonance – that is one's attempt to reduce any psychological state of tension between two cognitions which contradict each other – social dissonance is a phenomenon that describes social states of tension which contradict with each other. These social states of tension could be social paradoxes, dilemmas, social complexity or chaos, or even social despair. Authentic leaders, being inner-directed but others-driven, are motivated to reduce this social dissonance by aligning people's aspirations, easing the complexity, inspiring hope and optimism, resolving social paradoxes and dilemmas, and clarifying the purpose that is socially acceptable. This process of reducing the social dissonance in which authentic leaders engage is another form of reconciling the social realities.

> **Proposition 22**: As opposed to less authentic leaders, more authentic leaders engage in reducing social dissonance by aligning and empowering people, simplifying the complexity and uncertainty, and providing clear direction.

Reconciling social realities through constructive and creative politics

Many have argued that effective leaders have positive political skills which they use to reconcile issues within the organizations (Zanzi and O'Neill, 2001; Kan and Parry, 2004; Butcher and Clarke, 2006). Ammeter et al. (2002) describe politics in organizational leadership as creative management of shared meanings. Recognizing the negative connotation attached to organizational politics, Ammeter et al. perceive political leadership as an essential phenomenon in a positive sense. They argue that leaders must adopt political behaviors that align with the established political norms, because their behaviors must match the situational assessments of their followers. They note that social effectiveness is particularly relevant in organizational settings and that people rich in political skill not only know precisely what to do in different social situations but exactly how to create synergy among discrete behaviors that transcends the simple sum of the parts. Leaders undertake political tasks at the individual level, coalition level, and network level within the organizations. Ammeter et al. also explain two kinds of political behaviors: proactive and reactive. Proactive behaviors are those which leaders exhibit in response to a perceived opportunity for the organization whereas reactive behaviors are exhibited in response to a threat to the organization. By exhibiting such behaviors, leaders are able to reconcile between the competing demands of stakeholders and develop coalitions and alliances which are supportive to the leadership purpose.

Butcher and Clarke (2006) support the notion of political leadership in democratic systems. They note that democracy intends to realize the desires of individuals for meaningful control of their lives by treating them equally. This is natural in order to result in competing interests of people and diversity of opinions. However, political leadership is instrumental to reconcile the drive for cohesion and the productive exploitation of differences. They also reconcile between the exercise of bureaucratic politics and civic virtue in which they balance their personal interests with those of wider concern. They coalesce and distribute power in a micro-political process or in organizations. Therefore, they embrace political behaviors such as debate, lobbying, and coalition-building, and information management. Such micro-political behavior in organizations is far from being dysfunctional. On the other hand, it is beneficial for the achievement of managerial goals in the organizations (ibid.). Dynamic and creative reconciliation of social realities allows the authentic leaders to successfully and effectively operate within a wide variety of complex situations and circumstances.

Taking examples from project management, Bourne and Walker (2005) and others (Thomas and Mengel, 2008) emphasize the importance of political skills required by project leaders. Bourne and Walker (2005) note that the management of relationships and the engagement of stakeholders are key to success in project organizations. They suggest that project managers should be able to harness political skills to manage relationship and various competing forces. They should be vigilant and flexible to manage the hidden agendas as well as the actions of stakeholders. They note that it is important for the project manager to be able to control the project's time, cost, and scope parameters in addition to project relationships in an uncertain and unstable environment. Understanding the stakeholders' needs and employing appropriate political skills can help the project manager to overcome such project paradoxes.

> **Proposition 23**: As opposed to less authentic leaders, more authentic leaders influence their social contexts by engaging in positive and creative politics to garner support to achieve their purpose.

Reconciliation through organizational sense-making

The concept of sense-making was popularized by Weick (1995; 2001). Weick's (1995) book, entitled *Sense-making in Organizations*, encouraged an outburst of publications on the subject. The field is now well established, although Weick continues to be the most notable author who has written on sense-making. According to Weick, sense-making may be defined as an ongoing accomplishment through which people create their situations and actions and attempt to make them rationally accountable to themselves and others (Weick, 1993; 2001, p. 11). It involves the ongoing retrospective development of plausible images that rationalize what people are doing (Weick, 2001, p. 460). When people engage in acts of sense-making, it is more precise to think of them as accomplishing reality rather than discovering it (ibid., p. 460). It is an ongoing activity that never starts and never stops because it takes place in the context of pure duration (Weick, 1995). Sense-making activities involve the construction and bracketing of cues to be interpreted, linking them to a previous frame of reference that summarizes past experiences (such as traditions, ideologies, theories of actions or stories), and revising the interpretations that have thus developed as a result of actions, interactions, and their consequences (Weick, 1995, p. 8). The concept of sense-making fills important gaps in organizational theory because it is the primary site where meanings materialize that inform and constrain identity and action (Weick et al., 2005).

The conceptualizations of Weick show that sense-making is an inherent process in human interaction. Sense-making "is at the heart of organizing and is the soul of responsiveness" to cope with external demands in achieving sustainability of an organization (Seiling and Hinrichs, 2005). It is fundamentally a social process in which organizational members interpret their environment in and through interactions with others, construct accounts that allow them to comprehend the world, and act collectively (Maitlis, 2005). Seiling and Hinrichs (2005) emphasize that sense-making should not be confused with problem-solving. They note the distinction between sense-making and problem-solving in that sense-making is more than mere problem-solving. They argue that sense-making provides input into successful problem-solving and is informed by the decisions and outcomes that are, themselves, a result of solving problems. This way, sense-making happens before and after problem-solving. Sense-making encourages understanding the situation well enough to define the problem and take appropriate decision.

People engage in sense-making in all circumstances and establish their own interpretations of events. Based on the interpretations of events in the past, people formulate their future course of action and response to such events. Sense-making, in a way, is a mental technique that people use to deal with uncertainty, ambiguity, complexity, paradoxes, and dilemmas which are part of life. Sense-making is about labeling and categorizing to sta-bilize the streaming of experience. It strives to organize the chaos through appropriate action, and communication (Weick et al, 2005). Weick (1995) also notes that sense-making is a process of dealing with uncertainty. An underlying assumption is that if people interpret and attribute meaning correctly, they are able to make sense of the uncertainty and confusion being experienced. Action is taken which allows further understanding of the confusing environment.

With Weick's pioneering work on sense-making, other authors have attempted to relate sense-making with the complex social process of leadership. Reicher and Hopkins (2003) observe that, through an active sense-making process, leaders and followers construct joint identities that help define future possibilities as well as make sense out of current realities. Pye (2005) believes that leadership is a social process of influencing others in complex social and organizational settings. In this sense, leadership is also a sort of sense-making in which leaders synthesize the complex and dynamic environment, the paradoxical combination of traits, behaviors, principles, and relationships through a sense-making process. Colville and Murphy (2006) suggest that all leaders are tested by what they make out of certain events happening around them; how leaders interpret those events and what action they plan as a consequence of such events. They go on to express the view that once people begin to act (enactment), they generate outcomes (cues) in some context (social) and that helps them discover (retrospect) what is occurring (ongoing), what needs to be explained (plausibility) and what should be done next (identity enhancement).

In a study of executive sense-making, Parry (2003) produced seven themes: (1) reference to personal ontology in making sense (value-driven); (2) prevalence of the ethnicity, training, or professional background (culturally biased); (3) experience base of people (experiential); (4) the opportunity to reflect upon the strategic management role (reflection); (5) recounting of the stories about the organization (story-telling); (6) continuous action (action); and (7) the effect of gender (gender). To Allard-Poesi (2005), these processes are activated by, and situated in, the interactions that take place between members of organizations. In engaging in the sense-making process, organizations try to define their role in the present and the future, there is often a need to revise the past to be consistent with the way they currently see themselves (Gioia et al., 2002). The organizations also need to revise the past to be consistent with the way they want to be seen in the future.

Effective leadership is about sense-making, systems thinking, path-finding or orienting (Weick, 1995; Bourne and Walker, 2005). Collective sense-making is crucial for under-standing the decision-making process, actions, and change in organizations (Allard-Poesi, 2005). Seiling and Hinrichs (2005) note that leaders must comprehend their role as sense-managers to facilitate the sense-making process. However, in doing so, leaders and organi-zational members must remain mindful of, and engage in, creative accountability to each other to construct their new future. To Gioia et al. (2002), if leaders can imaginatively place themselves in the future and imagine as if the envisioned events have already occurred, and look back from that perspective, they can then influence the interpretation of the past when the imagined future actually emerges. Therefore, leaders play the role of sense-managers who encourage effective sense-making of organizational members to enable them to face common, unique, daunting challenges of everyday life (Seiling and Hinrichs, 2005).

Sense-making is typically considered to be a product of a perceived discrepancy between the current and expected state of the world (Weick, 1995; Colville and Murphy, 2006). In other words, sense-making begins when there is a lack of fit between what was expected and what is encountered. It starts in that early period of confusion, when someone asks, "What is going on here?" (Seiling and Hinrichs, 2005). In such circumstances, there is a shift from the experience of immersion in projects to a sense that the flow of action has become unintelligible in some way. To restore order and overcome the sense of the disruption, people look for reasons that will enable them to resume the interrupted activity and stay in action. This brings meaning into existence, meaning that they hope is stable enough for them to act into the future, and to have the sense that they remain in touch with the continuing flow of experience. Sense-making, is therefore, an invention, and an ongoing tensional process that leads to action.

Successful sense-making distinguishes the authentic leaders from their ordinary counterparts. Particularly in an age of uncertainty and change, authentic leaders look outside their organizations. By making sense of their environment, they are able to expand beyond their existing trails. By engaging in sense-making, authentic leaders foster better decisions in fluid environments and enable their followers to understand a situation and work together toward the same goal. Lord and Hall (2005) assert that the followers' understanding of a leader may be driven not just by cognitive factors such as implicit theories, but also by the emotional and motivational reactions that a leader elicits in followers through the sense-making process.

Du Toit (2005) views sense-making as a process that assists the individual in making sense of change and also to integrate new experiences into existing frames of reference. Organizational leaders frequently engage in sense-making. However, extraordinary circumstances such as turbulence, change, or crises give sense-making higher appeal. Du Toit (2005) advocates coaching as a catalyst to enhance the sense-making process. In coaching, the one being coached, or the sense-maker, gains feedback from the coach – who acts as a sounding board – resulting in free flow of ideas making sense-making visible and the emergence of clear alternatives for decision-making of the leader. In this process, the leader also engages in learning which is useful for future sense-making.

Particularly during extraordinary circumstances, sense-making is essential to redefine the situation out of the existing understanding of paradigms. As the leader engages in sense-making, ambiguity fades away and new horizons emerge. Environmental scanning and synthesis of situation equips the leader with the bases for preparing future action plans that can be useful for resolving the ambiguity or dilemma. Olson et al. (2006) present their perspective of sense-making in relation to aggression in the organization. In their view, employees' interpretations are culture-specific and their behaviors reflect this. However, leaders influence these interpretations by making sense of employees' sense-making process. For example, in managing workplace aggression, leaders can identify several points of intervention for managing sense-making, and in turn, managing aggression (ibid.).

Proposition 24: Authentic leaders engage in organizational sense-making to comprehend the existing organizational context so that they can shape it for the better.

Reconciling through sense-giving

Another form of sense-making is invoking the sense-making process in others. This is termed as sense-giving (Taylor, 1999) and is a process which is regarded as a fundamental leadership activity within organizational sense-making (Maitlis, 2005). Gioia and Chittipeddi (1991) argue that a critical leader behavior during strategic change is "sense-giving" – which they defined as the process of attempting to influence the sense-making and meaning construction of others toward a preferred redefinition of organizational reality.

Maitlis (2005) argues that the more leaders know about sense-making processes, the easier it will be to manage the sense-making process in their followers. He observes that the sense-making process is too complex for organizational leaders to manage all aspects of it. Therefore, it is more useful if they are able to inspire their followers in the sense-making process. Leaders should also understand how their organizational members make sense of the events (Taylor, 1999), as well as how they, as leaders, can engage in effective sense-giving to their immediate followers, organizational members, and other stakeholders of the organization. Through the sense-giving process, leaders are able to achieve alignment between the sense-making process of themselves and their organizational members. Sense-giving and sense-making enable leaders to have a holistic picture of their organizations.

> **Proposition 25**: Through their authentic leadership, authentic leaders employ sense-giving techniques in order to shape a positive organizational culture.

Reconciling social realities through management of meaning

Establishing a meaning of an event means that the individual is sticking to a certain frame to which some issues and facts are attached (Bean and Hamilton, 2006). The leader's fundamental job is to manage the meaning of the circumstances by articulating what is happening and the future goals they want to achieve with the help of their followers. To foster action toward a goal, leaders guide how others mind the world, helping followers to avoid chaos, and attempting to alleviate indeterminate interpretations (Weick, 1995). In order to convey their meanings to their followers, leaders communicate such that the followers understand a meaning similar to that of the leaders. Larsson et al. (2007), in a study of leadership in crises, noted that followers engage in sense-making of situational characteristics, organizational characteristics and the intent of the leader, and make decisions as a response to the sense-making process. However, this needs a high level of trust within the organization.

Fairhurst and Sarr (1996) also note that leaders spend a lot of time in communicating and their important tools are linguistic and symbolic, by which they make sense of the events happening around them and give sense to the others around them, influencing others' perceptions of the world, an understanding of the events, beliefs about the causes, consequences, and possible future. By managing the meanings, leaders influence their followers through framing and reframing of the events and situations. Fairhurst and Sarr suggest that leaders live their meaning by managing them effectively, conveying them in their small communications with others, and in actions they undertake. While leaders demonstrate consistency in their own frames, they are well cognizant of others' mental models which they want to change or seek to align with theirs (the leaders'). Through this framing process, leaders are able to simplify the ambiguity, uncertainty, or complexity that lie in a given situation as they strive to convey their meanings to others.

Another example of management of meaning is given by Berglas (2006) who discusses how the leaders can manage the expectations, insecurities, and vulnerabilities of "A players" – or what others might call high-performers, talented individuals, superstars, or high flyers – within the organizations. Berglas observes that

> [A p]layers are the people with the "right stuff." They are the most fiercely ambitious, wildly capable, and intelligent people in any organization. Yet despite their veneer of self-satisfaction, smugness, and even bluster, a significant number of your spectacular

performers suffer from a lack of confidence … because they spent their childhoods looking intently for clues about whether or not they had fulfilled parental expectations.

(ibid., p. 106)

However, to retain and nurture the "A players," the leaders must play a proactive role or be prepared to lose the talent. Scholars of human resource management also highlight the need for effort in attracting, motivating, and retaining knowledge workers (see, for example, Horwitz et al., 2003; Horwitz et al., 2006) through appropriate job design, high salaries, paid holidays, skills enhancement programs, increased personal autonomy, and flexible working hours (Stovel and Bontis, 2002).

Today, when most service-providing organizations are driven by knowledge, managing the expectations of "A players" or "knowledge workers" is a key issue that leaders have to deal with. The current research has shown that authentic leaders are adept at managing Berglas's (2006) "A players" or Horwitz et al.'s (2003) "knowledge workers." They deal with such individuals by appreciating their talent or addressing their inflated sense of superiority and underlying issues of poor self-worth. Authentic leaders manage their followers by understanding the strengths and weaknesses of each of them, and therefore treating them as individuals. They observe and understand their followers individually and give them a sense of confidence and self-worth that keeps them performing beyond their own expectations.

Reconciling social realities through the management of aesthetics

In a conceptualization of aesthetic leadership, Hansen et al. (2007) refer to leadership as "sensory knowledge and felt meaning of objects and experiences." They note that "aesthetics involves meanings we construct based on feelings about what we experience via our senses, as opposed to the meanings we can deduce in the absence of experience, such as mathematics or other realist ways of knowing" (ibid., p. 545). Hansen et al. also believe that aesthetic leadership lies at the conjunction of both "management of meaning" and "follower-centric models" of leadership. It is concerned with the experiential, but considers reality to be a subjective experience, a human-made interpretation based on awareness, perception, and the subjective materials used to make those interpretations. Aesthetic leadership also takes into account a holistic perspective by including the skills and competencies of people interacting in complex contexts instead of just relying on the cognitive faculties of leaders. It puts emphasis on direct experience and is constructed – made, shared, transformed, and transferred – in relationships among people through interactions.

This notion of "aesthetic leadership" is similar to what Ladkin (2008) describes as the "beauty" of leadership. Giving the example of a musician, Ladkin highlights three aspects that explain his conceptualization of "leading beautifully." These are:

1   mastery – understanding one's context and domain and mastery of the self;
2   coherence – expressing the self through forms which are congruent with one's overall message and purpose;
3   purpose – attending to the goal toward which one is leading.

In this sense, it can also be argued that aesthetic leaders, being fully self-aware, are managers of meanings and developers of followers and they reconcile with social realities around them in order to achieve social order and continuation of leadership.

Shelton and Darling (2001) proposed the quantum skills model which is similar to what Hansen et al. (2007) refer to as "aesthetics." The model incorporates seven skills:

1   quantum seeing – the ability to see intentionally;
2   quantum thinking – the ability to think paradoxically;
3   quantum feeling – the ability to feel vitally alive;
4   quantum knowing – the ability to know intuitively;
5   quantum acting – the ability to act responsibly;
6   quantum trusting – the ability to trust life's process;
7   quantum being – the ability to be in relationships.

Whereas quantum seeing, quantum thinking, and quantum feeling are psychological in nature, quantum knowing, quantum acting, and quantum trusting are grounded in "spiritual principles." Quantum being, as Shelton and Darling (2001) put it, is intricately connected to each of the other quantum skills. Proponents of the quantum skills model believe that it is key to enhancing leadership effectiveness.

> **Proposition 26**: As opposed to less authentic leaders, more authentic leaders are adept at understanding and managing the organizational meaning by managing the meanings of their relationships with their followers and outside stakeholders.
> **Proposition 27**: As opposed to less authentic leaders, more authentic leaders are better able to manage the organizational aesthetics by mastery of self, coherence, and purpose to achieve the goals of the organization.

Reconciling social realities through chaordic leadership

Reconciliation of social realities can also be well understood through another related concept, chaord – a concept derived from both chaos and order. Chaord refers to "any autocatalytic, self-regulating, adaptive, nonlinear, complex organism, organization, or system, whether physical, biological or social, the behavior of which harmoniously exhibits characteristics of both order and chaos" (Chaordic Commons, 2004). Chaordic refers to anything that is simultaneously orderly and chaotic, existing in the phase between order and chaos (ibid.).

Based on the work of Wilber (1996), Fitzgerald (1996) articulates five core characteristics of chaordic systems:

1   consciousness (mind ... not matter; thinking more than doing is the prime engine of a chaordic system);
2   connectivity (the system is one unbroken and unbreakable unity);
3   indeterminacy (the system is so dynamically complex and highly sensitive to initial conditions that any link between cause and effect is necessarily obscured, rendering its future indeterminate);
4   dissipation (the system is a dissipative structure perpetually cycling through a process of "falling apart" and "back together again" in a novel new form ungoverned by the past);
5   emergence (the inexorable thrust of the chaordic system is toward infinitely ascending levels of coherence and complexity).

These five properties illustrate the fact that humans taking initiatives are put in a central position in a chaotic system (Eijnatten et al., 2007). Therefore, processes such as dialogue, multilogue, and emergent leadership are important mechanisms.

Given the complexity of today's organizations (Boal and Schultz, 2007) that are intricately intertwined with individual and social demands, constraints, and choices (Stewart, 1982), leadership in organizations is even more complex and adaptive (Marion and Uhl-Bien, 2001). Its primary task is to establish a dynamic system where bottom-up structuration emerges and moves the system and its components to a more desirable level of fitness and order (Osborn and Hunt, 2007). Leadership capacity needed to reach a desired order is more intricate and complex and is more likely shared among managerial leadership role holders instead of being concentrated in a single individual in complex adaptive systems. In executive level leadership positions, leaders face more external pressures and less internal constraints while they develop, focus, and enable an organization's structural, human, and social capital and capabilities to meet real-time opportunities and threats (Boal, 2004). They engage in sense-making of environmental ambiguity, and sense-giving to their followers. They regularly operate on the edge of chaos and practice what can be termed chaordic leadership.

This view of management through the lens of complexity and chaos is emerging in project management studies as well, as discussed in Chapter 2. For example, Thomas and Mengel (2008) argue that projects and project environments are being recognized as influenced by complexity, chaos, and uncertainty, necessitating the understanding of project management education and development of project managers. Project challenges are increasing (Toor and Ofori, 2008a) and projects are being managed on the edge of chaos. The diverse range of challenges faced by organizations and projects in the modern business environment requires leaders to respond to each situation on its individual merits (Raiden and Dainty, 2006). This calls for leadership that fosters continuous change, creative and critical reflection, self-organized networking, virtual and cross-cultural communication, coping with uncertainty and various frames of reference, increasing self-knowledge and the ability to build and contribute to high-performance teams (Thomas and Mengel, 2008).

Such chaordic leadership is able to operate on the edge of chaos in the multiplicity of complex adaptive systems. It takes a systemic approach in which leaders are able to integrate ideas that previously were seen as unrelated or even as in contradiction with each other (Sternberg, 2007). What formerly were viewed as distinct ideas or disparate realities are now viewed as related and capable of being unified through the creative reconciliation of leaders. Hence, chaordic systems explain the dynamic and creative reconciliation of social realities that authentic leaders engage in while influencing their organizations, followers, and other social agents.

Proposition 28: Through successful political tactics, organizational sense-making, sense-giving, and the management of meanings and aesthetics, authentic leaders, as compared to less authentic leaders, are better able to lead on the edge of chaos and strive to take their organizations to higher levels of performance.

Proposition 29: As opposed to less authentic leaders, more authentic leaders make their followers and organizations unleash their best potential by effectively engaging in chaordic leadership.

Validation exercise

As noted in Chapter 5, a validation exercise was undertaken in which industry leaders were approached to seek their opinion on the resulting grounded theory model developed in this study. For this purpose, briefing sessions were conducted with nine of the leaders who had previously been interviewed (termed here "research participants"). The briefing sessions typically lasted for one hour, during which research design, general findings, and the resulting grounded theory model were explained to the executives who previously had been interviewed. Followed by the presentation, participants were asked to rate various questions about the model on a scale from 1 to 5 where 1 = Least Likely; and 5 = Most Likely. These questions, which are presented in Table 10.3, sought the participants' opinions on three major aspects: (1) do research participants see themselves in the theory and relate to the findings?; (2) do research participants think that the findings are comprehensive?; and (3) do research participants think that the findings are useful from the practical perspective and that their colleagues and subordinates can also benefit from the resulting frameworks? The ratings given to various questions by the nine participants are shown in Table 10.3.

The mean values obtained from this exercise show that the participants in the validation exercise were fairly satisfied with the rigor of the method used and the categories that emerged from the process. They also noted that the model was comprehensive and, from it, they could make sense of how they developed as leaders. The participants also noted that they could easily relate to the theory that emerged from the current research. They also found that it had a practical use and that they would recommend it to their colleagues to read.

In order to achieve the external validity of the findings, nine non-participant leaders were also approached. These non-participant leaders were nominated as authentic leaders by the participant leaders. Five of the non-participant leaders were directors in their organizations, whereas four of them were senior managers. The non-participant leaders were also given briefings on the research findings. These briefings typically lasted for one hour, during which the non-participants were told about the research objective, research process, and findings. After the briefings, the non-participants were requested to provide their ratings on the same scale on which the participants had given their ratings – a summary of these ratings is shown in Table 10.3. The average ratings given by the non-participant leaders were

Table 10.3 Mean values for validation questions

No.	Question	Mean	
		Participants (N = 9)	Non-participants (N = 9)
1.	Do you see yourself in the theory?	4*	4
2.	Have you gained any useful insight from this framework?	4	4
3.	How comprehensive are the categories noted in the framework?	5	4
4.	Do you think any process is missing in the framework?	2	2
5.	Do you think you can use any interpretation in your practical life?	4	4
6.	Can you interpret this framework in your own way?	4	3
7.	Would you want your project managers to know about this framework?	5	3
8.	Do you think the project managers will benefit from it?	5	4

Note:*5-Point Likert Scale:: 1 = Less Likely; 5 = Most Likely relatively lower than those given by the participant leaders. This is understandable as the participants were not interviewed and may not have been fully apprised about the details of the research that was explained to the participants during the interviews. Despite this, the non-participants did not think that any process was missing in the framework (see Question 4, in Table 10.3). Similar to participant leaders, non-participants also thought that they could relate to the research findings and had gained useful insights from the frameworks.

Many leaders who participated in the validation study also showed interest in the research being presented in open seminars and workshops in their firms so that their followers and colleagues could also learn from the findings and the resulting framework. Five of such seminars and workshops were held. They were attended by young professionals, management trainees, operational managers, and senior managers. The discussion sessions in these seminars and workshops were lively and attendees frequently mentioned that they could relate their own life story and leadership development to the GMALD. Although these seminars and workshops did not contribute to the study directly, feedback obtained from the participants was useful in the formulation of the "recommendations for future research."

Achieving validity

In the current research study, an attempt was made to achieve various forms of validity including:

- descriptive validity;
- interpretive validity;
- theoretical validity;
- generalizability (internal and external generalizability).

The efforts in these regards are now outlined.

Descriptive validity

Descriptive validity refers to the factual accuracy of the data and is concerned with the initial stage of the research study, usually involving data gathering. In order to achieve descriptive validity in the current research, all the interviews were recorded and transcribed into text. These transcriptions were again cross-checked with the audio recordings of the interviews by a third person. (Details regarding data collection are presented in Chapter 5.)

Interpretive validity

Interpretive validity refers to the match between the meaning attributed to participants' behaviors and the actual participants' perspective. This was achieved by using several "NVivo codes" (words which were used by the interviewees) that were established during the analysis of the textual data. Some examples of NVivo codes that were used in the study are: "seeing what others don't see"; "understanding their expectations"; "job matching exercise"; "leading without authority"; "connecting to people"; "honesty"; "integrity"; and "tenacity." Some of these Nvivo codes also emerged as major categories such as "sustainable leadership."

Theoretical validity

Theoretical validity refers to the ability of the findings to explain the phenomena studied, including its main concepts and the relationships among them. To achieve theoretical validity in this research, a core category, entitled "dynamic and creative reconciliation of self and social realities" was developed, grounded in the data, and supported with the extant literature. It was found that the core category not only has the explanatory power for the data, but also was adequately supported by other theories in the social sciences.

Internal generalizability

Generalizability means whether the same findings would be obtained with different data. Generalizability is usually divided into internal generalizability (or internal validity) and external generalizability (or external validity). Internal generalizability is whether generalizability exists within the community that has been studied. In the current research, internal generalizability was achieved by the validation exercise that was conducted. Nine participant leaders opined on the categories that had resulted from the data analysis. The participants leaders were generally very satisfied with the findings and could relate to the categories and processes that emerged from the data.

External generalizability

External generalizability is whether the findings are generalizable to the settings that were not studied. For this purpose, researchers are typically advised to study only those who have not participated in the research that generated original findings. External generalizability in the current research was achieved by asking the non-participant leaders to give their opinions about the research findings.

Summary

This chapter presents an integration of the concepts, processes, and categories which emerged in the research. The integration of processes under authentic leadership development and influence resulted in the paradigm model whereas the integration of conditional/interactional matrices resulted in the reflexive coding matrix and core-category representing all the other categories and processes. The grounded theory of "dynamic and creative reconciliation of self and social realities" showed that leaders constantly strive to optimize the outcomes of their leadership by creatively reconciling self and social realities. They reconcile with their inner realities, their outer context, people, relationships, roles, decisions, input, processes, and outcomes. This dynamic process of reconciliation results in authentic leadership development and influence.

The grounded theory of "dynamic and creative reconciliation of self and social realities" shows that authentic leaders understand their inner states and their social contexts and are able to creatively reconcile these realities to achieve authenticity in their own leadership and within their organizations. This is a new perspective to the understanding of how leaders develop and function within their social settings and achieve authenticity by aligning their purpose, goals, values, vision, and the legacy they aspire to leave behind. In doing so, they also align the aspirations of their followers and other people to their vision, and mobilize their contribution. This alignment helps them to manifest authentic leadership and inspire authenticity among their followers as well. By applying an established approach to validation, the results showed that the models developed, and the findings obtained, in this study satisfy all the elements of theoretical validity. The models were also found to be relevant and useful to practitioners.

11 Further development and application of the root construct

Introduction

This chapter presents a synthesis of the discussion in the book and considers the future of the subject of leadership in construction. It begins with a summary of the study and the findings of the research. The summary is followed by a comparison of authentic leadership with other forms of leadership, from the new light shone on the construct by the findings from the study. The research propositions outlined in the earlier chapters on the basis of the findings and inferences from the analysis are discussed. The contributions the study has made to knowledge are outlined. The implications for current leaders and aspiring leaders are considered. Those for human resource management units in firms are also outlined. For the benefit of aspiring leadership researchers, a critical evaluation of the grounded theory research methodology is presented. The limitations of the research are also discussed. Recommendations for future research are presented. Finally, the conclusion to the whole study ends the chapter.

Summary of findings

The research study explored how authentic leaders develop, remain authentic, manifest their leadership, and influence their followers and organizations. It considers leadership as a social process. It also recognizes that leadership development – the historical phenomenon of leader emergence and life experiences – is part of the leadership process. It recognizes that leadership can be fully understood only by studying all the related "where," "when," 'who," "how" and "what" of leadership emergence and influence.

Grounded theory methodology was adopted to accomplish the objectives of the study. This approach has been deemed suitable for uncovering complex social phenomena (see Conger, 1998; Parry, 1998; Bryman, 2004). Analysis of qualitative data collected from authentic leaders – chosen mainly through a peer nomination process – resulted in a number of concepts and categories. Integration of these conceptual categories provided a holistic understanding of authentic leadership development and influence.

It was found that authentic leadership development and influence are not two discrete aspects of authentic leadership. Instead, how authentic leaders manifest their leadership and how they influence others are deeply embedded in their life stories, values, purpose, goals, and motivations. The study found that authentic leaders develop in four phases: preparing, polishing and practicing, performing, and passing, through experiential learning under the influence of significant individuals and significant social institutions. Over the course of their development, the leaders learn from failures and successes, and discover their authentic self through transformative events. Discovery of their authentic self leads to further conscious development (self-regulation) on the part of the leaders. Self-transcendence, positive reframing, a systemic perspective, meta-reflection, and role modeling for followers are key processes in which leaders engage during the "performing" and "passing" phases of leadership development.

In this research, it was found that authentic leaders are primarily driven by self-leadership, self-transcendent leadership, and sustainable leadership. However, what makes them effective and successful is their ability to dynamically and creatively reconcile their self and social realities – a core social process of authentic leadership. In doing so, authentic leaders are able to bring harmony into their self-related knowledge by being cognizant of their values, ambitions, aspirations, goals, motivations, and purpose. They then seek an active alignment of their purpose with the goals and aspirations of their followers and various social agents around them. They are able to garner the support of others by inspiring mutual trust, commitment, and singularity of their goal. They are adept at harnessing followers' support due to their ability to creatively reconcile social realities – dilemmas, paradoxes, uncertainties, and ambiguities. This dynamic and creative reconciliation of self and social realities also makes them authentic and enables them to look authentic in the eyes of others. This congruence earns them authenticity of leadership, for which they receive support from followers and others.

This study has shown that leadership enactment, manifestation, or influence cannot be seen in isolation from leadership development. Authentic leadership is influenced by various trigger events which result in a progressive development of the leader's purpose and vision – that is grounded in the leader's self-awareness – that largely drives his or her subsequent actions – or self-regulation. Authentic leaders consciously engage in several cognitive and social processes over their life span. Experiential learning helps them to develop the competencies, higher-order attributes, skills, and dispositions to engage in self-leadership, self-transcendent leadership, and sustainable leadership – the three dimensions of authentic leadership found in this study.

Aware of their strengths and weaknesses, motivations and values, authentic leaders actively establish social networks to garner support from various social agents around them – followers, clients, outside stakeholders – to achieve their purpose and vision. This view also confirms that authentic leaders are not "perfect," "heroic," or "great man or woman" or "superman or superwoman" leaders. Their success lies in their ability to consider the big picture and balance the conflicting voices within their inner selves as well as in the environment they are part of. Their ability to creatively reconcile the complexities, paradoxes, dilemmas, uncertainties, ambiguities, and other challenges makes them stand out as effective and successful leaders. They do not have to be charismatic or flamboyant. They are leaders who lead themselves first, then go beyond themselves to lead others around them through self-transcendence, and finally they take appropriate actions to sustain the effect of their leadership.

In the process of dynamic and creative reconciliation, authentic leaders resolve inner and environmental paradoxes and dilemmas; understand, manage, and shape their inner and social states by reducing the socio-cognitive dissonance; frame and reframe their life stories to reconcile their inner self with, and shape their social context; manage the organizational meanings by sense-making and sense-giving; shape the working environment appropriately through positive and constructive politics; and are driven to seek to be effective in the interplay between order and chaos within their inner selves, as well as in their surroundings.

As many scholars assert, authentic leadership is a life-long journey (Avolio and Luthans, 2006; George and Sims, 2007), and not the destination. Authentic leaders realize this, and therefore continuously strive to enhance aspects of their leadership. Their values and purpose provide them the moral compass (George et al., 2007) to keep them on track so that they are not trapped in the famous six fallacies (Sternberg, 2002; 2007):

1   the unrealistic-optimism fallacy – that they are so smart and effective that they can do whatever they want;
2   the egocentrism fallacy – that they are the only ones that matter, and not the people;
3   the omniscience fallacy – that they know everything and lose sight of the limitations of their own knowledge;
4   the omnipotence fallacy – that they are all-powerful and can do whatever they like;
5   the invulnerability fallacy – that they can get away with anything because they are too clever to be caught;
6   the moral disengagement fallacy – that they cease to view their leadership in moral terms but rather only in terms of what is expedient.

Authentic leaders are not only driven to be true to themselves and others (Luthans and Avolio, 2003), but also they manage their authenticity within their social contexts by treating different people differently and making appropriate sense of varied situations and circumstances.

The most central quality of authentic leaders that emerged as the core category in this study is their ability to reconcile all three dimensions of authentic leadership: self-leadership, self-transcendent leadership, and sustainable leadership. They develop and influence others by simultaneously being self-leaders as well as self-transcendent leaders. Appreciating and inculcating these two crucial aspects, authentic leaders are able to sustain their leadership strategies and outcomes.

Authentic leadership: a "root construct"

Another finding in the current study is that authentic leadership is essentially a "root construct." Authentic leadership lies at the roots of positive forms of leadership (Avolio and Gardner, 2005; Gardner et al., 2005). As discussed in Chapters 7 and 8, authentic leaders are inherently inner-focused (self-leaders) and others-directed (self-transcendent) and manifest their leadership in various forms – such as servant, spiritual, transformational, or socially responsible leadership. In line with the views of George and Sims (2007), the current study shows that authentic leaders are able to motivate people based on mutual trust and connected relationships. George et al. (2007) assert that authentic leaders or "true north leaders" develop by knowing their authentic self, defining their values and leadership principles, understanding their motivations, building their support teams, and staying grounded by integrating all aspects of their lives. Through the wisdom they obtain from their experiences, they are able to reflect on situations, evaluating and making choices.

Authentic leaders are guided by a set of transcendent values which mediate their decisions about what is right and fair for all stakeholders (Luthans and Avolio, 2003; May et al., 2003; Sosik, 2005). They live with self-transcendent values such as benevolence—honesty, responsibility, and loyalty – and universalism – equality, social justice, and open-mindedness (Michie and Gooty, 2005). The findings in this study also show that authentic leaders genuinely desire to serve others and empower their followers and other people through their self-transcendent leadership. This notion of authentic leadership is somewhat similar to what Rock (2006) refers to as "quiet leadership." According to Rock, "quiet leaders" are masters in bringing out the best performance in others by noticing certain qualities in people's thinking, by helping them improve their thinking, and by helping them to make their own connections. They help the people to explore their real potential by giving them the power to think, by encouraging them to make their own decisions, and by letting them find solutions to their problems themselves. They validate, confirm, encourage, support, and believe in people's potential and empower their followers so that they can strive to realize their potential (ibid.).

The current study also shows that authentic leaders are not necessarily charismatic personalities or flamboyant in their rhetoric. Similar to Collins's (2001) Level-5 leaders, authentic leaders are timid and ferocious, shy and fearless, and modest with a commitment to high standards. They do have flaws and weaknesses, but they are not constrained by them. They take failures as opportunities to learn lessons. They challenge their followers by setting high performance standards through their own examples. By demonstrating commitment, devotion and dedication, they become the role models of veritable performance.

From the conceptual point of view, authentic leaders may manifest the characteristics of transformational, servant, spiritual, or ethical leadership. The findings from this study confirm this manifestation of different forms of positive leadership styles in the approaches taken by the authentic leaders. Thus, primarily, authentic leadership is the "root construct" that underlies all forms of positive leadership although it is distinct from other leadership styles in many respects (George, 2003; Avolio and Gardner, 2005; George et al., 2007).

Research propositions

As noted earlier, qualitative research is primarily driven to establish new models, descriptions, ideas, and propositions, instead of testing the already existing theories. The current research also sought to derive new concepts on, establish the frameworks of, and develop testable propositions about, authentic leadership development and influence. This is achieved in the analysis in Chapters 6–10 in which frameworks for authentic leadership development and influence are presented. Chapter 6 offers 13 propositions that pertain to how authentic leaders develop and remain authentic leaders.

These propositions are based on the concepts and categories related to authentic leadership development. The concepts and categories are integrated and grounded into the extant literature. These propositions address how significant others, social institutions, and transformative events intertwine to develop authentic leadership in leaders. They also address how various processes (such as reflection, self-awareness, self-regulation, positive reframing, and meta-reflection) and higher-order attributes are interlinked, resulting in authentic leadership development. Also, propositions 14–19 in Chapter 10 are presented in relation to the core category and grounded in the extant literature. These propositions are derived from the core category and pertain to how authentic leaders engage in creative and dynamic reconciliation of self-realities – while they are "being" and "becoming" authentic. Authentic leaders engage in reconciliation of self-realities by understanding and managing their inner states and inner sense-making, resolving inner dilemmas and paradoxes, and reframing their life stories.

Propositions 20–29 in Chapter 10 are also derived from the core category of this research. These propositions primarily pertain to how authentic leaders influence their followers and organizations by engaging in dynamic and creative reconciliation of their social realities. These address how authentic leaders influence their "existing" context by proactively "shaping" it through creative politics, organizational sense-making and sense-giving, management of organizational meaning, reducing the social dissonance, and chaordic leadership.

Based on the evidence in the data and as well as support in the literature, the propositions presented in the current study offer original contributions to the existing body of knowledge and suggest new directions for further empirical research. Future research should be undertaken to test and further validate these propositions.

Contributions of the study

Contributions to knowledge

The current research makes several contributions to the existing body of knowledge on leadership. The study addressed the main research question of how authentic leaders develop and influence their followers and organizations. The contributions to knowledge are now discussed.

First, the study supplements the understanding of authentic leadership – a construct that is still in its nascent stage. Through empirical qualitative data collected from authentic leaders, the study furthers the construct of authentic leadership by grounding the emerging concepts back into the data. The study offers various dimensions of authentic leadership – under the categories of self-transcendent leadership and sustainable leadership. As suggested by Cooper et al. (2005), the qualitative approach was particularly useful in identifying these dimensions as the construct under study is still in infancy. Researching authentic leadership development and influence in this manner was instrumental in the development of conceptual frameworks that related authentic leadership to its key antecedent, moderating, mediating, and dependent variables (ibid,).

Second, to the best of the authors' knowledge, the study is among the few which used the grounded theory methodology to explore authentic leadership development and influence. As noted above, the grounded theory approach was instrumental in enabling the unleashing of various dimensions of authentic leadership which would otherwise be difficult through qualitative methods. For example, in the absence of any validated objective measurements of authentic leadership, it would be difficult to relate it to its developmental antecedents and influence strategies. On the other hand, the paradigm model and reflexive coding matrix are very useful to relate how leadership development mediates the influence of authentic leaders. Examination of leadership development in conjunction with leadership influence integrates historical, proximal, and distal contexts of leadership in a single framework. In this regard, the study addresses the call to develop an integrated theory of leadership that has been made by many authors, such as Avolio (2007).

Third, the study offers a framework of authentic leadership development which illustrates different phases in which authentic leaders develop through various socio-cognitive processes. It divides leadership development into four phases and explains various development experiences that trigger authentic leadership development in individuals. Understanding these socio-cognitive processes can be useful to accelerate authentic leadership development in individuals. These socio-cognitive processes are also useful in making individuals appreciate their developmental path and take control of their leadership development.

Fourth, the study shows that authentic leadership is not merely about the leader's self-leadership. It is a complex interplay of self-leadership, self-transcendent leadership, and sustainable leadership. To develop and remain as authentic leaders, individuals reconcile the complexity of several worlds they are part of. The study recognizes leadership as a human endeavor and not a scientific one. The study appreciates that there is no single formula for effective leadership. However, being unique in their own way, individuals can be authentic leaders as long as they understand how their leadership is influenced by their life story, purpose, vision, motivations, desires, and legacy thinking. Therefore, to be authentic, leaders cannot be just "self-leaders"; they have to go beyond themselves (self-transcendent leadership) and then devise adequate strategies (sustainable leadership) to sustain the impact of their positive leadership.

Fifth, the study highlights that "dynamic and creative reconciliation of self and social realities" is a major socio-cognitive process in relation to authentic leadership development and influence. By explaining the concept in further detail and grounding it in the extant literature, this study takes the concept of "reconciliation" to a further level by incorporating the dimensions of sense-making and sense-giving, socio-cognitive dissonance, comprehension and management of inner states, framing and reframing of life stories, interplay between past, present, and possible self, management of social meaning and aesthetics, resolving the inner and social paradoxes and dilemmas, and chaordic leadership. The inclusion of "dynamics" and "creativity" also add notable dimensions to the construct of reconciliation. It shows that there should be reconciliation of the inner self with social realities if individuals are to develop and remain authentic. Hence, "being true to oneself" is not sufficient. Authentic leaders are true to themselves and others in such a way that others are aware of it, and appreciate it.

Sixth, the study offers a number of testable propositions based on the findings. These propositions relate to leadership development and influence and are also grounded in the literature. These propositions can be tested in future through empirical research. Thus, the propositions provide impetus for more research on authentic leadership.

Seventh, the study introduces several new concepts such as: the phases of authentic leadership development (preparing, polishing and practicing, performing, and passing); higher-order leadership attributes; the importance of meta-reflection; and the notion of social dissonance. It also distinguishes authentic leadership from other forms of leadership, and shows how authentic leadership can be considered to be at the root of the positive forms. For example, it breaks self-transcendent leadership into further concepts which are closely related to other forms of leadership, such as soulful leadership, servant leadership, spiritual leadership, shared leadership, service-oriented leadership, and socially responsible leadership. The concepts of self-leadership and sustainable leadership are explained. The concepts of vigilant outlook, creative problem-solving, change management, succession planning, sustainable human and social capital development, and building sustainable organizations have also been described.

Eighth, the study is the first of its kind in Singapore. Although, after a thorough search, a few other studies were found on leadership in Singapore (see, for example, Ji et al., 1997; Low and Lee, 1997; Low and May, 1997; Low and Leong, 2001; Weichun et al., 2005; Low and Chuan, 2006), these studies did not explore leadership development or authentic leadership. Nor did any study employ the grounded theory approach to study leadership. In this respect, the study provides impetus for further qualitative research on organizational leadership in the context of Singapore.

Contribution to practice: implications of leadership transformative events

The findings in this study offer the taxonomy of events in leadership development. The framework shows that leaders relate their development to their past experiences, their significant others, and social institutions. The findings show how the leader's self-awareness translates into self-regulation and desired impact of leadership. They also indicate that a greater and deeper insight into one's personal biography facilitates anchoring of one's values, ambitions, goals, convictions, and purpose. These findings are in line with earlier assertions that influential persons and positive transformative events serve as positive forces in developing leaders' self-awareness that determine their psychological well-being and help to generate more self-clarity and self-certainty (Gardner et al., 2005). Following this perspective, Cooper et al. (2005) believe that transformative or trigger events, if replicated appropriately through an intervention, can help to generate leadership authenticity. This is because the event may remind the individual of a similar past experience, create more self-awareness, generate disturbance in the conscience, and remind the person of something that is crucial. If employed with proven techniques, such as chaining and shaping (see B.F. Skinner, 1969, cited in Cooper et al., 2005), the effect of antecedents can be augmented to accelerate leadership development.

This seeking of insight from the past can be triggered through various interventions including workshops, seminars, and over the longer term, mentoring and coaching. Remembering and reconstructing such events can help to boost introspection and realization of self-related information, particularly about those events which brought a significant change in the individual's life. This information can then be used by the individuals not only to appreciate what contributes to their leadership development, but also for other purposes such as the positive reframing of negative events.

Luthans et al. (2006) argue that positive reframing of trigger events helps to develop positive psychological capacities (or what they call "PsyCap"). When individuals come across such planned or unplanned trigger events, they are in a better position to fully benefit from these episodes (Avolio and Luthans, 2006). In other words, it is a cyclic or iterative process in which trigger events result in the development of PsyCap; and with improved PsyCap, one is in a better position to identify and comprehend a certain trigger event. In this framing and reframing process, individuals can learn from their everyday experience and, thus, develop their leadership potential. Luthans and Avolio (2003) contend that unplanned and unpredictable trigger events involve the risk of exposing the individuals to negative perceptions. It is more beneficial for organizations to be proactive and expose the leaders to planned trigger events; this will challenge the leaders to positively react to such situations and will eventually result in the development of positive resilience (ibid.).

Therefore, in order to control the interventions and plan the replication of trigger events, individuals can be provided with mentoring and coaching to reframe their life stories into events of their self-development and episodes of significant change. Reframing of negative trigger events as positive self-development episodes by coaches and mentors can also help to boost the leadership development of emergent leaders. Moreover, it is also likely that once leadership trainees have been made aware of a repertoire of such events, they will react to similar events in a more positive way to accelerate their own leadership development. From reminders of negative events, trainees can also be helped to reframe these negative trigger episodes positively with the help of skilled coaching.

The description of trigger events by senior organizational leaders can also encourage the young professionals to frame and reframe their own life stories and appreciate the triggers that played a role in their leadership development. Such interventions can help to initiate reflection, self-awareness, and positive reframing. This can lead to self-regulation, meta-reflection, positive role modeling, and a systemic outlook, which are the necessary tools of authentic leaders.

Contribution to practice: implications for human resource management

Understanding the influences of leadership development is useful for human resource management in many practical ways. From knowledge of the individuals' background and events that have transformed them, human resource professionals can effectively plan their leadership development over their careers. The appraisal of leadership antecedents can help in identifying, recruiting, and promoting the individuals who have a better potential to develop as leaders (Cooper et al., 2005). Appraisal of antecedents will also help to differentiate between real and fake leaders (Chan and Chan, 2005). This is because real leaders are more likely to have consistency and harmony in their life stories and self-concepts, which can be found in their personal reports and accounts of their followers or colleagues; it would also be possible to deduce them in general terms from their behaviors as these should be consistent with their own life stories, and others' perceptions.

The opposite is true with fake leaders who are likely to have high inconsistency in their life stories, self-concept, and others' perceptions of them. The evaluation of antecedents can also help to grasp the temporal and logical interdependencies of events whose relationships to one another are not immediately evident (see Sparrowe, 2005). Such events can be drawn together through an assessment of the antecedents in an individual's life. Since leadership is temporal in nature (Avolio and Gardner, 2005; Chan and Chan, 2005), assessment of antecedents is likely to be useful as an indicator of the leader's self-related information such as purpose of leadership, personal values, and motivations of the leader. Appreciation of antecedents will also help to gauge and ultimately unleash the individual's potential for better performance (Gardner and Schermerhorn, 2004).

Contribution to practice: implications for practicing leaders

The study offers several ideas for practicing leaders in organizations. Practitioners in the industry can use these findings to prepare and take a personal assessment quiz in which they can score themselves and their organizations first, broadly in terms of leadership; and then score themselves on various dimensions such as self-leadership (self-awareness, self-regulations, role modeling, self-growth and competence, higher-order attributes). Leaders can also evaluate themselves on various dimensions of self-transcendent leadership (soulful leadership, servant leadership, spiritual leadership, shared leadership, service-oriented leadership, and socially-responsible leadership) and find out how their character fits with self-transcendence.

Finally, leaders can assess themselves on sustainable leadership (vigilant outlook, creative problem-solving, building high-performance and sustainable teams, sustainable human capital development, sustainable social capital development, leading and sustaining change, building sustainable organizations, and sustained success planning). They can engage in these self-assessment exercises to see the areas of their strength and identify the areas where they can improve their leadership. Leaders can also apply these assessment exercises to their followers and colleagues and help them identify their strengths and weaknesses. As Ofori and Toor (2008a) suggest, such evaluation exercises can also be carried out at the group, unit, and organizational levels to examine the prevalence of authentic leadership.

Also, by using the GMALD, practicing and emergent leaders can engage in discovering their own leadership journey; their transformative events, people who influenced them, and institutions where they learnt important lessons of leadership. Leaders can engage in: discovering their real leadership purpose, values, ambitions, and goals; assessing whether they have been able to find their true north (George and Sims, 2007); and taking steps to achieve the aspired leadership legacy they wish to leave. The leaders can also relate their stories of successes and failures to the narratives described in this study. Such assessments can enable the leaders to understand how they can take control of their own leadership journey.

Critical evaluation of the research methodology

A qualitative research method was adopted for this study. Guidelines provided by grounded theory methodology were used to examine how authentic leaders develop and influence their organizations and followers. Although Glaser and Strauss (1967) proposed grounded theory more than five decades ago, its use in leadership research has been limited. An added intricacy is that there are two schools of thought: the Straussian school (Strauss and Corbin, 1990; Corbin and Strauss, 2008), and the Glaserian school (Glaser, 1978; 1992) which significantly differ in their analysis techniques, such as coding methods and paradigm development. These differences were critical when the authors were finalizing the research methodology. Where Glaser (1978; 1992) stays closer to the original or "orthodox" version of grounded theory (Glaser and Strauss, 1967), the techniques offered by Strauss and Corbin (1990) are relatively more comprehensive and methodical in nature. Furthermore, the Straussian school dominates management research (Jones and Noble, 2007) and methodological techniques offered by Strauss and colleagues are useful in that they help the delineation of antecedents, and concurrent and consequent factors associated with authentic leadership development and influence. Therefore, this research adopted the Straussian school approach for data analysis. There is a common perception that qualitative grounded theory methodology is laborious and time-consuming. However, it is useful for examining a complex phenomenon such as leadership in detail (Parry, 1998; Toor and Ofori, 2008c).

During the research, there were challenges in finding the authentic leaders first for the interviews and then the validation study owing to the prevailing boom in the local construction industry. However, given that research on authentic leadership is still emerging, the use of grounded theory methodology helped to reveal various dimensions of the construct, and enabled various phenomena that take place in conjunction with authentic leadership development and influence to be described. In this sense, grounded theory helped to generate propositions instead of testing hypotheses. Although, some authors have used quantitative data in conjunction with qualitative data in their grounded theory analyses (see Kan and Parry, 2004), such examples are rare and more work is required to effectively reconcile a mix of both quantitative and qualitative data for grounded theory analysis.

Limitations of the research

Some limitations of this study are now outlined:

1   This study was undertaken on the construction industry of Singapore. Although the findings are well grounded in the extant literature, and were validated in accordance with the established conventions, it is still pertinent to note that the findings might not be universally generalizable.
2   The study was also a snapshot at a given time. It was cross-sectional in nature although the interviews sought to cover the leaders' experiences over their entire lives, as well as the future. A longitudinal design may be more useful to uncover further realities such as the effect of changing personal, organizational, and immediate contexts.

3   The study sought to explore authentic leadership development and influence. Although careful steps were incorporated into the research method and approach to avoid any personal bias and social desirability of the leaders interviewed, it still remains a limitation.

Recommendations for future research

Although Chapter 3 highlights some areas in which future research on authentic leadership could be focused, this section addresses some pertinent issues which demand deliberations among the research community in relation to authentic leadership.

Future research can focus on testing the propositions presented in this study through empirical studies on authentic leadership. In support of quantitative research on authentic leadership, it is necessary to develop objective scales through which authentic leadership can be measured at the individual and organizational levels. These scales can enable authentic leadership to be objectively related to leadership performance, effectiveness, organizational outcomes, employee well-being and satisfaction, personal growth of leaders, organizational and project performance, influence on managerial and technological innovation, effectiveness in terms of cost, time, and quality, management and satisfaction of stakeholders, decision-making and dispute resolution capabilities, and interface management. Objective measurement of leadership outcomes will help to estimate the return on the investment in leadership development programs.

There have been efforts to develop such scales. Walumbwa et al. (2008) offer a 16-item scale to measure higher-order constructs of authentic leadership. However, these scales have yet to go through extensive validation across many cultures and contexts. Therefore, as Cooper et al. (2005) suggest, "authentic leadership scholars must develop at least two ways of assessing each authentic leadership dimension or subdimension" (p. 479). One example of measuring authenticity is the scale developed by Kernis and Goldman (2005c; 2006). Such scales can be used to cross-validate the measures of authenticity in leaders. Once these scales have been validated, researchers can focus on combining the qualitative and quantitative methods to develop frameworks that can explain the larger number of variables.

Another subject for research on authentic leadership is the psychological and personality dimensions. A number of psychometric measures of personality, behavior, and motives are available in the literature and practice. These dimensions include the personality Big Five, emotional intelligence, psychological capital, stress management, attachment styles, motives, personal and interpersonal skills, personal well-being, and self-esteem. It would be valuable to ascertain relationships between authentic leadership and other forms of positive leadership such as servant leadership, transformational leadership, and ethical leadership, using these dimensions.

A promising avenue for future research is in-depth examination of environmental factors that impede authentic leadership development and influence within organizational settings. Such studies would help to identify the factors that should be eliminated to let authentic leadership flourish within organizational settings.

Another area which needs further deliberation is how authentic leaders avoid falling victims to cognitive fallacies, as noted by Sternberg (2007). Such studies would further the understanding of how leaders reconcile themselves with their weaknesses by focusing on their strengths while still appreciating that they are not perfect individuals.

More work on how leaders develop and master the higher-order leadership attributes would be useful. Particularly, research on practical wisdom and creativity (ibid.) can yield useful insights as to how authentic leaders gain practical wisdom and how wisdom intertwines with other higher-order attributes to enable the leaders to become and remain authentic. Another topic worth exploring is how authentic leaders assuage various uncertainties and self-limiting aspects which are related to their continuous learning. Understanding this would enable leaders to learn how to break out from their soft corners and exceed what they consider to be their limits.

There appears to be a scarcity of literature on leadership in times of emergency, crisis, or catastrophe. How authentic leaders remain authentic and how they utilize their past experience and learn from their prevailing experience when they are coping with crises can further the understanding of leadership.

In a review, Toor and Ofori (2008b) noted that few leadership studies in the construction literature focus on leadership development and almost no works have considered designing effective interventions for leadership training and development of construction professionals. In today's competitive business environment, there is a need to accelerate the authentic leadership development of capable individuals to maximize their leadership potential and benefit from their abilities. Researchers could consider developing effective interventions that can help accelerate authentic leadership development. Future studies should focus on customized, action learning-based interventions that are culturally specific, seek to address complex challenges, and structure, in order to develop leaders who can effectively function in the complex and dynamic business world of today.

Future studies can also examine the impact of authentic leaders on their subordinates, and on the success of the projects they lead. The influence of organizational culture, team environment; size, duration, and complexity of the construction project; number of stakeholders involved; and various other factors can also be considered mediating variables affecting the perception of leadership authenticity and performance.

While considering these research questions, due consideration has to be given to research designs. There is a need to employ multiple designs that may include life stories, idiographic, historiometric, psychometric, and psychobiographical approaches (Simonton, 1999) in leadership research in the construction industry. Applying methodological pluralism, researchers can use a quantitative approach to determine general patterns whereas qualitative data can assist to provide deeper insights into the quantitative findings (see Bryman, 2004).

Conclusion

Authentic leadership has become an important topic in leadership and organizational research. The current study fulfills a need for grounded empirical work on the subject. The study addressed several related issues such as authentic leadership development and influence. It shows that authentic leaders are profoundly influenced by their life stories. Their values, motivations, and purpose of leadership are a reflection of their past experiences. In this sense, the study shows that leadership influence cannot be seen in isolation from leadership development. What leaders do, why they do it, and how they do it are a manifestation of their values which they have gained in various social institutions and from interactions with significant individuals. The historical, proximal, and distal context of leaders are important dimensions to understand how they behave and act.

More importantly, the findings reveal that authentic leadership comprises three dimensions: self-leadership, self-transcendent leadership, and sustainable leadership. Authentic leaders engage in all three forms of leadership to achieve their leadership purpose and leave a lasting influence on their followers and in their organizations. Grounded theory analysis shows that authentic leaders engage in a basic social process that is labeled here "dynamic and creative reconciliation of self and social realities." In this process, the leaders make sense of their inner and outer realities and endeavor to reconcile their personal and social context.

Why do these findings matter? Much of previous research on leadership focused on preparing a list of traits, attributes, and behaviors that make successful and effective leaders. Leadership research has also considered how leaders act and behave but not on why they exhibit such behaviors and how they develop their leadership over time and what helps them gain the trust of their followers. However, the current study reports that leadership, or rather, authentic leadership is closely related to a person's humanity, character, values, and roots. Authentic leaders manifest their leadership in various positive forms; however, in each form, their leadership is inner-focused and others-directed. Their self-transcendent values and motivations translate into their purpose and the legacy they aspire to leave behind.

These findings open several avenues for research. The pertinent point here is that leadership without authenticity can be damaging for an organization. Given that humans intrinsically seek authenticity of self, the authentic leadership construct has several implications for research on leadership and leadership development. What is required now is more qualitative research in conjunction with quantitative research which can help devise tools to measure authentic leadership and establish its associations with other organizational constructs. An international consortium of researchers can take up the challenge to develop and validate a cross-cultural measurement tool for authentic leadership. The next step would then be to design interventions to develop authentic leaders.

Summary

This chapter presents a summary of the study and the findings of the research. The summary is followed by an attempt to compare authentic leadership with other forms of leadership, from the new light shone on the construct by the findings from the study. It is inferred that the findings from the study confirmed that authentic leadership is the root construct of all the positive forms of leadership, as examples of these forms of leadership emerged strongly in the analysis of the interviews of the authentic leaders.

The research propositions outlined in the earlier chapters of the book on the basis of the findings and inferences from the analysis are discussed. The contributions the study has made to knowledge are outlined. The implications for practice are considered on three dimensions. The first two dimensions comprise how current leaders and aspiring leaders respectively can use the book and the findings from the study. The third dimension is the set of implications for human resource management. It suggests that the model developed can be used to design interventions for leadership training for the staff members in their organizations. For the benefit of aspiring leadership researchers, a critical evaluation of the grounded theory research methodology used in the study is presented. The limitations of the research are also discussed. Recommendations for future research on authentic leadership in construction are presented. A conclusion to the whole study is presented at the end of the chapter.

12 A final reflection

Introduction

This chapter finally concludes the book. It begins with an analysis of a follow-up study which involved communication with a small number of the leaders who were interviewed in the main study. It ends with some personal reflections of the researchers.

Follow-up study

As outlined in Chapter 5, in May to July 2020, the researchers went back to some of the leaders, who had previously been interviewed for the study, to ask them some questions about their leadership. The researchers were interested in knowing how their leadership had evolved since they had been interviewed. Had anything changed them as leaders? What main lessons had they learned? How were they practicing their leadership in the Covid-19 era? What would they do to prepare their organizations for, and address, a similar situation in future? (A list of these questions is provided in Appendix 9.)

As indicated in Chapter 6, it was not easy to find the leaders, as many of them had retired, or had left their previous organizations. However, four of the original interviewees responded to the questions in detail. The four respondents were:

- Respondent A: former managing partner of a major quantity surveying consultancy firm; now a senior consultant in an international multi-disciplinary firm.
- Respondent B: former managing partner of a quantity surveying consultancy firm; now the regional head of a global consultancy conglomerate.
- Respondent C: former managing partner of a quantity surveying firm which has offices in many countries; now a senior adviser to the firm.
- Respondent D: former CEO of a quantity surveying firm; now a senior director in an international multi-disciplinary firm.

The responses of the leaders, organized under the themes of the questions, are now discussed.

The question themes

Evolution of leadership

The leaders were asked how their leadership had evolved since they had been interviewed in 2007–2008. All the leaders shared that their leadership had improved over the period.

Respondent D, the youngest of the four leaders, and a senior director of a consulting firm shared: "I feel that I am more well-rounded in business and I have also grown in social management – management of people. Catching up with technology and technical requirements is straightforward but the management of people and the environment takes experience."

Respondent B, the regional head of an international consultancy conglomerate, noted: "[In my current company], being a multi-disciplinary consultancy, my role has extended [from managing quantity surveyors] to managing architects, engineers, project managers, quantity surveyors and construction professionals. This has been very enriching."

Respondent C, the former managing partner of a quantity surveying firm who is now a senior adviser to it, highlighted two related dimensions of the changes in his leadership as the firm grew in the region: developing the required leaders; and adopting a leadership approach suitable for the current workforce:

> [M]y leadership style evolved towards a more participative "consensus-building" approach as I consciously encouraged the development and nurturing of the next generation of leaders within my company, which has grown considerably in terms of its presence in South East Asia and covers several countries ... New leaders had to be nurtured to lead and manage our new offices ... there was also an expansion of the resourcing and service consultancy scope of cost management, project management and cost advisory services offered.

Although the leaders held very senior positions at the time of the original interviews, and moved into other high positions, their leadership continued to evolve. This underlines the notion that authentic leadership is a lifetime phenomenon. The leaders had understood the dynamism of authenticity and developed their leadership to match their tasks, their people, and their operating environment. They had taken up leadership approaches, such as the consultative form, and strengthened the areas they felt they were lacking.

What has changed the leaders?

The authors were also interested to know what had changed the leaders in the evolution of their leadership. The leaders identified several factors, including the global economic situation, the nature of the organizations they worked in, their people, and technology.

Respondent C highlighted how the post-2007 financial crisis had changed him, and his approach to leadership. He noted: "The many challenges faced over 2007 to 2017 have made an impression – in particular, the global financial crisis, which had a significant impact on the local and regional construction markets."

Respondent A, who is a very senior leader in the construction industry, noted how technological advances had changed the construction business:

> The business environment and the factors that uphold its success have changed histrionically since our interview. Digitization, BIM and the ubiquitous smartphone have certainly shaped the way business has been conducted over recent years. Increase in productivity and sustainability in business will always remain the key considerations for any leader. However, my main focus now has been on social responsibility; placing greater emphasis on the welfare of the staff and their families in addition to looking at the need to fulfill the client's requirements.

Respondent A again noted the trends in the industry and how he and the firm are responding. This involved a change in his approach to leadership:

> Clients became more demanding, fees very competitive, and we faced increased costs in rentals, staff remuneration, and other overheads; all these within the global work environment facing disruptive developments of digitization, and media transformation. A more participative approach indeed helped to build up a unified purpose and objectivity, the proliferation of ideas and solutions, and more vigorous engagement of problem-solving.

Respondent A again indicated that the nature of the company he was working for has changed his leadership approach:

> Given the new corporate environment of a multi-discipline consultancy establishment, it has provided the fillip for me to be more patient in my interactions with my colleagues and subordinates; adopting a more approachable disposition and hearing out others' points of view before making a decision or giving a direction. Patience is a virtue one must espouse while an affable temperament will create an environment of pleasantness and cordiality.

Respondent D, the former CEO, who is now a senior director in a larger firm, shared that the things which had changed his leadership were more personal, and related to human issues: "The type of people coming into the office and industry, economics in the industry as well as personal experiences as I volunteer in the industry. I have to say the Bible has also influenced my leadership decision."

Respondent A again noted:

> With the ever-changing and capricious nature of the global economic landscape, it is important to have a workplace environment that is conducive to healthy staff and productive work, which would in turn engender good quality and reliable work. Both the company and the client will then be happy with the outcome and this would foster an enduring business relationship that is mutually beneficial to both parties.

The authenticity of the leaders has deepened over time. Their self-transcendence and keenness to ensure that their followers and organizations succeed are evident.

Three most important lessons for the leaders

The leaders were asked about the three most important leadership lessons they had learnt since 2007–2008. This is where the importance of authenticity of leadership was even clearer. The leaders underscored:

- the importance of integrity;
- reliability;
- social awareness and responsibility;
- leading by example;
- having the courage to do the right thing;
- staying positive in the face of challenges;

- making sacrifices to fulfill leadership responsibilities;
- making a difference in the lives of people;
- being open to new ideas;
- finding completeness in imperfection.

Respondent A expanded on the three lessons in leadership:

1   *Integrity* – this is fundamental and central to our core values in any business relationship; without which one may gain in the short term but will lose both reputation and business in the long term.
2   *Reliability* – the success or otherwise of our service delivery is inextricably dependent on reliability. Service that lacks reliability is destined to failure and eventually the demise of one's business undertaking.
3   *Social awareness and responsibility* – given the multifarious nature of our business, it is inevitable some ... units will need help on various matters and I take it upon myself to extend help where required. Such social awareness or responsibility and a spontaneity to render assistance will create a culture of teamwork that can only serve to enhance our reputation, and (1) leadership has the responsibility of making a difference to the people around you; (2) finding completeness in imperfection; as solutions are all inadvertently imperfect, and therefore perfection cannot be an all-consuming ambition; and (3) openness to new ideas, innovation, hybridity of opinions and views, are essential as a leader.

The lessons highlighted by Respondent B were: "Courage to do what is right and equitable for the business; Making sacrifices for the responsibilities entrusted in me; Being positive in every challenge." The lessons of Respondent C again included the importance of his religious belief: "(1) Lead by example; (2) extreme ownership (Navy Seals), and (3) biblical principles in leadership."

Elements of leadership in the Covid-19 pandemic

The leaders were asked to relate some of the most important elements of their leadership which is helping them during the prevailing Covid-19 pandemic. This was most apt because the pandemic has had a deep impact on the construction industry in Singapore, and in other countries. Singapore, which relies heavily on foreign construction workers, has been hit by a rampant spread of the virus in the dormitories of the workers. This upended the country's early success in keeping the virus under control, and leaders at various levels of the industry had to act.

Respondent D highlighted: "Super networking. Able to connect to clients and vendors in this time to still generate work and to spur my team on." Respondent B shared: "Remaining agile and adaptable." Respondent C noted that: "Important elements of leadership to deal with the Covid-19 crisis include empathy and compassion, foresight and risk management, and mitigation."

Respondent A again gave a full statement:

Undoubtedly, it's the sense of social responsibility to our fellow workers that provides the impetus for me to focus on ways that would help to contain the spread of the virus and minimize the risks to their health and safety. This would include arrangements such as 'Working from Home" and having virtual meetings. Such arrangements have a profound impact on the well-being and health of the workers. In fact, they seem to produce more productive work and I am definitely happy with the positive impact on their work life.

The leaders' single most critical Covid-19 focus

The leaders were asked about their most critical focus during the ongoing Covid-19 crisis. Respondent A noted:

> The single most critical focus is ... planning for the future. Post Covid-19 will be an era with a new normal of living and working, which inevitably raises concern about the ability to secure jobs from the depressed market and more intense competition for reduced opportunities ... While business continuity plans are put in place ... the need to deal with the effects of Covid-19 and contain its spread remains the most critical focus for any company to ensure not only the success of its business but also the well-being and livelihood of its staff.

Respondent D stated: "Keep plowing the field to get the team going, keeping the flame of passion going to spur the troops on but at all times, taking note of the BCP [business continuity planning] requirements by the company and government." Respondent B's concern was: "Well-being of my staff." Respondent C shared:

> My thoughts would be focused on business continuity, as most staff worked from home, and there would be inefficiencies in collaboration, communication, technology, etc. affecting work productivity, client liaison, marketing, billing, and meeting client deadlines. Telecommuting gives the health of staff priority but it has its drawbacks in work efficacy, social interaction, and technological/ICT limitations.

Preparing the organizations for crises in the future

The leaders were asked what they would do to better prepare their companies and their followers for a similar crisis in future. The responses ranged from long-term planning and business continuity to flexibility in the variety of projects.

Respondent A gave a full statement:

> According to experts, COVID-19 will last for a long time ... possibly ... even beyond 2021. It's time to reset our expectations and to confront a new way of life and human behavior. Foremost would be the need to have contingency plans for safe working and safe traveling ... In addition, there must be plans for office seating to ensure safe distancing and plans to minimize the number of people having face-to-face meetings.
>
> Lessons learnt from the current crisis will prepare us for the next similar crisis, but what is important is the willingness to abide by rules and measures ... being implemented by the authorities to curb the spread ... The message must be conveyed that no amount of preparation will succeed if there are blatant acts of ... not complying with the measures ... in place ...

Respondent D suggested: "Diversify to do more resiliency and digital work. Even in this down time, those sectors are still very busy." Respondent B indicated: "To focus on implementing business continuity plans." Respondent C noted:

Future-proofing of the practice against the effects of another pandemic in the future – review of business continuity plans, safe distancing protocols; managing client and staff expectations while in telecommuting mode; how to manage a "new normal" in practice, the respective roles and responsibilities of staff and management.

Summary of follow-up study

The authentic leaders who were interviewed more than a decade ago continued to learn, grow and become better leaders by adhering to their values. Their leadership was still all about nurturing future leaders, and serving people and making a positive difference in their lives. The lessons they had learned as leaders included: their personal integrity and that of their businesses; reliability in business; social awareness and responsibility; making a difference in their leadership; openness – to new ideas, innovation and hybridity of opinions and views; and realism, such as in finding completeness in imperfection because all solutions are inherently imperfect. They adhered to their authenticity as leaders, embraced change, and stayed humble. They did not have to say or do any heroic things. Their clarity about their own leadership and the simplicity with which they described what changed them or how their leadership evolved are inspiring. That is what authentic leadership is about.

Reflections of the researchers

When the main part of this research study was undertaken a few years ago, authentic leadership was an emerging stream of scholarship and most of the work in the field was conceptual and theoretical. As in all studies, we, the researchers, faced numerous challenges during the empirical study. The study was undertaken during a construction boom in Singapore which was coming on the end of many years of continuous decline in demand and output. We would make appointments with potential interviewees and have to wait for several days to obtain mutually convenient time slots. However, the interviewees were generous with their time, and their enthusiasm for the study, once they got to know more about it and realized how they could enhance their own leadership; build a better business organization; and help to change the situation in the industry for the better, meant they freely shared intimate information about themselves. All the interviewees allowed the sessions to be audio recorded. This was immensely useful to us in the research.

There were challenges in, or relating to, many other aspects of the field study. They included: preparing transcriptions of the interviews, analyzing the interviews, writing memos, drawing relationships diagrams, integrating different concepts, going back to the field for theoretical sampling, going through the analysis cycle all over again, and finally coming up with a holistic theoretical framework. Theoretical sampling was instrumental in revealing important concepts when some research participants made points which were similar to, or related to, some specific points or ideas which had previously been mentioned by another participant. In such instances, the effort which we had made in building a relationship of trust was helpful in overcoming the challenge to seek out more information from one or more of the interviewees. We spent several hours individually grappling with, and jointly discussing concepts, abstraction, and integration of relevant ideas. Often, we had to make conscious attempts to distance ourselves from the data to write memos and conceptualize the data. The challenge was always to record ideas which came to us, and test them later in the light of data.

During this whole process of studying authentic leadership, knowing authentic leaders, analyzing their leadership development, and discovering how they manifest their leadership and achieve their purpose, we went through a novel experience of discovering leadership at work. The leaders interviewed showed a remarkable willingness to share some very personal incidents, sometimes with emotion. They were also ready to share about their values and principles as well as what undergirded them, such as their religious beliefs; their weaknesses and vulnerabilities; and their successes and failures, as well as the consequences. They openly related the way in which many other people had touched their lives and how they were making the effort to influence others' lives in return. Indeed, this authenticity of the leaders who participated in the study as interviewees drove the spirit of this study and enhanced our passion as researchers.

After completing this book, we are convinced, more than ever, that a person's journey of authenticity only begins and never ends. Authentic leaders seek satisfaction in the process of discovery of their inner self and social realities. Only through understanding that they have to lead themselves before leading others do they come to realize and start to enjoy their leadership. Once they are willing to lead a self-aware and self-regulated life, they are ready to embark on the kind of leadership that can benefit others. This inner strength that leaders gain from leading themselves then translates into the joy and satisfaction they gain from leading others with authenticity.

Finally, every authentic leader has to engage in a struggle within the self and within society. Since authentic leaders have gone through transformative events, defining episodes, and crucibles to gain authenticity, they are very well aware of its value and worth. This keeps them anchored to their values, principles, purpose, visions, aspirations, and motivations. To become and remain authentic, leaders creatively reconcile with the realities of self and the social system of which they are a part. This dynamic process helps them adhere to their authenticity.

It is difficult to imagine a world without leaders. But, at the same time, there are those who are authentic and those who are not. The world is growing more complex in many ways. It is also increasingly vulnerable to crises, epidemics, natural disasters, and socio-political unrest. Leadership has always mattered. It matters more now than it ever did in the past. The development of more authentic leaders in all spheres of life and at all levels of society deserves to be given priority.

Appendix 1: Leadership interview questions

1. Introduction

- Would you briefly describe your career to date, beginning with your education and then when you first entered a management position?
- Would you please describe your own philosophy of leadership (what do you think is your leadership approach all about)?
- In your opinion, what do you think is the difference between leadership and management?
- In general, what do managers in construction firms do, more of leadership or more of management?
- What are your major strengths with respect to your functioning as a CEO/GM/MD/Director/Vice President in your current position?
- What do you think are some issues (weaknesses) you struggle with while performing your routine leadership role?
- Would you agree that leading a construction organization is unique from leading other businesses? If so, what is so unique about it?

2. Impact of key people, experiences, and turning points

- How would you briefly describe your early life? (Was it comfortable, easy-going, difficult, struggling, challenging, or fearful?)
- During your early years (before university life), which people had the greatest impact on you?

  a   parents, siblings, teachers, peers, or any other person

- In your later years (university and professional life), which people had the greatest impact on you?

  a   teachers, your own family, bosses or colleagues, or any other person

- What is your opinion on how leaders are developed, by choice or by chance? What was it in your own case?
- Today, who inspires you the most when it comes to leadership (who is your leadership role model)?
- From your early life, can you remember two such events when you felt you were really at your best and it boosted your leadership development?

- Can you remember a single most important event/incident/turning point that transformed you as a person and inspired your leadership style? And would you call this moment the defining moment for your leadership development?
- Can you identify an instance where you were dissatisfied with your leadership and you thought you had to revise your leadership style?

3. Discovering authentic leadership

- Would you agree that a leader learns more from failures rather than successes? Can you remember some examples from your career and what lessons you have learnt from these examples?
- We all come across failures and disappointments. However, some learn from them and others suffer from them. Do the failures or disappointments you experienced earlier in your life constrain you, even today, or have you been able to reframe them as learning and positive experiences?
- What are your three strongest desires, motivations, or goals you really want to achieve as a leader?
- What has been your greatest professional challenge in career and how did you overcome that?

4. Approaches to leadership development

- In your opinion, what are the most important attributes of an authentic leader? And how did you develop and polish those attributes?
- In practicing your leadership authenticity, what kind of problems/challenges do you really come across and how do you overcome these challenges?
- As an authentic leader, what is the secret to develop your followers into authentic followers?
- What would you advise the young professionals to do in order to develop as authentic leaders in their careers?

5. Leadership purpose and legacy

- What would you really like to achieve from your leadership role in this organization and further?
- As a leader, what kind of changes would you really like to bring in your organization and construction industry at large?
- What would you like to leave behind which you would regard as your own leadership legacy?

Appendix 2: List of interviewees

No.	Name	Company	Category
1.	Philip Ng	Fareast Developers	Developers
2.	Eddie Wong	City Developments Limited	Developers
3.	Christina Lim-Chia+	City Developments Limited	Developers
4.	Stephen Choo	Keppel Land	Developers
5.	Lee Pek Yan	Mappletree	Developers
6.	Siew Heh Lau	Mappletree	Developers
7.	Peter Chua	Mappletree	Developers
8.	Goh Chun Kiat	Mappletree	Developers
9.	Lim Bok Ngam	Land Transport Authority	Developers
10.	Irene Meta+	Overseas Union Enterprise Ltd	Developers
11.	Tian Thai Hong	Surbana	Consultants and Developers
12.	Seah Choo Meng*	Davis Langdon & Seah	Quantity Surveyors
13.	Eugene Seah	Davis Langdon & Seah	Quantity Surveyors
14.	Teoh Wooi Sin	Davis Langdon & Seah	Quantity Surveyors
15.	Iris Teng+	Davis Langdon & Seah	Quantity Surveyors
16.	Billy Wong	KPK	Quantity Surveyors
17.	Winston Hauw	Rider, Hunt and Levitt	Quantity Surveyors
18.	Ashvin Kumar*	Architect Team 61	Architects
19.	Vikas Gore	DP Architects	Architects
20.	Goh Peng Tong	Alfred Wong Partnership	Architects
21.	Jeffery Allan	Alfred Wong Partnership	Architects
22.	Rita Soh*+	RDC Architects	Architects
23.	Lai Huen Poh	RSP Architects Planners & Engineers	Architects, Planners and Engineers
24.	Lee Kut Cheung	RSP Architects Planners & Engineers	Architects, Planners and Engineers
25.	Matthew Hon	P&T Group	Architects and Consultants
26.	Kingsley Ng	P&T Group	Architects and Consultants
27.	Allan Low	P&T Group	Architects and Consultants
28.	Lee Chuan Seng	Square Mech	Engineers
29.	Kong Mung Kwong*	Lereno Bio-Chemical	Engineers
30.	Shehzad Nassim	Meinhardt	Engineering Consultants

(Continued)

(Cont.)

No.	Name	Company	Category
31.	Foo See Lim	Meinhardt	Engineering Consultants
32.	Tan Siew Chin	Meinhardt	Engineering Consultants
33.	Dr. Junaid Qureshi	Meinhardt	Engineering Consultants
34.	Lee Bee Wah+	LBW Consultants	Civil, Structural, and Project Management
35.	Low Yut Peng	LBW Consultants	Civil, Structural, and Project Management
36.	Bonaventure Lek	Shimizu Corporation	Consultants
37.	Whey Ying Mao+	Jurong Consultants	Consultants
38.	Ruby Lai+	CPG Consultants	Consultants
39.	Chan Shelt Tsong	Obayashi Corporation	Contractors
40.	Desmond Hill*	Penta Ocean	Contractors
41.	Eugene Yong	Woh Hup	Contractors
42.	Eddie Lam	Woh Hup	Contractors
43.	Chu Chiang Yong	Woh Hup	Contractors
44.	Jimmy Koh*	Antara Koh	Contractors
45.	Allen Yeo	Thompson Medical	Practitioner

+Woman leader.

Appendix 3: List of questions for validation exercise

1 Do you see yourself in the theory?
2 Have you gained any useful insight from this framework?
3 How comprehensive are the categories noted in the framework?
4 Do you think any process is missing in the framework?
5 Do you think you can use any interpretation in your practical life?
6 Can you interpret this framework in your own way?
7 Would you want your project managers to know about this framework?
8 Do you think the project managers will benefit from it?

Appendix 4: Results of ANOVA for various demographic groups

Demographic factor	Gender		Age		Working experience		Experience as a leader		Nationality	
Variable code	F	Sig.	F	Sig.	F	Sig.	F	Sig.	F	Sig.
SI 1	.089	.766	3.512	.021*	.580	.887	.423	.830	.927	.494
SI 2	.099	.754	3.322	.026	.689	.794	1.242	.303	1.695	.132
SI 3	.045	.832	.981	.409	1.230	.287	.478	.791	2.148	.055
SI 4	.059	.809	.860	.468	.680	.803	1.104	.370	.557	.787
SI 5	1.041	.312	2.261	.092	.748	.737	.233	.947	.842	.558
SI 6	.061	.806	.683	.567	.897	.581	.079	.995	.652	.711
SI 7	.841	.363	.469	.705	.705	.779	.320	.899	1.603	.157
SI 8	.530	.470	.525	.667	.638	.841	.321	.898	1.555	.171
SI 9	.033	.856	1.643	.190	.467	.954	.844	.525	1.176	.333
SI 10	5.708	.020*	.282	.838	.293	.996	1.871	.115	1.043	.414
SI 11	2.602	.112	1.861	.147	.797	.686	.393	.851	1.965	.079
SE 1	1.232	.272	.066	.978	.765	.719	1.739	.142	1.026	.425
SE 2	2.262	.138	.695	.559	.806	.677	1.179	.332	.749	.632
SE 3	5.525	.022*	3.474	.022*	1.001	.477	.930	.469	.991	.448
SE 4	.547	.463	.160	.923	.720	.764	1.982	.097	.728	.649
SE 5	.664	.419	.397	.756	1.034	.446	1.558	.189	.641	.719
SE 6	.814	.371	.450	.719	.939	.538	1.220	.313	.653	.710
SE 7	.104	.748	1.106	.355	.529	.921	1.146	.348	.645	.717
SE 8	1.753	.191	1.451	.238	.693	.790	1.407	.237	.449	.866
SE 9	.012	.914	.255	.857	1.668	.092	1.291	.282	1.011	.435
SE 10	1.274	.264	.941	.427	1.569	.120	.561	.729	.937	.487
SE 11	2.020	.161	.242	.867	1.297	.244	2.671	.032*	1.100	.378
SE 12	1.388	.244	.906	.444	.592	.878	1.363	.253	.387	.906
SE 13	3.549	.065	.551	.650	.598	.874	2.043	.088	2.208	.049*
SE 14	1.111	.296	3.99	.012*	.848	.632	.671	.647	1.300	.270
SE 15	.530	.470	1.276	.292	1.638	.100	1.357	.256	.859	.545
SE 16	2.821	.099	.036	.991	1.113	.376	3.362	.010*	1.325	.258
SE 17	1.772	.189	.381	.767	1.031	.448	1.964	.100	.884	.526

* Significant difference (p<.05).

Appendix 5: Frequencies for significant individuals

Variable code	Significant individual	Frequencies							
		N/A	Extremely Negative Influence	Somewhat Negative Influence	Somewhat Negative Influence	Neutral	Positive Influence	Somewhat Positive Influence	Extremely Positive Influence
		0	1	2	3	4	5	6	7
SI* 1	Mother					15	12	13	18
SI 2	Father					17	8	13	20
SI 3	Spouse	15	1			13	9	8	12
SI 4	Siblings	7			2	17	9	13	10
SI 5	Teachers (at school, college, or university)	3			2	7	15	14	17
SI 6	Mentor/s (in personal, organizational, social life)	1				10	18	16	13
SI 7	Peers (friends)	1			2	12	19	17	7
SI 8	Colleagues	2			2	12	21	14	7
SI 9	Bosses (in organizations)	5				15	20	7	11
SI 10	Some ideal personality (political, organizational, social, or other leaders)	6				25	7	8	12
SI 11	Any other significant individual	16	1			29	3	6	3

* SI denotes significant individual.

Appendix 6: Ranking of significant individuals

Significant individual	N	Mean	Rank
Teachers	55	5.67	1
Father	58	5.62	2
Mother	58	5.59	3
Mentors	57	5.54	4
Spouse	43	5.35	5
Bosses	53	5.26	6
Peers	57	5.26	7
Siblings	51	5.24	8
Colleagues	56	5.21	9
Some ideal personalities	52	5.13	10
Any other significant individual	43	4.40	11

Appendix 7: Frequencies for significant experiences

Variable code	Significant experience	Frequencies							
		N/A	Extremely negative influence	Somewhat negative influence	Somewhat negative influence	Neutral	Positive influence	Extremely positive influence	Extremely positive influence
		0	1	2	3	4	5	6	7
SE 1	Early childhood experiences within family (such as care/or lack of care/love/guidance within family)			1		7	15	18	17
SE 2	Early childhood experiences outside family (such as acceptance among outside people, attention, fame)			1		13	15	14	15
SE 3	Loss or death of someone special	11	4	1	2	35		3	2
SE 4	Experiences during school/college education				2	9	13	25	9
SE 5	Experiences during university education					6	9	27	16
SE 6	Experiences during organizational work			1		7	8	21	21
SE 7	Experiences during stay in home country	1		1		3	14	25	14

(Continued)

(Cont.)

Variable code	Significant experience	Frequencies							
		N/A	Extremely negative influence	Somewhat negative influence	Somewhat negative influence	Neutral	Positive influence	Extremely positive influence	Extremely positive influence
		0	1	2	3	4	5	6	7
SE 8	Experiences during stay abroad		2	2		17	9	12	16
SE 9	Experience of adversities in life (such as financial crisis, lack of facilities, lack of privileges)		1	2	3	21	9	13	9
SE 10	Experience of comfortable life (such as rich parents, nice home, excellent school, and other socio-economic facilities)		2		3	21	14	11	7
SE 11	Experience of meeting someone (loved one/ideal figure/some famous leader, and so on)	2		1	2	10	11	20	12
SE 12	Experience of reading some book/article/watching a movie	2	1	1		15	8	19	12
SE 13	Experience of surrounding social environment (positive or negative condition of societal situation)	2	1		3	9	18	13	12
SE 14	Some special incident	14	1		2	26	6	5	4
SE 15	Participation in extracurricular activities during education	1		5		14	18	19	11

(Continued)

(Cont.)

Variable code	Significant experience	Frequencies							
		N/A	Extremely negative influence	Somewhat negative influence	Somewhat negative influence	Neutral	Positive influence	Extremely positive influence	Extremely positive influence
		0	1	2	3	4	5	6	7
SE 16	Participation in leadership activities during education and professional career	1		1		14	17	15	10
SE 17	Participation in other social activities	1		3		18	16	9	11

Appendix 8: Ranking of significant experiences

Significant experience	N	Mean	Rank
Experiences during university education (SE5)	58	5.91	1
Experiences during organizational work (SE6)	58	5.91	2
Experiences during stay in home country (SE7)	57	5.84	3
Early childhood experiences within family (such as care/or lack of care/ love/guidance within family) (SE1)	58	5.74	4
Experiences during school/college education (SE4)	58	5.52	5
Early childhood experiences outside family (such as acceptance among outside people, attention, fame) (SE2)	58	5.50	6
Experience of meeting someone (loved one/ideal figure/ some famous leader, and so on) (SE11)	56	5.48	7
Experience of reading some book/article or watching a movie (SE12)	56	5.39	8
Experience of surrounding social environment (SE13)	56	5.34	9
Participation in leadership activities during education and professional career (SE16)	57	5.33	10
Experiences during stay abroad (SE8)	58	5.26	11
Participation in extracurricular activities during education (SE15)	57	5.12	12
Participation in other social activities (SE17)	57	5.12	13
Experience of adversities in life (such as financial crisis, lack of facilities, lack of privileges) (SE9)	58	4.90	14
Experience of comfortable life (such as rich parents, nice home, excellent school, and other socio-economic facilities) (SE10)	58	4.83	15
Some special incident (SE14)	44	4.52	16
Loss or death of someone special (SE3)	47	3.91	17

Appendix 9: List of questions for follow-up study in 2020

1 How has your leadership evolved over the years, since we first interviewed you in 2007?
2 What has changed you as a leader?
3 What are the three most important leadership lessons you have learnt in the past few years?
4 What are some of the most important elements of your leadership that you think are helping you during the on-going COVID-19 pandemic crisis?
5 As a leader, what is your single most critical focus during the on-going COVID-19 crisis?
6 What will you do to better prepare your company and your followers for a similar crisis in future?

References

Adler, P.S. and Kwon, S. (2002) Social capital: Prospects for a new concept. *Academy of Management Review*, 27(1), 17–40.

Aguinaldo, J.P. (2004) Rethinking validity in qualitative research from a social constructionist perspective: From "Is this valid research?" to "What is this research valid for?" *The Qualitative Report*, 9(1), 127–136.

Akintoye, A., Beck, M. and Kumaraswamy, J. (Eds.) (2015) *Public Private Partnerships: A Global Review*. Abingdon: Routledge.

Aksenova, G., Kiviniemi, A., Kocaturk, T., and Lejeune, A. (2018) From Finnish AEC knowledge ecosystem to business ecosystem: lessons learned from the national deployment of BIM. *Construction Management and Economics*, 37(6), 317–335, doi:10.1080/01446193.2018.1481985.

Allard-Poesi, F. (2005) The paradox of sensemaking in organizational analysis. *Organization*, 12(2), 69–96.

Allport, G.W. (1961) *Pattern and Growth in Personality*. New York: Holt, Rinehart and Winston.

Alvesson, M. (1996) Leadership studies: From procedure and abstraction to reflexivity and. situation. *The Leadership Quarterly*, 7(4), 455–485.

American Heritage Dictionary of English Language (2020) Reconciliation. Online. Available at: http s://ahdictionary.com/ (accessed February 2008).

American Society of Civil Engineers (ASCE) (2007) Combating corruption in engineering and construction: An engineer's charter. Reston, VA: ASCE.

Ammeter, A.P., Douglas, C., Gardner, W.L., Hochwarter, W.A., and Ferris, G.R. (2002) Toward a political theory of leadership. *The Leadership Quarterly*, 13, 751–796.

Andeweg, R.B. and Van Den Berg, S.B. (2003) Linking birth order to political leadership: The impact of parents or sibling interaction. *Political Psychology*, 20(3), 605–623.

Antonakis, J., Ashkanasy, N.M., and Dasborough, M.T. (2009) Does leadership need emotional intelligence? *The Leadership Quarterly*, 20, 247–261.

Antonakis, J. and Dalgas, O. (2009) Predicting elections: Child's play! *Science*, 323(5918), 1183.

Antonakis, J. and Day, D.V. (2018) Leadership: past, present, and future. In J. Antonakis and D.V. Day (Eds.), *The Nature of Leadership* (3rd ed.) (pp. 3–26). Los Angeles: Sage.

Antonakis, J. and Eubanks, D.L. (2017) Looking leadership in the face. *Current Directions in Psychological Science*, 26(3), 270–275.

Antwi, S.K. and Hamza, K. (2015) Qualitative and quantitative research paradigms in business research: A philosophical reflection. *European Journal of Business and Management*, 7(3), 217–225.

Arup (2018) *Construction Industry Capacity Framework*. London: Department for International Development.

Arvey, R.D., Rotundo, M., Johnson, W., Zhang, Z., and McGue, M. (2006) The determinants of leadership role occupancy: Genetic and personality factors. *The Leadership Quarterly*, 17(1), 1–20.

Arvey, R.D., Zhang, Z., Avolio, B.J., and Krueger, R. (2007) Developmental and genetic determinants of leadership role occupancy among women. *Journal of Applied Psychology*, 92(3), 693–706.

Asch, S.E. and Zukier, H. (1984) Thinking about persons. *Journal of Personality and Social Psychology*, 46, 1230–1240.

Ashforth, B.E. (1994) Petty tyranny in organizations. *Human Relations*, 47, 755–778.

Ashforth, B.E. (1997) Petty tyranny in organizations: A preliminary examination of antecedents and consequences. *Canadian Journal of Administrative Sciences*, 14, 126–140.

Ashmos, D.P. and Duchon, D. (2000) Spirituality at work: Conceptualization and measure. *Journal of Management Enquiry*, 9, 134–145.

Association for Consultancy and Engineering, Build UK, Chartered Institute of Building, Civil Engineering Contractors Association, Construction Plant-hire Association, Construction Products Association, Federation of Master Builders, Highways Term Maintenance Association and National Federation of Builders (2019) Shortage occupations in construction: A cross-industry research report. London: Build UK. Available at: https://builduk.org/wp-content/uploads/2019/01/Shortage-Occupations-in-Construction-A-cross-industry-research-report-January-2019.pdf

Atkin, B. (1998) *Unravelling the value chain in construction*. Paper presented at 6th Conference, International Group for Lean Construction, Cuaruja, Brazil.

Avolio, B.J. (1999) *Full Leadership Development: Building the Vital Forces in Organizations*. Thousand Oaks, CA: Sage.

Avolio, B.J. (2005) *Leadership Development in Balance: Made/Born*. Mahwah, NJ: Lawrence Erlbaum Associates.

Avolio, B.J. (2007) Promoting more integrative strategies for leadership theory building. *American Psychologist*, 62, 25–33.

Avolio, B.J. (2010) Pursuing authentic leadership development. In N. Nohria and R. Khurana (Eds.), *Handbook of Leadership Theory and Practice: A Harvard Business School Centennial Colloquium* (pp. 739–768). Boston, MA: Harvard Business School Publishing.

Avolio, B.J. and Bass, B.M. (1991) *The Full Range Leadership Development Programs: Basic and Advanced Manuals*. New York: Bass, Avolio Associates.

Avolio, B.J. and Bass, B.M. (1995) Individual consideration viewed at multiple levels of analysis: A multi-level framework for examining the diffusion of transformational leadership. *The Leadership Quarterly*, 6(2), 199–218.

Avolio, B.J. and Bass, B.M. (2004) *Multifactor Leadership Questionnaire: Manual and Sampler Set* (3rd ed.). Redwood City, CA: Mind Garden.

Avolio, B.J. and Gardner, W.L. (2005) Authentic leadership development: Getting to the root of positive forms of leadership. *The Leadership Quarterly*, 16(3), 315–338.

Avolio, B.J., Gardner, W.L., and Walumbwa, F.O. (2005) *Authentic Leadership and Practice: Origins, Effects and Development*. Amsterdam: Elsevier JAI.

Avolio, B.J., Gardner, W.L., Walumbwa, F.O., Luthans, F., and May, D.R. (2004) Unlocking the mask: A look at the process by which authentic leaders impact follower attitudes and behaviors. *The Leadership Quarterly*, 15(6), 801–823. Avolio, B.J. and Gibbons, T.C. (1988) Developing transformational leaders: A life span approach. In J.A. Conger, R.N. Kanungo and Associates (Eds.), *Charismatic Leadership: The Elusive Factor in Organizational Effectiveness* (pp. 276–308). San Francisco, CA: Jossey-Bass.

Avolio, B.J. and Luthans, F.L. (2006) *The High Impact Leader: Moments Matter in Authentic Leadership Development*. New York: McGraw-Hill.

Avolio, B.J., Sosik, J.J., Jung, D.I., and Berson, Y. (2003) Leadership models, methods and applications: Small steps and giant leaps. In W.C. Borman, R.J. Klimoski, D.J. Ilgen, and I.B. Weiner (Eds.), *Handbook of Psychology* (pp. 277–307). New York: John Wiley & Sons, Inc.

Bandura, A. (1977) Self-efficacy: Toward a unifying theory of behavioral change. *Psychological Review*, 84(2), 191–215.

Barber, E. and Warn, J.R.(2005)Leadership in project management: From firefighter to fire lighter. *Management Decisions*, 43(7/8), 1032–1039.

Barbosa, F., Woetzel, J., Mischke, J., Ribeirinho, M.J., Sridhar, M., Parsons, M., Bertram, N. and Brown, S. (2017) Reinventing construction through a productivity revolution. McKinsey Global Institute. Available at: www.mckinsey.com/industries/capital-projects-and-infrastructure/our-insights/reinventing-construction-through-a-productivity-revolution

Barker, R.A (1997) How can we train leaders if we do not know what leadership is? *Human Relations*, 50, 343–362.

Barnes, L.B. (1981) Managing the paradox of organizational trust. *Harvard Business Review*, 59(2), 107–116.

Bass, B.M. (1960) *Leadership, Psychology, and Organizational Behavior*. New York: Harper.

Bass, B.M. (1985) *Leadership and Performance Beyond Expectations*. New York: The Free Press,

Bass, B.M. (1990) *Bass and Stogdill's Handbook of Leadership* (3rd ed.). New York: The Free Press.

Bass, B.M. (1996) A new paradigm of leadership: An inquiry into transformational leadership. Fort Belvoir, VA: US Army Research Institute for the Behavioral and Social Sciences.

Bass, B.M. (1997) Personal selling and transactional/transformational leadership. *Journal of Personal Selling and Sales Management*, 17(3), 19–28.

Bass, B.M. (1998) *Transformational Leadership: Industrial, Military, and Educational Impact*. Mahwah, NJ: Lawrence Erlbaum Associates.

Bass, B.M. and Avolio, B.J. (1990) *Transformational Leadership Development: Manual for the Multifactor Leadership Questionnaire*. Palo Alto, CA: Consulting Psychologist Press.

Bass, B.M. and Avolio, B.J. (1995) *Multifactor Leadership Questionnaire: Technical Report*. Redwood City, CA:Mind Garden Inc.

Bass, B.M., Avolio, B., and Atwater L. (1996) The transformational and transactional leadership of men and women. *Applied Psychology*, 45(1), 5–34.

Bass, B.M. and Steidlmeier, P. (1999) Ethics, character, and authentic transformational leadership behaviour. *The Leadership Quarterly*, 10, 181–217.

Baumgardner, A.H. (1990) To know oneself is to like oneself: Self-certainty and self-affect. *Journal of Personality and Social Psychology*, 58(6), 1062–1072.

Bazeley, P. (2006) *Qualitative Data Analysis with NVivo*. London: Sage.

Bean, C.J. and Hamilton, F.E. (2006) Leader framing and follower sensemaking: Response to downsizing in the brave new workplace. *Human Relations*, 59(3), 321–349.

Bedell-Avers, K.E., Hunter, S.T., and Mumford, M.D. (2008) Conditions of problem-solving and the performance of charismatic, ideological, and pragmatic leaders: A comparative experimental study. *The Leadership Quarterly*, 19, 89–106.

Bendahan, S., Zehnder, C., Pralong, F.P., and Antonakis, J. (2015) Leader corruption depends on power and testosterone. *The Leadership Quarterly*, 26, 101–122.

Bennis, W.G. (1989) *On Becoming a Leader*. Reading, MA: Addison-Wesley.

Bennis, W.G. (1996) The leader as storyteller. *Harvard Business Review*. Jan–Feb.

Bennis, W.G. (2004) The seven ages of the leader. *Harvard Business Review*. January.

Bennis, W.G. (2007) The challenges of leadership in the modern world: Introduction to the special issue. *American Psychologist*, 62, 2–5.

Bennis, W.G. and Nanus, B. (1985) *Leaders: The Strategies for Taking Charge*. New York: Harper and Row.

Bennis, W.G. and Thomas, R.J. (2002) Crucibles of leadership. *HBR at Large*, September.

Bentz, V. (1987) Explorations of scope and scale: The critical determinant of high-level executive effectiveness. Technical Report 31. Greensboro, NC: Center for Creative Leadership.

Berglas, S. (2006) How to keep A players productive. *Harvard Business Review*, September, 104–112.

Bernerth, J.B., Cole, M.S., Taylor, E.C., and Walker, H.J. (2017) Control variables in leadership research: A qualitative and quantitative review. *Journal of Management*, 44(1), 131–160. doi:10.1177/0149206317690586.

Blake, R.R. and Mouton, J.S. (1964) *The Managerial Grid: The Key to Leadership Excellence*. Houston, TX: Gulf Publishing Co.

Blake, R.R. and Mouton, J.S. (1978) *The New Managerial Grid*. Houston, TX: Gulf Publishing Co.

Blanchard, K., Zigarmi, P., and Zigarmi, D. (2015) *Leadership and the One Minute Manager*. New York: Harper Thorsons.

Blismas, N.G. and Dainty, A.R.J. (2003) Computer-aided qualitative data analysis: Panacea or paradox? *Building Research and Information*, 31(6), 455–463.

Bluck, S. and Habermas, T. (2000) The life story schema. *Motivation and Emotion*, 24, 121–147.

Boal, K.B. (2004) Strategic leadership. In G.R. Goethals, G.J. Sorenson, and J.M. Burns (Eds.), *Encyclopedia of Leadership* (pp. 1497–1504). Thousand Oaks, CA: Sage.

Boal, K.B. and Hooijberg, R. (2000) Strategic leadership research: Moving on. *The Leadership Quarterly*, 11(4), 515–550.

Boal, K.B. and Schultz, P.L. (2007) Storytelling, time and evolution: The role of strategic leadership in complex adaptive systems. *The Leadership Quarterly*, 18(4), 411–428.

Bono, J.E. and Judge, T.A. (2004) Personality and transformational and transactional leadership: A meta-analysis. *Journal of Applied Psychology*, 89, 901–910.

Bourne, L. and Walker, D.H.T. (2005) The paradox of project control. *Team Performance Management*, 11(5/6), 157–178.

Boyatzis, R.E. (1982) *The Competent Manager: A Model for Effective Performance*. New York: John Wiley & Sons, Inc.

Bradbury, H. (2003) Webs not Kevlar: Designing sustainable organizations. *Reflections: The Society for Organizational Learning Journal*, 4(4), 65–68.

Brandenburg, S.G., Haas, C.T., and Byrom, K. (2006) Strategic management of human resources in construction. *Journal of Management in Engineering*, 22(2), 89–96.

Brent, E. (1984) Qualitative computing: Issues and answers. *Qualitative Sociology*, 7(1/2), 34–59.

Bresnen, M.J. (1986) The leader orientation of construction site managers. *Construction Engineering and Management*, 112(3), 370–386.

Brewer, M.B. (1991) The social self: On being the same and different at the same time. *Personality and Social Psychology Bulletin*, 17, 475–482.

Briner, W., Geddes, M., and Hastings, C. (1990) *Project Leadership*. New York: Van Nostrand Reinhold.

Briner, W., Hastings, C., and Geddes, M. (1996) *Project Leadership* (2nd ed.). Farnham: Gower.

Bronfenbrenner, U. (1961) Some familial antecedents of responsibility and leadership in adolescents. In L. Petrullo and B.M. Bass (Eds.), *Leadership and Interpersonal Behavior* (pp. 239–271). New York: Holt, Rinehart and Winston.

Broussine, M. and Miller, C. (2005) Leadership, ethical dilemmas and "good" authority in public service partnership working. *Business Ethics: A European Review*, 14(4), 379–391.

Brown, J.A. and Gardner, W.L. (2005) *Exemplification and authentic leadership: An exploratory study*. Paper presented at 1st Gallup Leadership Institute Biennial Leadership Summit, Omaha, USA.

Brown, L.M. (2019) Women in the future of construction. *Innovate UK* blog, February 11. Available at: https://innovateuk.blog.gov.uk/2019/02/11/women-in-the-future-of-construction/

Brown, M.E. and Treviño, L.K. (2006) Ethical leadership: A review and future directions. *The Leadership Quarterly*, 17, 595–616.

Brown, M.E., Treviño, L.K., and Harrison, D. (2005) Ethical leadership: A social learning perspective for construct development and testing. *Organizational Behavior and Human Decision Processes*, 97, 117–134.

Brungardt, C. (1996) The making of leaders: A review of the research in leadership development and education. *Journal of Leadership Studies*, 3(3), 81–95.

Bryant, A. (2017) *Grounded Theory and Grounded Theorizing*. Oxford: Oxford University Press.

Bryant, A. and Kazan, A. (2012) *Self-Leadership: How to Become a More Successful, Efficient, and Effective Leader from the Inside Out*. New York: McGraw-Hill.

Bryman, A. (1984) The debate about qualitative and quantitative methods: A question of method or epistemology? *British Journal of Sociology*, 35, 75–92.

Bryman, A. (1993) Charismatic leadership in business organizations: Some neglected issues. *The Leadership Quarterly*, 4(3/4), 289–304.

Bryman, A. (1996) Leadership in organizations. In S.R. Clegg, C. Hardy, and W.R. Nord (Eds.), *Handbook of Organization Studies* (pp. 276–292). London: Sage.

Bryman, A. (2004) Qualitative research on leadership: A critical but appreciative review. *The Leadership Quarterly*, 15(6), 729–769.

Bryman, A. (2007) The research question in social research: What is its role? *International Journal of Social Research Methodology*, 10, 5–20.

Bryman, A., Bresnen, M., Beardsworth, A., and Keil, T. (1988) Qualitative research and the study of leadership. *Human Relations*, 41(1), 13–30. Bryman, A., Bresnen, M., Ford, J., Beardsworth, A., and Keil, T. (1987) Leader orientation and organizational transience: An investigation using Fiedler's LPC Scale. *Journal of Occupational Psychology*, 60, 13–19.

Bryman, A., Stephens, M., and Campo, C. (1996) The importance of context: Qualitative research and the study of leadership. *The Leadership Quarterly*, 7(3), 353–370.

Buchanan, D. and Bryman, A. (2007) Contextualizing methods choice in organizational research. *Organizational Research Methods*, 10, 483–501.

Building and Construction Authority (BCA) (2014) 3rd Green Building Master Plan. Singapore: BCA.

Building and Construction Authority (BCA) (2017) Annex A, Appendix: Background of Industry Transformation Maps (ITM): Integrated roadmaps to drive industry transformation. Available at: www.bca.gov.sg/newsroom/others/Fact_Sheet_Construction_ITM_launch_231017.pdf

Building and Construction Authority (BCA) (2020) About BCA. Available at: www.bca.gov.sg/about-us/about-bca

Bull, R.C. (2010) *Moving from Project Management to Project Leadership*. Boca Raton, FL: CRC Books.

Bureau of Labor Statistics (2020a) Labor force statistics from the current population survey: Employed persons in nonagricultural industries by age, sex, race, and Hispanic or Latino ethnicity. Available at: www.bls.gov/cps/cpsaat14.pdf

Bureau of Labor Statistics (2020b) Labor force statistics from the current population survey: Employed persons by detailed occupation, sex, race, and Hispanic or Latino ethnicity. Available at: www.bls.gov/cps/cpsaat11.pdf

Burkitt, I. (1991) *Social Selves: Theories of the Social Formation of Personality*. London: Sage.

Burns, J.M. (1978) *Leadership*. New York: Harper and Row.

Burns, J.M. (2003) *Transformational Leadership*. New York: Atlantic Monthly Press.

Butcher, D. and Clarke, M. (2006) Political leadership in democracies: Some lessons for business? *Management Decision*, 44(8), 985–1001.

Cameron, K.S., Dutton, J.E., and Quinn, R.E. (Eds.) (2003) *Positive Organizational Scholarship: Foundation of a New Discipline*. San Francisco: Berrett-Koehler.

Campbell, F. (2006) *Occupational Stress in the Construction Industry*, Berkshire: Chartered Institute of Building.

Campbell, J.P., Dunnette, M.D., Lawler, E.E., and Weick, K.E., Jr. (1970) *Managerial Behavior, Performance, and Effectiveness*. New York: McGraw-Hill.

Cardona, P. (2000) Transcendent leadership. *The Leadership and Organization Development Journal*, 21(4), 201–206.

Carlyle, T. (1841) *Heroes and Hero-worship*. London: James Fraser.

Carver, C.S., and Scheier, M.F. (1998) *On the Self-regulation of Behavior*. Cambridge: Cambridge University Press.

Casey, B. and Long, A. (2002) Reconciling voices. *Journal of Psychiatric and Mental Health Nursing*, 9, 603–610.

Cashman, K. (1999) *Leadership from the Inside Out*. London: Executive Excellence Publishing.

Centre for Excellence in Management and Leadership (2002) *Managers and Leaders: Raising Our Game*. London: Council for Excellence in Management and Leadership.

Centre for Liveable Cities (2015) *Built by Singapore: From Slums to a Sustainable Built Environment*. Singapore: Centre for Liveable Cities.

Chan, A. (2005) Authentic leadership measurement and development: Challenges and suggestions. In W.L. Gardner, B.J. Avolio, and F.O. Walumbwa (Eds.), *Authentic Leadership Theory and Practice: Origins, Effects and Development* (pp. 227–250). Greenwood, CT: Elsevier JAI.

Chan, A.P.C., Ho, D.C.K., and Tam, C.M. (2001) Design and Build project success factors; Multivariate analysis. *Journal of Construction Engineering Management*, 127(2), 93–100.

Chan, A.T. and Chan, E.H. (2005) Impact of perceived leadership styles on work outcomes: Case of building professionals. *Construction Engineering and Management*, 131(4), 413–422.

Chan, E.H.W., Chan, M.W., Scott, D., and Chan, A.T.S. (2002) Educating the 21st century construction professionals. *Journal of Professional Issues in Engineering Education and Practice*, 128(1), 44–51.

Chaordic Commons (2004) Available at: www.chaordic.org

Charmaz, K. (2006) *Constructing Grounded Theory: A Practical Guide Through Qualitative Analysis*. Thousand Oaks, CA: Sage.

Charmaz, K. and Bryant, A. (2010) Grounded theory: A overview. In A. Bryant and K. Charmaz (Eds.), *International Encyclopedia of Education* (3rd ed.). Oxford: Academic Press.

Cheetham, G. and Chivers, G. (2001) How professionals learn in practice: What the empirical research found. *Journal of European Industrial Training*, 25(5), 270–292.

Chemers, M.M. (1997) *An Integrative Theory of Leadership*. Mahwah, NJ: Lawrence Erlbaum Publishers.

Chemers, M.M. (2000) Leadership research and theory: A functional integration. *Group Dynamics: Theory, Research, and Practice*, 4(1), 27–43.

Chinowsky, P.S. and Taylor, J. (2007) Leadership and team development in construction. In CIB, *Revaluing Construction: A W065 'Organisation and Management of Construction' Perspective* (pp. 54–65). London: CIB.

Cho, J. and Trent, A. (2006) Validity in qualitative research revisited. *The Journal of Qualitative Research*, 6(3), 319–340.

Chong, L.M.A. and Thomas, D.C. (1997) Leadership perceptions in cross-cultural context: Pakeha and Pacific Islands of New Zealand. *The Leadership Quarterly*, 8(3), 275–293.

Christodoulou, S. (2004) Educating the civil engineering professionals of tomorrow. *Journal of Professional Issues in Engineering Education and Management*, 130(2), 90–94.

Cialdini, R.B. and Trost, M.R. (1998) Social influence: Social norms, conformity, and compliance. In D. Gilbert, S. Fiske, and G. Lindzey (Eds.), *The Handbook of Social Psychology* (4th ed.) (pp. 151–192). New York: McGraw-Hill.

CIB (2006) TG64: Leadership in construction: Introducing new CIB Task Group. *CIB Newsletter*, December. Available at: www.cibworld.nl/website/newsletter/0608/tg64.html

Cilia, J. (2019) The construction labor shortage: Will developers deploy robotics? *Forbes*. Available at: www.forbes.com/sites/columbiabusinessschool/2019/07/31/the-construction-labor-shortage-will-developers-deploy-robotics/#65e8b48b7198

Ciulla, J.B. (1995) Leadership ethics: Mapping the territory. *Business Ethics Quarterly*, 5(1), 5–28.

Ciulla, J.B. (2004) Ethics and leadership effectiveness. In J. Antonakis, A.T. Cianciolo, and R.J. Sternberg (Eds.), *The Nature of Leadership* (pp. 302–327). Thousand Oaks, CA: Sage.

Cleland, D.I. (1995) Leadership and the project management body of knowledge. *International Journal of Project Management*, 13(2), 82–88.

Coleman, S. and Bourne, M. (2018) *Project Leadership: Skills, Behaviours, Knowledge and Values*. Princes Risborough, Buckinghamshire: Association for Project Management.

Coleman, S. and MacNicol, D. (2015) *Project Leadership*. Farnham: Gower.

Collier, J. and Esteban, R (2000) Systemic leadership: Ethical and effective. *The Leadership and Organizational Development Journal*, 21(4): 207–215.

Collins, J. (2001) *Good to Great*. New York: HarperCollins.

Collinson, D. (2003) Identities and insecurities: Selves at work. *Organization*, 10(3), 527–547.

Colville, I.D. and Murphy, A.J. (2006) Leadership as the enabler of strategizing and organizing. *Long Range Planning*, 39(6), 663–677.

Committee on the Future Economy (2017) *Report of the Committee on the Future Economy: Pioneers of the Next Generation*. Singapore:Ministry of Finance and Ministry of Trade and Industry.

Conger, J A.(1989) *The Charismatic Leader*. San Francisco: Jossey-Bass.

Conger, J.A. (1990) The dark side of leadership. *Organizational Dynamics*, 19(2), 44–55.

Conger, J.A. (1998) Qualitative research as the cornerstone methodology for understanding leadership. *The Leadership Quarterly*, 9(1), 107–121.

Conger, J.A. (2004) Developing leadership capability: What's inside the black box? *Academy of Management Executive*, 18(3), 136–139.

Conger, J.A. and Fulmer, R.M. (2003) Developing your leadership pipeline. *Harvard Business Review*, 81(12), 56–64.

Conger, J.A. and Kanungo, R.N. (1987) Towards a behavioural theory of charismatic leadership in organizational setting. *Academy of Management Review*, 12, 637–647.

Construction 21 Steering Committee (1999) *Reinventing Construction*. Singapore: Ministry of Manpower and Ministry of National Development.

Construction Industry Development Board (CIDB) (2015) *Construction Industry Transformation Programme 2015–2020: Driving Construction Excellence Together*. Kuala Lumpur: CIDB.

Construction Industry Joint Committee (2016) Purpose of CIJC. Available at: www.cijc.sg/about

Cooper, C.D., Scandura, T.A., and Schriesheim, C.A. (2005) Looking forward but learning from our past: Potential challenges to developing authentic leadership theory and authentic leaders. *The Leadership Quarterly*, 16, 475–493.

Corbin, J.M. and Strauss, A.L (2008) *Basics of Qualitative Research*. London: Sage.

Crigger, N.J. and Meek, V.L. (2007) Toward a theory of self-reconciliation following mistakes in nursing practice. *Journal of Nursing Scholarship*, 39(2), 177–183.

Cross, S.E. and Markus, H. (1991) Possible selves across the lifespan. *Human Development*, 34, 230–255.

Dahles, H. and Wels, H. (Eds.) (2002) *Culture, Organization and Management in East Asia: Doing Business in China*. New York: Nova Science Publishers, Inc.

Dainty, A.R.J. (1998) A grounded theory of the determinants of women's under-achievement in large construction companies. Unpublished PhD dissertation, Loughborough University.

Dainty, A.R.J. (2007) *A review and critique of construction management research methods*. Paper presented at Construction Management and Economics, 25th Anniversary Conference on 'Past, Present and Future,' July 16–18, University of Reading, Reading, United Kingdom.

Dainty, A.R.J., Bagilhole, B.M., and Neale, R.H. (1999) Career dynamics of professional women in large construction companies. In *Proceedings of the CIB W55/W65/W92/W99 Joint Triennial Symposium* (pp. 121–130).

Dainty, A.R.J., Bagilhole, B.M., and Neale, R.H. (2000) A grounded theory of women's career under-achievement in large construction companies. *Construction Management and Economics*, 18, 239–250.

Dainty, A.R.J., Bagilhole, B.M. *et al.* (2004) Creating equality in the construction industry: An agenda for change for women and ethnic minorities. *Journal of Construction Research*, 5(1), 75–86.

Dainty, A.R.J., Cheng, M., and Moore, D. (2005) A comparison of the behavioural competencies of client-focused and production-focused project managers in the construction sector. *Project Management Journal*, 36(1), 39–48.

Dainty, A.R.J., Cheng, M., and Moore, D. (2006) What makes an effective project manager? Findings of a four-year program of research. In A. Songer, P. Chinowsky, and P. Carrillo (Eds.), *Proceedings of the ASCE/CIB 2nd Speciality Conference in Leadership and Management in Construction* (pp. 116–123).

Davidson, E.S., Mitchell, A., Beverly, C., Brown, L.M., Rettiganti, M., Walden, M., and Wright, P. (2018) Psychometric properties of the authentic leadership inventory in the nursing profession. *Journal of Nursing Measurement*, 26(2), 364–377.

Dawes, J. (2008) Do data characteristics change according to the number of scale points used? *International Journal of Market Research*, 50(1), 61–77.

Day, D. (2000) Leadership development: A review in context. *The Leadership Quarterly*, 11(4), 581–613.

Dearlove, D. and Coomber, S. (2005) A leadership miscellany. *Business Strategy Review*, 16(3), 32–64.

Debrah, Y.A. and Ofori, G. (1997) Flexibility, labour subcontracting and HRM in the construction industry in Singapore: Can the system be refined? *International Journal of Human Resource Management*, 8(5), 690–709.

Debrah, Y.A. and Ofori, G. (2005) Emerging managerial competencies of professionals in the Tanzanian construction industry. *Human Resource Management*, 16(8), 1399–1414.

De Cremer, D. (2007) Which type of leader do I support in step-level public good dilemmas? The roles of level of threshold and trust. *Scandinavian Journal of Psychology*, 48(1), 51–59.

De Hoogh, A.H.B. and Den Dartog, D.N. (2008) Ethical and despotic leadership, relationships with leader's social responsibility, top management team effectiveness and subordinates' optimism: A multi-method study. *The Leadership Quarterly*, 19, 297–311.

Deloitte Australia and the Victorian Equal Opportunity and Human Rights Commission (2013) Waiter, is that inclusion in my soup? A new recipe to improve business performance. Report. Canberra: Deloitte.

De Neve, J.-E., Mikhaylov, S., Dawes, C.T., Christakis, N.A., and Fowler, J.H. (2013) Born to lead? A twin design and genetic association study of leadership role occupancy. *The Leadership Quarterly*, 24(1), 45–60.

Den Hartog, D.N., House, R.J., Hanges, P.J., and Ruiz-Quintanilla, S.A. (1999) Cultural specific and cross-culturally generalizable implicit leadership theories: Are attributes of charismatic/transformational leadership universally endorsed? *The Leadership Quarterly*, 10, 219–256.

Den Otter, A.F.H.J. and Prins, M. (2003) Architectural design management within the digital design team. *Engineering, Construction and Architectural Management*, 9(3), 162–173.

Dent, E.B., Higgins, M.E., and Wharff, D.M. (2005) Spirituality and leadership: An empirical review of definitions, distinctions, and embedded assumptions. *The Leadership Quarterly*, 16, 625–653.

Department for Business, Innovation and Skills Leadership and Management Network Group (2012) *Leadership and Management in the UK: The Key to Sustainable Growth*. London: DBIS.

Department of Statistics (2018a) *Yearbook of Statistics 2018*. Singapore. Available at: www.singstat.gov.sg

Department of Statistics (2018b) *Singapore in Figures 2018*. Singapore. Available at: www.singstat.gov.sg

Designing Buildings Wiki (2018) Fragmentation of the construction industry. Available at: www.designingbuildings.co.uk/wiki/Fragmentation_of_the_UK_construction_industry

Dettbarn, J., Ibbs, C.W., and Murphree, E.L. (2005) Capital project portfolio management for federal real property. *Journal of Management in Engineering*, 21(1), 4453.

Devanna, M. and Tichy, N. (1990) Creating the competitive organization of the 21st century: The boundary-less corporation. *Human Resource Management*, 29, 445–471.

De Waal, F.B.M. (2000) Primates: A natural heritage of conflict resolution. *Science*, 289, 586–590.

Dey, I. (1999) *Grounding Grounded Theory: Guidelines for Qualitative Research*. San Diego, CA: Academic Press.

Dickson, M.W., Den Hartog, D.N., and Mitchelson, J.K. (2003) Research on leadership in a cross-cultural context: Making progress, and raising new questions. *The Leadership Quarterly*, 14, 729–768.

Diebig, M., Bormann, K.C., and Rowold, J. (2016) A double-edged sword: Relationship between full-range leadership behaviors and followers' hair cortisol level. *The Leadership Quarterly*, 27(6), 684–696.

Dinh, J.E., Lord, R.G., Gardner, W.L., Meuser, J.D., Liden, R.C., and Hu, J. (2014) Leadership theory and research in the new millennium: Current theoretical trends and changing perspectives. *The Leadership Quarterly*, 25, 36–62.

Drath, W.H. and Palus, C.J. (1994) *Making Common Sense: Leadership as Meaning-Making in a Community of Practice*. Greensboro, NC: Center for Creative Leadership.

Drewer, S. (2001) A perspective of the international construction system. *Habitat International*, 25, 69–79.

Drucker, P. (1996) The executive in action: Managing for results, innovation and entrepreneurship. In P. Drucker, *The Effective Executive*. New York: Harper Business. DuBrin, A.J. (1995) *Leadership: Research Findings, Practice, and Skills*. Boston, MA: Houghton Mifflin.

Dulaimi, M.F. (2005) The influence of academic education and formal training on the project manager's behaviour. *Journal of Construction Research*, 6(1), 179–193.

Dulaimi, M.F. and Langford, D.A. (1999) Job behaviour of construction project managers: Determinants and assessment. *Construction Engineering and Management*, 125(4), 256–264.

Dunkel, C.S. (2000) Possible selves as a mechanism for identity exploration. *Journal of Adolescence*, 23, 519–529.

Du Toit, A. (2005) A guide to executive coaching: Advice to managers and their organizations. *Development and Learning in Organizations: An International Journal*, 19(2), 11–12.

Eakin, J.M. and Gladstone, B. (2020) "Value-adding" analysis: Doing more with qualitative data. *International Journal of Qualitative Methods*, 19, 1–13.

Earley, P.C. and Ang, S. (2003) *Cultural Intelligence: Individual Interactions across Cultures.* Palo Alto, CA: Stanford University Press.

Early, P. (2006) Leading cultural research in the future: A matter of paradigms and taste. *Journal of International Business Studies*, 37, 922–931.

Economic Development Board (2017) Singapore business environment: Singapore today. Singapore: Economic Development Board.

Economic Strategies Committee (2010) Report of the Economic Strategies Committee: High skilled people, innovative economy, distinctive global city. Singapore: Ministry of Finance.

Edelman (2019) Edelman Trust Barometer: Expectations for CEOs. Available at: www.edelman.com/research/trust-barometer-expectations-for-ceos-2019

Edgeman, R.L. and Scherer, F. (1999) Systemic leadership via core value deployment. *Leadership and Organization Development Journal*, 20(2), 94–98. https://doi.org/10.1108/01437739910259190.

Egbu, C. (1999) Skills, knowledge and competencies for managing construction refurbishment works. *Construction Management and Economics*, 17(1), 29–43.

Egbu, C. and Ofori, G. (Eds.) (2018) *Proceedings of the International Conference on Professionalism and Ethics in Construction.* London, November 21–22, 2018.

Egri, C.P., and Herman, S. (2000) Leadership in the North American environmental sector: Values, leadership styles, and contexts of environmental leaders and their organizations. *Academy of Management Journal*, 43, 571–604.

Einarsen, S., Aasland, M.S., and Skogstad, A. (2007) Destructive leadership behavior: A definition and conceptual model. *The Leadership Quarterly*, 18, 207–216.

Eisner, E.W. and Peshkin, A. (Eds.) (1990) *Qualitative Inquiry in Education: The Continuing Debate.* New York: Teachers College Press.

Elder, G.H. (1974) *Children of the Great Depression: Social Change of Life Experience.* Chicago: University of Chicago Press.

Elder, G.H. and Rockwell, R.C. (1979) The life-course and human development: An ecological perspective. *International Journal of Behavioral Development*, 2(1), 1–21.

Elliott, J. (2005) *Using Narrative in Social Research: Qualitative and Quantitative Approaches.* London: Sage.

Emden, C. and Sandelowski, M. (1999) The good, the bad and the relative, part two: Goodness and the criterion problem in qualitative research. *International Journal of Nursing Practice*, 5, 2–7.

Erickson, R.J. (1995) The importance of authenticity for self and society. *Symbolic Interaction*, 18(2), 121–144.

Erikson, E.H. (Ed.) (1963) *Youth: Change and Challenge.* New York: Basic Books.

Erikson, E.H. (1968) *Identity: Youth and Crisis.* New York: W.W. Norton and Company.

Erikson, E.H. (1980) *Identity and the Life Cycle.* New York: W.W. Norton and Company.

Fairhurst, G.T. and Sarr, R.A. (1996) *The Art of Framing: Managing the Language of Leadership.* San Francisco: Jossey-Bass.

Fan, L., Ho, C., and Ng, V. (2001) A study of quantity surveyors' ethical behaviour. *Journal of Construction Management and Economics*, 19(1), 19–36.

Fawcett, W., Ellingham, I., and Platt, S. (2008) Reconciling the architectural preferences of architects and the public: The ordered preference model. *Environment and Behavior*, doi:10.1177/0013916507304695.

Fellows, R., Liu, A., and Fong, C.M. (2003) Leadership style and power relations in quantity surveying in Hong Kong. *Construction Management and Economics*, 21, 809–818.

Ferdig, M. (2007) Sustainability leadership: Co-creating a sustainable future. *Journal of Change Management*, 7(2), 25–35.

Fernando, M., and Jackson, B. (2006) The influence of religion-based workplace spirituality on leaders' decision-making: An inter-faith study. *Journal of Management and Organization*, 12, 23–39.

Festinger, L. (1957) *A Theory of Cognitive Dissonance.* Stanford, CA: Stanford University Press.

Fiedler, F.E. (1963) *A Contingency Model for the Prediction of Leadership Effectiveness.* Urbana, IL: University of Illinois, Group Effectiveness Research Laboratory.

Fiedler, F.E. (1967) *A Theory of Leadership Effectiveness.* New York: McGraw-Hill.

Fiedler, F.E. and Garcia, J.E. (1987) *New Approaches to Effective Leadership: Cognitive Resources and Organizational Performance.* Chichester: John Wiley & Sons, Ltd.

Fielden, S.L., Davidson, M.J., Gale, A.W., and Davey, C.L. (2000) Women in construction: The untapped resource. *Construction Management and Economics,* 18(1), 113–121.

Fielding, N.G. and Lee, R.M. (1998) *Computer Analysis and Qualitative Research.* Thousand Oaks, CA: Sage.

Fisher, T.F. and Ranasinghe, M. (2001) Culture and foreign companies' choice of entry mode: The case of the Singapore building and construction industry. *Construction Management and Economics,* 19(4), 343–353.

Fisher, W.R. (1987) *Human Communication as a Narration: Toward a Philosophy of Reason, Value, and Action.* Columbia, SC: University of South Carolina Press.

Fisher-Yoshida, B. and Geller, K. (2008) Developing transnational leaders: Five paradoxes for success. *Industrial and Commercial Training,* 40(1), 42–50.

Fitzgerald, L.A. (1996) *Chaordic System Properties, Classical Assumptions and the Chaos Principle.* Denver, CO: The Consultancy.

Flannery, B. and May, D. (1994) Prominent factors influencing environmental leadership: Application of a theoretical model from the waste management industry. *The Leadership Quarterly,* 5, 201–222.

Flick, U. (2002) Qualitative research: State of the art. *Social Science Information,* 41(1), 5–24.

Flood, R.L. and Jackson, M.C. (1991) *Creative Problem Solving: Total Systems Intervention.* Chichester: John Wiley & Sons, Ltd.

Floricel, S. and Miller, R. (2001) Strategizing for anticipated risks and turbulence in large-scale engineering projects. *International Journal of Project Management,* 19, 445–455.

Fox, P. (1999) *Construction industry development: Exploring values and other factors from a grounded theory approach.* Paper presented at CIB W55 and W65 Joint Triennial Symposium, Cape Town, South Africa.

Fox, P. (2003) Construction industry development: Analysis and synthesis of contributing factors. Unpublished PhD dissertation, Queensland University of Technology, Brisbane, Australia.

Fox, P. and Skitmore, M. (2007) Factors facilitating construction industry development. *Building Research and Information,* 35(2), 178–188.

Foxell, S. (2018) *Professionalism for the Built Environment.* London: Routledge.

Fraser, C. (2000) The influence of personal characteristics on effectiveness of construction site managers. *Construction Management and Economics,* 18, 29–36.

Freeman, M. (2006) Life "on holiday"? In defense of big stories. *Narrative Inquiry,* 16, 131–138.

Frost, P.J. (2004) Handling toxic emotions: New challenges for leaders and their organization. *Organizational Dynamics,* 33, 111–127.

Fry, D.P. (2000) Conflict management in cross-cultural perspective. In F. Aureli and F.B.M. de Waal (Eds.), *Natural Conflict Resolution* (pp. 334–351). Berkeley, CA: University of California Press.

Fry, L.W. (2003) Toward a theory of spiritual leadership. *The Leadership Quarterly,* 14, 693–727.

Fry, L.W., Vitucci, S., and Cedillo, M. (2005) Spiritual leadership and army transformation: Theory, measurement, and establishing a baseline. *The Leadership Quarterly,* 16, 835–862.

Fusco, T., O'Riordan, S., and Palmer, S. (2016) Assessing the efficacy of authentic leadership group-coaching. *International Coaching Psychology Review,* 11(2), 6–16.

Gale, A.W. (1994) Women in non-traditional occupations: The construction industry. *Women in Management Review,* 9(2): 3–14.

Gardner, H. (2007) The ethical mind: A conversation with psychologist Howard Gardner. *Harvard Business Review,* March, 51–56.

Gardner, J.W. (1990) *On Leadership.* New York: Free Press.

Gardner, W.L. and Avolio, B.J. (1998) The charismatic relationship: A dramaturgical perspective. *Academy of Management Review,* 23(1), 32–58.

Gardner, W.L., Avolio, B.J., Luthans, F., May, D.R., and Walumba, F.O. (2005) Can you see the real me? A self-based model of authentic leader and follower development. *The Leadership Quarterly,* 16, 343–372.

Gardner, W.L., Lowe, K.B., Moss, T.W., Mahoney, K.T., and Cogliser, C.C. (2010) Scholarly leadership of the study of leadership: A review of *The Leadership Quarterly*'s second decade, 2000–2009. *The Leadership Quarterly,* 21, 922–958.

Gardner, W.L. and Schermerhorn, J. (2004) Unleashing individual potential: Performance gains through positive organizational behavior and authentic leadership. *Organizational Dynamics*, 33(3), 270–281.

Gelade, G.A., Dobson, P., and Gilbert, P. (2006) National differences in organizational commitment: Effect of economy, product of personality, or consequence of culture? *Journal of Cross-Cultural Psychology*, 37, 542–556.

George, B. (2003) *Authentic Leadership: Rediscovering the Secrets to Creating Lasting Value*. San Francisco: Jossey-Bass.

George, B. and Sims, P. (2007) *True North: Discover Your Authentic Leadership*. San Francisco: Wiley.

George, B., Sims, P., McLean, A.N., and Mayer, D. (2007) Discovering your authentic leadership. *Harvard Business Review*, 85, 129–138.

Gerstner, C.R. and Day, D.V. (1994) Cross-cultural comparison of leadership prototypes. *The Leadership Quarterly*, 5, 121–134.

Gherardi, S. and Turner, B. (2002) Real men don't collect soft data. In A.M. Huberman and M.B. Miles (Eds.), *The Qualitative Researcher's Companion* (pp. 81–100). Thousand Oaks, CA: Sage.

Gibbons, T.C. (1986) Revisiting the question of born vs. made: Toward a theory of development of transformational leadership. Unpublished doctoral dissertation, Fielding Institute, Santa Barbara, CA.

Giddens, A. (1991) *Modernity and Self-Identity: Self and Society in the Late Modern Age*. Cambridge: Polity.

Gilbert, P. (1983) Styles of project management. *Project Management Journal*, 1(4), 189–193.

Gioia, D. and Chittipeddi, K. (1991) Sensemaking and sensegiving in strategic change initiation. *Strategic Management Journal*, 12, 433–448.

Gioia, D., Corley, K.G., and Fabbri, T. (2002) Revising the past while thinking about the future perfect tense. *Journal of Organizational Change Management*, 15(6), 622–634.

Giritli, H. and Topcu Oraz, G. (2004) Leadership styles: Some evidence from the Turkish construction industry. *Construction Management and Economics*, 22, 253–262.

Glaser, B.G. (1978) *Theoretical Sensitivity*. Mill Valley, CA: Sociology Press.

Glaser, B.G. (1992) *Basics of Grounded Theory Analysis: Emergence Vs. Forcing*. Mill Valley, CA: Sociology Press.

Glaser, B.G. and Strauss, A.L. (1967) *The Discovery of Grounded Theory: Strategies for Qualitative Research*. Chicago: Aldine Press.

Global Construction Perspectives and Oxford Economics (2015) Global Construction 2030: A global forecast for the construction industry to 2030. Available at: https://policy.ciob.org/wp-content/uploads/2016/06/GlobalConstruction2030_ExecutiveSummary_CIOB.pdf

Goethals, G.R. and Sorenson, G.J. (2007) *The Quest for a General Theory of Leadership*. Cheltenham: Edward Elgar Publishing.

Goethals, G.R., Sorenson, G.J., and Burns, J.M. (Eds.) (2004) *Encyclopedia of Leadership*, Vols 1–4. Thousand Oaks, CA: Sage.

Goffee, R. and Jones, G. (2000) Why should anyone be led by you? *Harvard Business Review*, September/October, 63–70.

Goffee, R. and Jones, G. (2005) Managing authenticity: The paradox of great leadership. *Harvard Business Review*, 83(12), 87–94.

Goldman, B.M. (2006) Making diamonds out of coal: The role of authenticity in healthy (optimal) self-esteem and psychological functioning. In M.H. Kernis (Ed.), *Self-Esteem Issues and Answers: A Sourcebook of Current Perspectives* (pp. 133–139). New York: Psychology Press.

Goldman, B.M. and Kernis M. (2002) The role of authenticity in healthy psychological functioning and subjective well-being. *Annals of the American Psychotherapy Association*, 5, 18–20.

Goldman, B.M. and Kernis, M. (2004) The development of the authenticity inventory, V.3. Unpublished data.

Goulding, C. (1999) Grounded theory: Some reflections on paradigm, procedures and misconceptions. Technical Working Paper, No. WP006/99, University of Wolverhampton, UK.

Goulding, C. (2002) *Grounded Theory: A Practical Guide for Management, Business and Market Researchers*. Birmingham: University of Birmingham.

Goulding, C. (2003) Issues in representing the postmodern consumer. *Qualitative Market Research: An International Journal*, 6(3), 152–159.

Graen, G.B., Hui, C., and Taylor, E. (2006) Experience-based learning about LMX leadership and fairness in project teams: A dyadic directional approach. *The Academy of Management Learning and Education*, 5(4), 448–460.

Granovetter, M. (1985) Economic action and social structure: The problem of embeddedness. *American Journal of Sociology*, 91(3), 481–510.

Gray, P.B. and Campbell, B.C. (2009) Human male testosterone, pair-bonding, and fatherhood. In P. T. Ellison and P.B. Gray (Eds.), *Endocrinology of Social Relationships* (pp. 270–293). Cambridge, MA: Harvard University Press.

Green, S. (2011) *Making Sense of Construction Improvement*. Chichester: Wiley-Blackwell.

Green, S.D., Larsen, G.D., and Kao, C.C. (2008) Competitive strategy revisited: Contested concepts and dynamic capabilities. *Construction Management and Economics*, 26(1), 63–78.

Greenleaf, R.K. (1977) *Servant-Leadership: A Journey into the Nature of Legitimate Power and Greatness*. Mahwah, NJ: Paulist Press.

Gronn, P. (2000) Distributed properties: A new architecture for leadership. *Educational Management and Administration*, 28(3), 317–338.

Guba, E.G. and Lincoln, Y.S. (1994) Competing paradigms in qualitative research. In N.K. Denzin and Y.S. Lincoln (Eds.), *Handbook of Qualitative Research*. Thousand Oaks, CA: Sage.

Guerrero, S., Lapalme, M.E., and Séguin, M. (2015) Board chair authentic leadership and non-executives' motivation and commitment. *Journal of Leadership and Organizational Studies*, 22, 88–101.

Hackman, J.R. and Wageman, R. (2007) Asking the right questions about leadership. *American Psychologist*, 62, 43–47.

Hackman, M.Z. and Johnson, C.E. (1996) *Leadership: A Communication Perspective* (2nd ed.) . Prospect Heights, IL: Waveland.

Hall, D. (2006) *Project member behaviour and client service expectations:A case study in implementation*. Paper presented at Cooperative Research Centre (CRC) for Construction Innovation, Clients Driving Innovation: Moving Ideas into Practice,March 12–14, 2006.

Halpin, A.W. and Winer, B.J. (1957) A factorial study of the leader behaviour descriptions. In R.M. Stogdill and A.E. Coons (Eds.), *Leader Behavior: Its Description and Measurement*. Columbus, OH: Bureau of Business Research, Ohio State University.

Hampson, S.E. (1998) When is an inconsistency not an inconsistency? Trait reconciliation in personality description and impression formation. *Journal of Personality and Social Psychology*, 74, 102–117.

Hannah, S.T., Lester, P.B., and Vogelgesang, G.R. (2005) Moral leadership: Explicating the moral component of authentic leadership. In W.L. Gardner, B.J. Avolio, and F.O. Walumbwa (Eds.), *Authentic Leadership Theory and Practice: Origins, Effects and Development* (pp. 43–82). Greenwood, CT: Elsevier JAI.

Hansen, H., Ropo, A., and Sauer, E. (2007) Aesthetic leadership. *The Leadership Quarterly*, 18, 544–560.

Harchar, R.L. and Hyle, A.E. (1996) Collaborative power: A grounded theory of administrative instructional leadership in the elementary school. *Journal of Educational Administration*, 34(3), 15–29.

Hargreaves, A. and Fink, D (2003) The seven principles of sustainable leadership. *Educational Leadership*, 61(7), 1–12.

Hari, S. (2006) Facilitating knowledge management initiatives for improved competitiveness in small and medium enterprises in construction. Unpublished PhD dissertation. Glasgow Caledonian University, Glasgow, UK.

Harter, S. (1999) *Distinguished Contributions in Psychology: The Construction of the Self: A Developmental Perspective*. New York: Guilford Press.

Harter, S. (2002) Authenticity. In C.R. Snyder and S.J. Lopez (Eds.), *Handbook of Positive Psychology* (pp. 382–394). Oxford: Oxford University Press.

Hartman, S.J. and Harris, O.J. (1992) The role of parental influence on leadership. *Journal of Social Psychology*, 132(2), 153–167.

Harvey, P., Stoner, J., Hochwarter, W.A., and >Kacmar, C. (2007) Coping with abusive supervision: The neutralizing effects of ingratiation and positive affect on negative employee outcomes. *The Leadership Quarterly*, 18(3), 264–280.

Hawkins, J. and Prado, M.G.F.A. (2020) CoST – the Infrastructure Transparency Initiative: from disclosed data to sector reform. *Civil Engineering*, 173,(1), 3–9.

Hay, A. and Hodgkinson, H. (2006) Rethinking leadership: A way forward for teaching leadership? *Leadership and Organization Development Journal*, 27(2), 144–158.

Hay, R., Samuel, F., Watson, K.J., and Bradbury, S. (2018) Post-occupancy evaluation in architecture: Experiences and perspectives from UK practice. *Building Research and Information*, 46(6), 698–710.

Hayden, W.M., Jr. (2013) How many oceans do we have? *Leadership and Management in Engineering*, 13(4). https://doi.org/10.1061/(ASCE)LM.1943-5630.0000251.

Health and Safety Executive (2020) *Construction Statistics in Great Britain, 2019*. London: HSE.

Hemphill, J.K. (1949) *Situational Factors in Leadership*. Columbus, OH: The Ohio State University Press.

Hersey, P. and Blanchard, K. (1982) *Management of Organizational Behavior* (4th ed.). Englewood Cliffs, NJ: Prentice-Hall.

Herzberg, F. (1966) *Work and the Nature of Man*. New York: Thomas Y. Crowell Publishers.

Hicks, D.A. (2002) Spiritual and religious diversity in the workplace: Implications for leadership. *The Leadership Quarterly*, 13, 379–396.

Higgins, E.T. (2005) Value from regulatory fit. *Current Directions in Psychological Science*, 14, 209–213.

Higgins, E.T. and Pittman, T. (2008) Motives of the human animal: Comprehending, managing, and sharing inner states. *Annual Review of Psychology*, 59, 361–385.

Higgs, M. (2003) How can we make sense of leadership in the 21st century? *Leadership and Organizational Development Journal*, 24(5), 273–284.

Hillebrandt, P.M. (2000) *Economic Theory and the Construction Industry* (3rd ed.). Basingstoke: Macmillan.

Hirst, G., Mann, L., Bain, P., Pirola-Merlo, A., and Richver, A. (2004) Learning to lead: The development and testing of a model of leadership learning. *The Leadership Quarterly*, 15(3), 311–327.

Hitt, M.A., Keats, B.W., and DeMarie, S.M. (1998) Navigating in the new competitive landscape: Building strategic flexibility and competitive advantage in the 21st century. *Academy of Management Executive*, 12(4), 22–42.

Hixon, J.G. and Swann, W.B. (1993) When does introspection bear fruit? Self-reflection, self-insight, and interpersonal choices. *Journal of Personality and Social Psychology*, 64(1), 35–43.

HM Government (2013) *Construction 2025: Joint Strategy from Government and Industry for the Future of the UK Construction Industry*. London: TSO.

HM Government (2018) *Industrial Strategy: Construction Sector Deal*. London: TSO.

Hmieleski, K.M., Cole, M.S., and Baron, R.A. (2012) Shared authentic leadership and new venture performance. *Journal of Management*, 38(5), 1476–1499.

Hock, D. (1999) *Birth of the Chaordic Age*. San Francisco: Berrett-Koehler Publishers.

Hofstede, G. (1980) *Culture's Consequences: International Differences in Work-Related Values*. Beverly Hills, CA: Sage.

Hofstede, G. (1991) *Cultures and Organizations: Software of the Mind*. London: McGraw-Hill.

Hogan, R., Raskin, R., and Fazzini, D. (1990) The dark side of charisma. In K. Clark and M. Clark (Eds.), *Measures of Leadership* (pp. 343–354). West Orange, NJ: Leadership Library of America.

Hollander, E.P. (2008) *Inclusive Leadership: The Essential Leader-Follower Relationship*. New York: Routledge.

Hooper, D.T. and Martin, R. (2007) Beyond personal Leader Member Exchange (LMX) quality: The effects of perceived LMX variability on employee reactions. *The Leadership Quarterly*, 19, 20–30.

Hopkins, M.M. and O'Neil, D.A. (2015) Authentic leadership: Application to women leaders. *Frontiers in Psychology*, 6, 959.

Horna, J. (1994) *The Study of Leisure*. Oxford: Oxford University Press.

Horwitz, F.M., Chan, T.H., Quazi, H., Nonkwelo, C., Roditi, D., and van Eck, P. (2006) Human resource strategies for managing knowledge workers: An Afro-Asian comparative analysis. *International Journal of Human Resource Management*, 17(5), 775–811.

Horwitz, F.M., Heng, C.T., and Quazi, H.A. (2003) Finders, keepers? Attracting, motivating and retaining knowledge workers. *Human Resource Management Journal*, 13(4), 23–44.

House, R.J. (1971) A path-goal theory of leader effectiveness. *Administrative Science Leadership Review*, 16, 321–339.

House, R.J. (1977) A 1976 theory of charismatic leadership. In J.G. Hunt and L.L. Larson (Eds.), *Leadership: The Cutting Edge* (pp. 189–207). Carbondale, IL: Southern Illinois University Press.

House, R.J. and Aditya, R.N. (1997) The social scientific study of leadership: Quo vadis? *Journal of Management*, 23(3), 409–473.

House, R.J. and Dessler, G. (1974) The path-goal theory of leadership: Some post hoc and a priori tests. In J.G. Hunt and L.L. Larson (Eds.), *Contingency Approaches to Leadership* (pp. 29–55). Carbondale, IL: Southern Illinois University Press.

House, R.J., Hanges, P.J., Javidan, M., Dorfman, P., and Gupta, V. (Eds.) (2004) *GLOBE, Cultures, Leadership, and Organizations: GLOBE Study of 62 Societies*. Newbury Park, CA: Sage.

House, R.J. and Howell, J.M. (1992) Personality and charismatic leadership. *The Leadership Quarterly*, 3, 81–108.

House, R.J. and Mitchell, T.R. (1974) Path-goal theory of leadership. *Contemporary Business*, 3, 81–98.

House, R.J., Spangler, W.D., and Woycke, J. (1991) Personality and charisma in the U.S. presidency: A psychological theory of leadership effectiveness. *Administrative Science Quarterly*, 36, 364–396.

House, R.J., Woycke, J., and Fodor, E.M. (1988) Charismatic and non-charismatic leaders: Differences in behavior and effectiveness. In J.A. Conger and R.N. Kanungo (Eds.), *Charismatic Leadership: The Elusive Factor in Organizational Effectiveness* (pp. 98–121). San Francisco: Jossey-Bass.

House, R.J., Wright, N.S., and Aditya, R.N. (1997) Cross-cultural research on organizational leadership: A critical analysis and a proposed theory. In P.C. Earley and M. Erez (Eds.), *New Perspectives in International Industrial Organizational Psychology* (pp. 535–625). San Francisco: New Lexington.

Howell, J.M. and Avolio, B.J. (1992) The ethics of charismatic leadership: Submission or liberation? *Academy of Management Executive*, 6, 43–54. Howell, J.P. (1997) Substitutes for leadership: Their meaning and measurement – an historical assessment. *The Leadership Quarterly*, 8(2), 113–116.

Hoy, W.K. and Henderson, J.E. (1983) Principal authenticity, school climate, and pupil-control orientation. *Alberta Journal of Educational Research*, 2, 123–130.

Hoyle, R.H., Kernis, M.H., Leary, M.R., and Baldwin, M.W. (1999) *Selfhood: Identity, Esteem, Regulation*. Boulder, CO: Westview Press.

Hung, S-C. (2004) Explaining the process of innovation: The dynamic reconciliation of action and structure. *Human Relations*, 57(11), 1479–1497.

Hunt, J.G. (1991) *Leadership: A New Synthesis*. Newbury Park, CA: Sage.

Hunt, J.G., and Ropo, A. (1995) Multi-level leadership: Grounded theory and mainstream theory applied to the case of general motors. *The Leadership Quarterly*, 6(3), 379–412.

Hunt, V., Layton, D., and Prince, S. (2015) *Why Diversity Matters*. London: McKinsey and Company.

Hunt, V., Prince, S., Dixon-Fyle, S., and Yee, L. (2018) *Delivering through Diversity*. London: McKinsey and Company.

Hunter, K. (2006) A link between project value management and best value in the public sector. Unpublished PhD dissertation. Glasgow Caledonian University, Glasgow, UK. Hunter, K., Hari, S., Egbu, C., and Kelly, J. (2005) Grounded theory: Its diversification and application through two examples from research studies on knowledge and value management. *The Electronic Journal of Business Research Methodology*, 3(1), 57–68.

Ilies, R., Gerhardt, M., and Le, H. (2004) Individual differences in leadership emergence: Integrating meta-analytic findings and behavioral genetics estimates. *International Journal of Selection and Assessment*, 12, 207–219.

Ilies, R., Morgeson, F.P., and Nahrgang, J.D. (2005) Authentic leadership and eudaemonic well-being: Understanding leader-follower outcomes. *The Leadership Quarterly*, 16(3), 373–394.

Infrastructure Transparency Initiative (CoST) (2020) *Business Plan 2020–2025: Strengthening Economies and Improving Lives*. London: CoST.

International Centre for Complex Project Management (2018) Project leadership: The game changer in large-scale complex projects. Roundtable Series. Available at: www.iccpm.com

International Labour Organisation (2001) *The Construction Industry in the Twenty-first Century: Its Image, Employment Prospects and Skill Requirements.* Geneva: ILO.

Iqbal, S., Farid, T., Ma, J., Khattak, A., and Nurunnabi, M. (2018) The impact of authentic leadership on organizational citizenship behaviours and the mediating role of corporate social responsibility in the banking sector of Pakistan. *Sustainability*, 10(7), 2170.

Ireland, R.D. and Hitt, M.A. (2005) Achieving and maintaining strategic competitiveness in the 21st century: The role of strategic leadership. *Academy of Management Executive Journal*, 19(4), 63.

Irurita, V.F. (1990) Optimizing as a leadership process: A grounded theory study of nurse leaders in Western Australia. Unpublished PhD Dissertation, The University of Western Australia, Perth.

Irurita, V.F. (1996) Optimizing: A leadership process for transforming mediocrity to excellence. In K. Parry (Ed.), *Leadership Research and Practice: Emerging Themes and New Challenges.* Melbourne: Pitman Publishing.

Irving, J.A. and Longbotham, G.J. (2007) Team effectiveness and six essential servant leadership themes: A regression model based on items in the organizational leadership assessment. *International Journal of Leadership Studies*, 2(2), 98–113.

Javidan, M. (2010) Bringing the global mindset to leadership. *Harvard Business Review*, May, 19.

Jensen, S.M. and Luthans, F. (2006) Relationship between entrepreneurs' psychological capital and their authentic leadership. *Journal of Managerial Issues*, 18(2), 254–273.

Ji, L., Koh, W.K.X., and Heng, S.H. (1997) The effects of interactive leadership on human resource management in Singapore's banking industry. *The International Journal of Human Resource Management*, 8(5), 710–719.

Jia, A.Y., Rowlinson, S., Loosemore, M., Xu, M., Li, B., and Gibb, A. (2017) Institutions and institutional logics in construction safety management: The case of climatic heat stress. *Construction Management and Economics*, 35:6, 338–367, doi:10.1080/01446193.2017.1296171.

Jones, I. (1997) Mixing qualitative and quantitative methods in sports fan research. *The Qualitative Report*, 3(4).

Jones, R. and Kriflik, G. (2006) Subordinate expectations of leadership within a cleaned-up bureaucracy: A grounded theory study. *Journal of Organizational Change Management*, 19(2), 154–172.

Jones, R. and Noble, G. (2007) Grounded theory and management research: A lack of integrity? *Qualitative Research in Organizations and Management*, 2(2), 84–103.

Judge, T., Bono, J., Ilies, R., and Gerhardt, M.W. (2002) Personality and leadership: A qualitative and quantitative review. *Journal of Applied Psychology*, 87, 765–780.

Jung, C.G. (1959) Conscious, unconscious, and individuation. In *Collected Works*, vol. 9. Princeton, NJ: Princeton University Press.

Kan, M.M. and Parry, K.W. (2004) Identifying paradox: A grounded theory of leadership in overcoming resistance to change. *The Leadership Quarterly*, 15(4), 467–491.

Kangis, P. and Lee-Kelley, L. (2000) Project leadership in clinical research organizations. *International Journal of Project Management*, 18(6), 393–401.

Kanungo, R.N. (2001) Ethical values of transactional and transformational leaders. *Canadian Journal of Administrative Sciences*, 18, 257–265.

Katz, D. and Kahn, R.L. (1952) Some recent findings in human relations research. In E. Swanson, T. Newcombe, and E. Hartley (Eds.), *Readings in Social Psychology*. New York: Holt, Rinehart and Winston.

Kegan, R. (1982) *The Evolving Self: Problem and Process in Human Development.* Cambridge, MA: Harvard University Press.

Kelle, U. and Laurie, H. (1995) Computer use in qualitative research and issues of validity. In U. Kelle (Ed.), *Computer-Aided Qualitative Data Analysis: Theory, Methods and Practice.* London: Sage.

Keller, T. (1999) Images of the familiar: Individual differences and implicit leadership theories. *The Leadership Quarterly*, 10(4), 589–607.

Keller, T. (2003) Parental images as a guide to leadership sensemaking: An attachment perspective on implicit leadership theories. *The Leadership Quarterly*, 14, 141–160.

Kellerman, B. (2004) *Bad Leadership: What It Is, How It Happens, Why It Matters.* Boston: Harvard Business Publishing.

Kellerman, B. and Webster, S.W. (2001) The recent literature on public leadership reviewed and considered. *The Leadership Quarterly*, 12, 485–514.

Kellett, J.B., Humphrey, R.H., and Sleeth, R.G. (2006) Empathy and the emergence of task and relations leaders. *The Leadership Quarterly*, 17(2), 146–162. doi:10.1016/j.leaqua.2005.12.003.

Kelman, H.C. (2006) Interests, relationships, identities: Three central issues for individuals and groups in negotiating their social environment. In S.T. Fiske, A.E. Kazdin, and D.L. Schacter (Eds.), *Annual Review of Psychology* (pp. 1–26). Palo Alto, CA: Annual Reviews.

Kempster, S. (2006) Leadership learning through lived experience: A process of apprenticeship? *Journal of Management and Organization*, 12(1), 4–22.

Kempster, S. and Parry, K. (2011) Grounded theory and leadership research: A critical realist perspective. *The Leadership Quarterly*, 22(1), 106–120.

Kendall, J. (1999) Axial coding and the grounded theory controversy. *Western Journal of Nursing Research*, 21(6), 743–757.

Kernis, M.H. (2003) Toward a conceptualization of optimal self-esteem. *Psychological Inquiry*, 14, 1–26.

Kernis M.H. and Goldman, B.M. (2005a) Authenticity, social motivation, and wellbeing. In J.P. Forgas, K.D. Williams, and S.M. Laham (Eds.), *Social Motivation: Conscious and Unconscious Processes* (pp. 210–227). Cambridge: Cambridge University Press.

Kernis, M.H. and Goldman, B.M. (2005b) Authenticity: A multicomponent perspective. In A. Tesser, J. Wood, and D. Stapel (Eds.), *On Building, Defending, and Regulating the Self: A Psychological Perspective* (pp. 31–52). New York: Psychology Press.

Kernis, M.H. and Goldman, B.M. (2005c) From thought and experience to behavior and interpersonal relationships: A multicomponent conceptualization of authenticity. In A. Tesser, J.V. Wood, and D. Stapel (Eds.), *On Building, Defending and Regulating the Self: A Psychological Perspective* (pp. 31–52). New York: Psychology Press.

Kernis, M.H. and Goldman, B.M. (2006) A multicomponent conceptualization of authenticity: Research and theory. In M.P. Zanna (Ed.), *Advances in Experimental Social Psychology* (pp. 284–357). San Diego, CA: Academic Press.

Kerr, S. and Jermier, J.M. (1978) Substitutes for leadership: Their meaning and measurement. *Organizational Behavior and Human Performance*, 22, 375–403.

Kets de Vries, M.F.R. (1996) Leaders who make a difference. *European Management Journal*, 14(5), 486–493.

Kets de Vries, M.F.R. (2003) The retirement syndrome: The psychology of letting go. *European Management Journal*, 21(6), 707–716.

Khan, S.N. (2010) Impact of authentic leaders on organization performance. *International Journal of Business and Management*, 5(12), 167–172.

King, S. (1994) What is the latest on leadership? *Management Development Review*, 7, 7–9.

Kirkman, B.L. and Shapiro, D.L. (2001) The impact of team members' cultural values on productivity, cooperation, and empowerment in self-managing work teams. *Journal of Cross-Cultural Psychology*, 32, 597–617.

Kirkpatrick, S.A. and Locke, E.A. (1991) Leadership: Do traits really matter? *Academy of Management Executive*, 5, 48–60.

Kloppenborg, T.J., Shriberg, A., and Venkatraman, J. (2003) *Project Leadership.* Vienna, VA: Management Concepts,

Kodish, S. (2006) The paradoxes of leadership: The contribution of Aristotle. *Leadership*, 2(4), 451–468.

Kolb, J.A. (1995) Leader behaviors affecting team performance: Similarities and differences between leader/member assessments. *Journal of Business Communication*, 32(3), 233–248.

Komives, S.R., Longerbeam, S.D., Owen, J.E., Mainella, F.C., and Osteen, L. (2006) A leadership identity development model: Applications from a grounded theory. *Journal of College Student Development*, 47(4), 401–418.

Komives, S.R., Owen, J.E., Longerbeam, S.D., Mainella, F.C., and Osteen, L. (2005) Developing a leadership identity: A grounded theory. *Journal of College Student Development*, 46(6), 593–611.

Kondo, D.K. (1990) *Crafting Selves: Power, Gender and Discourses of Identity in a Japanese Work-Place*. Chicago: University of Chicago Press.

Kotre, J. and Hall, E. (1997) *Seasons of Life: The Dramatic Journey from Birth to Death*. Ann Arbor, MI: University of Michigan Press.

Kotter, J.P. (1982) What effective general managers really do. *Harvard Business Review*, 60(6), 156–168.

Kotter, J.P. (1990) What leaders really do. *Harvard Business Review*, 5(3), 3–11.

Kouzes, J.M. and Posner, B.Z. (1987) *The Leadership Challenge: How to Get Extraordinary Things Done in Organizations*. San Francisco: Jossey-Bass.

Kramer, R.S.S., Arend, I., and Ward, R. (2010) Perceived health from biological motion predicts voting behaviour. *Quarterly Journal of Experimental Psychology*, 63(4), 625–632.

Kriger, M. and Seng, Y. (2005) Leadership with inner meaning: A contingency theory of leadership based on the worldviews of five religions. *The Leadership Quarterly*, 16, 771–806.

Kuhl, J. and Beckmann, J. (Eds.) (1985) *Action Control: From Cognition to Behavior*. Berlin: Springer.

Kumaraswamy, M. and Dulaimi, M. (2002) Empowering innovative improvements through creative construction procurement. *Engineering, Construction and Architectural Management*, 8(5–6), 325–334. https://doi.org/10.1046/j.1365-232X.2001.00215.x.

Labaree, R.V. (2002) The risk of 'going observationalist': Negotiating the hidden dilemmas of being an insider participant observer. *Qualitative Research*, 2(1), 97–122.

Ladkin, D. (2008) Leading beautifully: How mastery, congruence and purpose create the aesthetic of embodied leadership practice. *The Leadership Quarterly*, 19, 31–41.

Ladkin, D. and Spiller, C. (2013) *Authentic Leadership: Clashes, Convergences and Coalescences*. Cheltenham: Edward Elgar Publishing.

Lakey, C.E., Kernis, M.H., Heppner, W.L., and Lance, C.E. (2008) Individual differences in authenticity and mindfulness as predictors of verbal defensiveness. *Journal of Research in Personality*, 42, 230–238.

Lakshman, C. (2007) Organizational knowledge leadership: A grounded theory approach. *Leadership and Organization Development Journal*, 28(1), 51–75.

Larsson, G., Hærem, T., Sjöberg, M., Alvinius, A., and Bakken, B. (2007) Indirect leadership under severe stress: A qualitative inquiry into the 2004 Kosovo riots. *International Journal of Organizational Analysis*, 15(1), 23–34.

Larsson, G., Sjöberg, M., Vrbanjac, A., and Björkman, T. (2005) Indirect leadership in a military context: A qualitative study on how to do it. *Leadership and Organization Development Journal*, 26(3), 215–227.

Latham, M. (1994) *Constructing the Team*. London: HMSO.

LeCompte, M.D. and Goetz, J.P. (1982) Problems of reliability and validity in ethnographic research. *Review of Educational Research*, 52(2), 31–60.

Lee, A.Y. and Hutchison, L. (1998) Improving learning from examples through reflection. *Journal of Experimental Psychology: Applied*, 4(3), 187–210.

Lee-Kelley, L. and Leong, K. (2003) Turner's five functions of project-based management and situational leadership in IT services projects. *International Journal of Project Management*, 21(8), 583–591.

Levine, K.J. (2008) Trait theory. In A. Marturano and J. Gosling (Eds.), *Leadership: The Key Concepts* (pp. 163–166). London: Routledge.

Levinson, D.J. (1977) The mid-life transition: A period in adult psychosocial development. *Psychiatry*, 40(2), 99–112.

Levinson, D.J. (1981) Explorations in biography. In A.L. Rabin, J. Aronoff, A.M. Barclay, and R.A. Zucker (Eds.), *Further Explorations in Personality* (pp. 44–79). New York: Wiley.

Levinson, D.J., with Darrow, C.N., Klein, E.B., Levinson, M.H., and McKee, B. (1978) *The Seasons of a Man's Life*. New York: Knopf.

Lewin, K., Lippitt, R., and White, R.K. (1939) Patterns of aggressive behavior in experimentally created social climates. *Journal of Social Psychology*, 10, 271–301.

Lewis, J. (2003) *Project Leadership*. New York: McGraw-Hill.

Lewis, M.W. (2000) Exploring paradox: Toward a more comprehensive guide. *Academy of Management Review*, 25(4), 760–776.

Lewis, T.M. (2007) Impact of globalization on the construction sector in developing countries. *Construction Management and Economics*, 25(1), 7–23.

Li, H., Cheng, E.W.L., Love, P.E.D., and Irani, Z. (2001) Co-operative benchmarking: A tool for partnering excellence in construction. *International Journal of Project Management*, 19(3), 171–179.

Liden, R.C., Wayne, S.J., Zhao, H., and Henderson, D. (2008) Servant leadership: Development of a multidimensional measure and multi-level assessment. *The Leadership Quarterly*, 19, 161–177.

Likert, R. (1967) *The Human Organization*. New York: McGraw-Hill.

Lim, G.S. and Daft, R.L. (2004) *The Leadership Experience in Asia*. Singapore: Thomson Learning.

Lim, L.L.C. (Ed.) (2016) *Singapore's Economic Development: Retrospection and Reflections*. Singapore: World Scientific.

Lincoln, Y.S and Guba, E.G. (1985) *Naturalistic Inquiry*. Beverly Hills, CA: Sage.

Lingard, H. and Rowlinson, S.M. (2006) *Occupational Health and Safety in Construction Project Management*. London: Routledge.

Lipman-Blumen, J. (2005) *The Allure of Toxic Leaders*. New York: Oxford University Press.

Liu, A., Fellows, R., and Fang, Z. (2003) The power paradigm of project leadership. *Construction Management and Economics*, 21(8), 819–829.

Liu, Y., Fuller, B., Hester, K., Bennett, R.J., and Dickerson, M.S. (2018) Linking authentic leadership to subordinate behaviors. *Leadership and Organization Development Journal*, 39(2), 218–233.

Locke, E. (2003) Leadership: Starting at the top. In C.L. Pearce and J.A. Conger (Eds.), *Shared Leadership: Reframing the Hows and Whys of Leadership* (pp. 271–284). Thousand Oaks, CA: Sage.

Locke, K. (2007) *Grounded Theory in Management Research*. London: Sage.

Loftus, E.F. and Loftus, G.R. (1980) On the permanence of stored information in the human brain. *American Psychologist*, 35, 409–420.

Lombardo, M.M., Ruderman, M.N., and McCauley, C.D. (1988) Explanations of success and derailment in upper-level management positions. *Journal of Business Psychology*, 2, 199–216.

London, K. and Chen, J. (2007) Role of cultural capital towards the development of a sustainable business model for design firm internationalization. In T.C. Haupt and R. Milford (Eds.), *CIB World Building Congress: Construction for Development* (pp. 4–17). Cape Town, South Africa: CIB.

Loosemore, M. (1999) International construction management research: Cultural sensitivity in methodological design. *Construction Management and Economics*, 17, 553–561.

Loosemore, M., Dainty, A., and Lingard, H. (2003) *Human Resource Management in Construction Projects: Strategic and Operational Approaches*. London: Spon Press.

Loosemore, M. and Lee, P. (2002) Communication problems with ethnic minorities in the. construction industry. *International Journal of Project Management*, 20, 517–524.

Lopata, H.Z. and Levy, J.A. (2003) *Social Problems across the Life Course*. Lanham, MD: Rowman and Littlefield.

Lopes, J. (2012) Construction in the economy and its role in socio-economic development. In G. Ofori (Ed.), *New Perspectives on Construction in Developing Countries* (pp. 40–71). Abingdon: Routledge.

Lord, R.G. and Brown, D.J. (2004). *Leadership Processes and Follower Identity*. Mahwah, NJ: Lawrence Erlbaum Associates.

Lord, R.G., Brown, D.J., and Freiberg, S.J. (1999) Understanding the dynamics of leadership: The role of follower self-concepts in the leader/follower relationship. *Organizational Behavior and Human Decision Processes*, 78, 167–203.

Lord, R. and Hall, R. (2005) Identity, deep structure and the development of leadership skill. *The Leadership Quarterly*, 16(4), 591–615.

Love, P.E.D., Holt, G.D., and Li, H. (2002) Triangulation in construction management research. *Engineering, Construction and Architectural Management*, 9(4), 294–303.

Low, S.P. and Chuan, Q.T. (2006) Environmental factors and work performance of project managers. *International Journal of Project Management*, 21(1), 24–37.

Low, S.P. and Lee, B.S. (1997) East meets West: Leadership development for construction project management. *Journal of Managerial Psychology*, 12(6), 383–400.

Low S.P. and Leong, C.H. (2001) Asian management style versus Western management theories: A Singapore case study in construction project management. *Journal of Managerial Psychology*, 16(2), 127–141.

Low, S.P. and May, C.F. (1997) Quality management systems: A study of authority and empowerment. *Building Research and Information*, 25(3), 158–169.

Lowe, K.B. and Gardner, W.L. (2000) Ten years of *The Leadership Quarterly*: Contributions and challenges for the future. *The Leadership Quarterly*, 11, 459–514.

Lowe, K.B., Kroeck, K.G., and Sivasubramaniam, N. (1996) Effectiveness correlates of transformational and transactional leadership: A meta-analytic review of the MLQ literature. *The Leadership Quarterly*, 7, 385–425.

Lowy, A. (2008) The leader's dilemma agenda. *Strategy and Leadership*, 36(1), 33–38.

Lowy, A. and Hood, P. (2004) Leaders manage dilemmas. *Strategy and Leadership*, 32(3), 21–26.

Lozano, R. (2008) Developing collaborative and sustainable organizations. *Journal of Cleaner Production*, 16, 499–509.

Luthans, F. (2002) Positive organizational behavior: Developing and maintaining psychological strengths. *Academy of Management Executive*, 16, 57–72.

Luthans, F. (2003) Positive organizational behavior: Implications for leadership and HR development and motivation. In L.W. Porter, G.A. Bigley, and R.M. Steers (Eds.), *Motivation and Work Behavior* (pp. 178–195). New York: McGraw-Hill/Irwin.

Luthans, F. and Avolio, B.J. (2003) Authentic leadership development. In K.S. Cameron, J.E. Dutton, and R.E. Quinn (Eds.), *Positive Organizational Scholarship: Foundations of a New Discipline* (pp. 241–258). San Francisco, CA: Berrett-Koehler.

Luthans, F. and Youssef, C.M. (2004) Human, social, and now positive psychological capital management: Investing in people for competitive advantage. *Organizational Dynamics*, 33, 143–160.

Luthans, F., Youssef, C.M., and Avolio, B.J. (2006) Psychological capital: Investing and developing positive organizational behavior. In C.L. Cooper and D. Nelson (Eds.), *Positive Organizational Behavior: Accentuating the Positive at Work*. Thousand Oaks, CA: Sage.

Luthans, F., Youssef, C.M., and Avolio, B.J. (2007) *Psychological Capital: Developing the Human Competitive Edge*. Oxford: Oxford University Press.

Lyubovnikova, J., Legood, A., Turner, N., and Mamakouka, A. (2017) How authentic leadership influences team performance: The mediating role of team reflexivity. *Journal of Business Ethics*, 141, 59–70.

MacDonald, D.A. (2000) Spirituality: Description, measurement and relation to the five factor model of personality. *Journal of Personality*, 68(1), 153–197.

MacDonald, K., Rezania, D., and Baker, R. (2020) A grounded theory examination of project managers' accountability. *International Journal of Project Management*, 38(1), 27–35.

MacPherson, S.J., Kelly, J.R., and Webb, R.S. (1993) How designs develop: Insights from case studies in building engineering services. *Construction Management and Economics*, 11, 475–485.

Madsen, S. (2015) *The Power of Project Leadership*. London: Kogan Page.

Maitlis, S. (2005) The social processes of organizational sensemaking. *Academy of Management Journal*, 48(1), 21–49.

Makilouko, M. (2004) Coping with multicultural projects: The leadership styles of Finnish project managers. *International Journal of Project Management*, 22, 387–396.

Malhotra, N. and Peterson, M. (2006) *Basic Marketing Research: A Decision-Making Approach* (2nd ed.). Englewood Cliffs, NJ: Prentice Hall.

Maloney, W.F. (1997) Strategic planning for human resource management in construction. *Journal of Management in Engineering (ASCE)*, 13(3).

Manring, S.L. (2007) Creating and managing interorganizational learning networks to achieve sustainable ecosystem management. *Organization and Environment*, 20(3), 325–346.

Manz, C.C. (1986) Self-leadership: Toward an expanded theory of self-influence processes in organizations. *Academy of Management Review*, 11(3), 585–600.

Manz, C.C. (1992) *Mastering Self Leadership: Empowering Yourself for Personal Excellence*. Englewood Cliffs, NJ: Prentice-Hall.

Manz, C.C., Anand, V., Joshi, M., and Manz, K.P. (2008) Emerging paradoxes in executive leadership: A theoretical interpretation of the tensions between corruption and virtuous values. *The Leadership Quarterly*, 19, 385–392.

Manz, C.C. and Sims, H.P. (1982) The potential for "groupthink" in autonomous work groups. *Human Relations*. 35(9), 773–784.

Manz, C.C. and Sims, H.P. (1987) Leading workers to lead themselves: The external leadership of self-managing work teams. *Administrative Science Quarterly*, 32, 106–128.

Manz, C.C. and Sims, H.P. (1989) *Super Leadership: Leading Others to Lead Themselves*. Englewood Cliffs, NJ: Prentice-Hall.

Marion, R. and Uhl-Bien, M. (2001) Leadership in complex organizations. *The Leadership Quarterly*, 12, 389–418.

Markus, H. and Nurius, P. (1986) Possible selves. *American Psychologist*, 41, 954–969.

Martin, M. (1986) *Self-Deception and Morality*. Lawrence, KS: University Press of Kansas.

Martin, P.Y. and Turner, B.A. (1986) Grounded theory and organizational research. *Journal of Applied Behavioral Science*, 22, 141–157.

Maslow, A.H. (1954) *Motivation and Personality*. New York: Harper and Row.

Maslow, A.H. (1968) *Toward a Psychology of Being*. New York: D. Van Nostrand Company.

Maslow, A.H. (1971) *The Farther Reaches of Human Nature*. New York: Viking.

Masters, C., Carlson, D.S., and Pfadt, E. (2006) Winging it through research: An innovative approach to a basic understanding of research methodology. *Journal of Emergency Nursing*, 32(5), 382–384.

May, D.R., Chan, A., Hodges, T., and Avolio, B.J. (2003) Developing the moral component of authentic leadership. *Organizational Dynamics*, 32, 247–260.

Mayo, E. (1933) *The Human Problems of an Industrial Civilization*. New York: Macmillan.

Mays, N. and Pope, C. (1995) Qualitative research: Rigor and qualitative research. *British Medical Journal*, 311, 109–112.

McCaffer, R. (1995) Editorial. *Engineering, Construction and Architectural Management*, 15(4), 2008.

McCall, M.W. Jr. (2004) Leadership development through experience. *Academy of Management Executive*, 18(3), 127–130.

McCall, M.W., Jr., Lombardo, M., and Morrison, A.M. (1988) *The Lessons of Experience*. Lexington, MA: Lexington Books.

McCaslin, M.L. (1993) The nature of leadership within rural communities: A grounded theory. Unpublished PhD dissertation, University of Nebraska, Lincoln.

McClelland, D.C., Atkinson, J.W., Clark, R.A., and Lowell, E.L. (1953) *The Achievement Motive*. New York: Appleton-Century-Crofts.

McGregor, D. (1960) *The Human Side of Enterprise*. New York: McGraw-Hill.

Mehta, S. (2003) MCI: Is being good good enough? *Fortune*, 27, 117–124.

Meindl, J.R., Ehrlich, S.B., and Dukerich, J.M. (1985) The romance of leadership. *Administrative Science Quarterly*, 30, 78–102.

Merriam, S.B., *et al.* (2002) *Qualitative Research in Practice*. San Francisco: Jossey-Bass.

Merriam-Webster Online Dictionary (2020) Reconciliation. Available at: www.merriam-webster.com (accessed February 2008).

Meyer, T. and Xu, Y. (2005) Academic and clinical dissonance in nursing education: Are we guilty of failure to rescue? *Nurse Educator*, 30(2), 76–79.

Mezher, T.M. and Tawil, W. (1998) Causes of delays in the construction industry in Lebanon. *Engineering, Construction and Architectural Management*, 5(3), 252–260.

Miao, C., Humphrey, R., and Qian, S. (2018) Emotional intelligence and authentic leadership: A meta-analysis. *Leadership and Organization Development Journal*, 39(5), 679–690.

Michaels, E.D., Handfield-Jones, H., and Axelrod, A. (2001) *The War for Talent*. Boston, MA: Harvard Business School Press.

Michie, S. and Gooty, J. (2005) Values, emotions, and authenticity: Will the real leader please stand up? *The Leadership Quarterly*, 16(3), 441–457.

Miles, M.B. and Huberman, A.M. (1994) *Qualitative Data Analysis* (2nd ed.). Thousand Oaks, CA: Sage.

Mintzberg, H. (1979) An emerging strategy of "Direct Research." *Administrative Science Quarterly*, 24(4), 582–590.

Mirvis, P. and Ayas, K. (2003) Reflective dialogue, life stories and leadership development. *Reflections: The SoL Journal*, 4(4), 39–48.

Mishler, E. (1986) *Research Interviewing: Context and Narrative*. Cambridge, MA: Harvard University Press.

Mitchell, M. (2002) My journey to reconcile religious beliefs with reality. *Rehabilitation Counseling Bulletin*, 46(1), 51–53.

Mitroff, I. and Denton, E.A. (1999) *A Spiritual Audit of Corporate America: A Hard Look at Spirituality, Religion, and Values in the Workplace*. San Francisco: Jossey-Bass.

Monaghan T.J. (1981) An investigation of leadership styles and organizational structure, and their influence on the conduct of construction projects. Unpublished MSc dissertation, Heriot-Watt University.

Monetary Authority of Singapore (2019) Exchange rates. Available at: https://secure.mas.gov.sg/msb/ExchangeRates.aspx.

Moore, J.D. (2006) Women in construction management: Creating a theory of career choice and development. Unpublished PhD Dissertation, Colorado State University, Fort Collins, CO.

Morgan, D.L. and Nica, A. (2020) Iterative thematic inquiry: A new method for analyzing qualitative data. *International Journal of Qualitative Methods*, 19, 1–11.

Morrell, P. (Ed.) (2015) *The Edge Commission Report on the Future of Professionalism: Collaboration for Change*. London: The Edge.

Morse, J.M. (1998) Validity by committee. [Editorial] *Qualitative Health Research*, 8, 443–445.

Morse, J.M., Barrett, M., Mayan, M., Olson, K., and Spiers, J. (2002) Verification strategies for establishing reliability and validity in qualitative research. *International Journal of Qualitative Methods*, 1(2), 1–19.

Morse, J.M., Hutchinson, S.A., and Penrod, J. (1998) From theory to practice: The development of assessment guides from qualitatively derived theory. *Qualitative Health Research Journal*, 8(3), 329–340.

Muir, I. and Langford, D. (1994) Managerial behaviour in two small construction organizations. *International Journal of Project Management*, 12(4), 244–253.

Mumford, M.D., Espejo, J., Hunter, S.T., Bedell-Avers, K.E., Eubanks, D.I., and Connelly, S. (2007) The sources of leader violence: A comparison of ideological and non-ideological leaders. *The Leadership Quarterly*, 18, 217–223.

Mumford, M.D., Gessner, T.L., Connelly, M.S., O'Connor, J.A., and Clifton, T.C. (1993) Leadership and destructive acts: Individual and situational influences. *The Leadership Quarterly*, 4(2), 115–147.

Mustapha, F.H. and Naoum, S. (1998) Factors influencing the effectiveness of construction site managers. *International Journal of Project Management*, 16(1), 1–8.

Nadler, A. (2002) Social-psychological analysis of reconciliation: Instrumental and socio- emotional routes to reconciliation. In G. Salomon and B. Nevo (Eds.), *Peace Education Worldwide: The Concept, Underlying Principles, the Research*. Mahwah, NJ: Erlbaum.

Nadler, A. and Liviatan, I. (2006) Intergroup reconciliation: Effects of adversary's expressions of empathy, responsibility, and recipients' trust. *Personality and Social Psychology Bulletin*, 32, 459–470.

Nathan, E., Mulyadi, R., Sen, S., Dirkvan, D., and Robert, C.L. (2019) Servant leadership: A systematic review and call for future research. *The Leadership Quarterly*, 30(1), 111–132.

Nathaniel, A. (2020) From the Editor's desk. *The Grounded Theory Review*, 19(1). Available at: http://groundedtheoryreview.com/2020/08/12/from-the-editors-desk-3/

Navin, T.R. (1971) Passing on the mantle: Management succession in industry. *Business Horizons*, 14(5), 83–93.

Neider, L.L. and Schriesheim, C.A. (2011) The Authentic Leadership Inventory (ALI): Development and empirical tests. *The Leadership Quarterly*, 22(6), 1146–1164.

Newport, F. and Harter, J. (2016) Presidential candidates as leaders: The public's view. *Gallup*, 29 April.

Newton, S., Ruddock, L., and Gruneberg, S. (2016) W55 Construction Industry Economics Research Roadmap. Available at: www.cibworld.nl/app/attach/tLFKJEfc/20152487/9e3dab6fb020f23bfe51e13bd07309c6/Les_Ruddock.pdf

Nguyen, Q.T., Preece, C.N., and Male, S.P. (2006) *Strategic management practice and tendency in Vietnamese small and medium-size construction firms.* Paper presented at CIB W107 Construction in Developing Countries International Symposium "Construction in Developing Economies: New Issues and Challenges,"18–20 January 2006,Santiago, Chile.

Noordegraaf, M. and Stewart, R. (2000) Managerial behaviour research in private and public sectors: distinctiveness, disputes and directions. *Journal of Management Studies*, 37(3), 427–443.

Norrie, J. and Walker D.H.T. (2004) A balanced scorecard approach to project management leadership. *Journal of Project Management*, 35(4), 47–56,.

Northouse, P.G. (2016) *Leadership Theory and Practice.* Thousand Oaks, CA: Sage.

Norwood, S.R. and Mansfield, N.R. (1999) Joint venture issues concerning European and Asian construction markets of the 1990's. *International Journal of Project Management*, 17(2), 89–93.

O'Connor, J., Mumford, M.D., Clifton, T.C., Gessner, T.L., and Connelly, M.S. (1995) Charismatic leaders and destructiveness: An historiometric study. *The Leadership Quarterly*, 6(4), 529–555.

Odusami, K.T. (2002) Perceptions of construction professionals concerning important skills of effective project leaders. *Journal of Management in Engineering*, 18(2), 61–67.

Odusami, K.T. and Ameh, O.J. (2006) *The leadership profile of Nigerian construction project leaders.* Paper presented at the 2nd Specialty Conference on Leadership and Management in Construction, Grand Bahama Island, Bahamas.

Odusami, K.T., Iyagba, R.R., and Omirin, M.M. (2003) The relationship between project leadership, team composition and construction project performance in Nigeria. *International Journal of Project Management*, 21, 519–527.

Ofori, G. (1993) Research on construction industry development at the crossroads. *Construction Management and Economics*, 12, 295–306.

Ofori, G. (2003) Frameworks for analysing international construction. *Construction Management and Economics*, 21(4), 379–391.

Ofori, G. (2019) *Professionalism of the built environment researcher.* Paper presented at Future Trends in Civil Engineering 2019, October 17, 2019, Zagreb, Croatia.

Ofori, G. and Toor, S.R. (2008a) Leadership: A pivotal factor for sustainable development. *Construction Information Quarterly*, Special Issue on Leadership, 10(2), 67–72.

Ofori, G. and Toor, S.R. (2008b) *Leading the development of sustainable organizations.* Paper presented at CIB Joint International Symposium: Transformation through Construction,November 17–19, 2008,Dubai, UAE.

Ofori, G. and Toor, S.R. (2008c) *Leadership in sustainability in construction.* Paper presented at CIOB International Construction Conference on "Construction: Regional outlook and Sustainability," February 20, 2008, Singapore.

Ofori, G. and Toor, S.R. (2009) Research on cross-cultural leadership and management in construction: A review and directions for future research. *Construction Management and Economics*, 27(2), 119–133.

Ofori-Dankwa, J. and Julian, D. (2004) Conceptualizing social science paradoxes using the diversity and similarity curves model: Illustrations from the work/play and theory novelty/continuity paradoxes. *Human Relations*, 57(11), 1449–1477.

Ogunlana, S.O., Niwawate, C., Quang, T., and Thang, L.C. (2002a) Effect of humour usage by engineers at construction sites. *Journal of Management in Engineering*, 22(2), 81–88.

Ogunlana, S.O., Siddiqui, Z., Yisa, S., and Olomolaiye, P. (2002b) Factors and procedures used in matching project managers to construction projects in Bangkok. *International Journal of Project Management*, 20(5), 385–400.

O'Leary, B.S., Lindholm, M.L., Whitford, R.A., and Freeman, S.E. (2002) Selecting the best and brightest: Leveraging human capital. *Human Resource Management*, 41(3), 325–340.

O'Leary, Z. (2004) *The Essential Guide to Doing Research.* London: Sage.

Olson, B.J., Nelson, D.L., and Parayitam, S. (2006) Managing aggression in organizations: What leaders must know. *Leadership and Organization Development Journal*, 27(5), 384–398.

Oluokun, C.O. (2008) A grounded theory study of younger and older construction workers' perceptions of each other in the work place. Unpublished PhD dissertation, The George Washington University.

Orton, J.D. and Weick, K.E. (1990) Loosely coupled systems: A reconceptualization. *Academy of Management Review*, 15, 203–223.

Osborn, R.N. and Hunt, J.G. (2007) Leadership and the choice of order: Complexity and hierarchical perspectives near the edge of chaos. *The Leadership Quarterly*, 18(4), 319–340.

Osborn, R.N., Hunt, J.G., and Jauch, LR. (2002) Toward a contextual theory of leadership. *The Leadership Quarterly*, 13(6), 797–837.

Ouchi, W.G (1981) *Theory Z: How American Business Can Meet the Japanese Challenge*. Reading, MA: Addison-Wesley.

Oxford English Dictionary (2008) Reconciliation. Online. Available at: www.oed.com/

Oyserman, D. and Markus, H. (1990) Possible selves in balance: Implications for delinquency. *Journal of Social Issues*, 46, 141–157.

Padilla, A., Hogan, R., and Kaiser, R.B. (2007) The toxic triangle: Destructive leaders, susceptible followers, and conducive environments. *The Leadership Quarterly*, 18, 176–194.

Palanski, M. and Yammarino, F. (2007) Integrity and leadership: Clearing the conceptual confusion. *European Management Journal*, 25, 171–184.

Pandit, N.R. (1996) The creation of theory: A recent application of the grounded theory method. *The Qualitative Report*, 2(4), 1–14.

Parameshwar, S. (2005) Spiritual leadership through ego-transcendence: Exceptional responses to challenging circumstances. *The Leadership Quarterly*, 16, 689–722.

Parameshwar, S. (2006) Inventing a higher purpose through suffering: The transformation of the transformational leader. *The Leadership Quarterly*, 17(5), 454–474.

Parry, J. (2003) Making sense of executive sensemaking. *Journal of Health Organization and Management*, 17(4), 240–263.

Parry, K.W. (1998) Grounded theory and social process: A new direction for leadership research. *The Leadership Quarterly*, 9(1), 85–105.

Parry, K.W. (1999) Enhancing adaptability: Leadership strategies to accommodate change in local government settings. *Journal of Organizational Change Management*, 12, 134–156.

Parry, K.W. (2002) Four phenomenologically-determined social processes of organizational leadership: Further support for the construct of transformational leadership. In B.J. Avolio, and F.J. Yammarino (Eds.), *Transformational and Charismatic Leadership: Monographs in Leadership and Management*, vol. 2 (pp. 339–372). Oxford: Elsevier Science. Parry, K.W. (2004) Comparative modeling of the social processes of leadership in work units. *Journal of the Australian and New Zealand Academy of Management*, 10(2), 69–80.

Parry, K.W. (2006) Qualitative method for leadership research: Now there's a novel idea! *Compliance and Regulatory Journal: The Journal of the Australasian Compliance Institute*, 1(1), 24–25.

Parry, K.W. and Meindl, J.R. (Eds.) (2002) *Grounding Leadership Theory and Research: Issues, Challenges, and Perspectives*. Charlotte, NC: Information Age Publishing.

Paulhus, D.L. (1982) Individual differences, cognitive dissonance, and self-presentation: Their concurrent operation in forced compliance. *Journal of Personality and Social Psychology*, 43, 838–852.

Pearce, C.L. and Conger, J.A. (Eds.) (2003) *Shared Leadership: Reframing the Hows and Whys of Leadership*. Thousand Oaks, CA: Sage.

Pearce, C.L., Conger, J.A., and Locke, E.A. (2008) Shared leadership theory. *The Leadership Quarterly*, 19, 622–628.

Pearce, C.L. and Sims, H.P. (2002) Vertical versus shared leadership as predictors of the effectiveness of change management teams: An examination of aversive, directive, transactional, transformational, and empowering leader behaviors. *Group Dynamics: Theory, Research, and Practice*, 6(2), 172–197.

Peiffer, E. (2017) 10 construction industry trends to watch in 2017. Blog. Available at : https://futureofconstruction.org/blog/10-construction-industry-trends-to-watch-in-2017/.

Phillips, D.L. (1973) *Abandoning Method*. San Francisco: Jossey-Bass.

Phua, F.T.T. (2004) Modeling the determinants of multi-firm project success: A grounded exploration of different participant perspectives. *Construction Management and Economics*, 22(5), 451–459.

Pietersen, W. (2002) The Mark Twain dilemma: The theory and practice of change leadership. *Journal of Business Strategy*, Sept/Oct, 32–37.

Platt, D.G. (1996) Building process models for design management. *Journal of Computing in Civil Engineering*, 10(3), 194–203. Podsakoff, P.M. and MacKenzie, S.B. (1997) Kerr and Jermier's substitutes for leadership model: Background, empirical assessment, and suggestions for future research. *The Leadership Quarterly*, 8(2), 117–132.

Popper, M. and Mayseless, O. (2003) Back to basics: Applying a parenting perspective to transformational leadership. *The Leadership Quarterly*, 14(1), 41–65.

Popper, M., Mayseless, O., and Castelnovo, O. (2000) Transformational leadership and attachment. *The Leadership Quarterly*, 11, 267–289.

Portes, A. 1998. Social capital: Its origins and applications in modern sociology. *Annual Review of Sociology*, 24: 1–25.

Portugal, E. and Yukl, G. (1994) Perspectives on environmental leadership. *The Leadership Quarterly*, 5, 271–276.

Posner, B.Z. and Kouzes, J.M. (1988) Development and validation of the leadership practices inventory. *Educational and Psychological Measurement*, 48(2), 483–496.

Price, T.L. (2003) The ethics of authentic transformational leadership. *The Leadership Quarterly*, 14, 67–81.

PwC (2014) Fighting corruption and bribery in the construction industry. Available at: www.transparency. org/news/pressrelease/a_world_built_on_bribes_corruption_in_construction_bankrupts_countries_and

Pye, A.J. (2005) Leadership and organizing: Sensemaking in action. *Leadership*, 1(1), 31–49.

Quinnell, K. (2018) The gap between CEO and worker compensation continues to grow. *AFL-CIO Executive Paywatch*, May 22, 2018.

Raftery, J., Pasadilla, B., Chiang, Y.H., Hui, E.C.M., and Tang, B.S. (1998) Globalization and construction industry development: Implications of recent developments in the construction sector in Asia. *Building Journal*, February, 73–77.

Raiden, A.B. and Dainty, A.R.J. (2006) Human resource development (HRD) in construction organisations: An example of a 'chaordic' learning organisation? *Learning Organization*, 13(1), 63–79.

Ralph, N., Birks, M., and Chapman, Y. (2015) The methodological dynamism of grounded theory. *International Journal of Qualitative Methods*, 14: 1–6.

Rameezdeen, R. (2007) Image of the construction industry. In M. Sexton, K. Kähkönen, and S. Lu (Eds.), *CIB Priority Theme: Revaluing Construction: A W065 'Organization and Management of Construction' Perspective*. Ottawa, Canada: CIB.

Reave, L. (2005) Spiritual values and practices related to leadership effectiveness. *The Leadership Quarterly*, 16, 655–687.

Rego, A., Sousa, F., Marques, C., and Cunha, M.P.E. (2012) Authentic leadership promoting employees' psychological capital and creativity. *Journal of Business Research*, 65(3), 429–437.

Regumyamheto, J.A. and Batatia, C. (1994) *Consultancy Report: A Study of Human Resources in the Construction Sector*. Dar es Salaam: National Construction Council.

Reicher, S. and Hopkins, N. (2003) On the science and art of leadership. In D. van Knippenberg and M.A. Hogg (Eds.), *Leadership and Power: Identity Processes in Groups and Organizations* (pp. 197–209). London: Sage.

Remington, K. (2011) *Leading Complex Projects*. Farnham: Gower.

Revell, J. (2003) The fires that won't go out. *Fortune*, 13, 139.

Rhodes, C. (2019) Construction industry: Statistics and policy. Briefing Paper No. 01432. London: House of Commons Library.

Ricoeur, P. (1992) *Oneself as Another*. Trans. K. Blamey. Chicago: University of Chicago Press.

Rubin, H.J. and Rubin, I.S. (2005) *Qualitative Interviewing: The Art of Hearing Data* (2nd ed.). Thousand Oaks, CA: Sage.

Rock, D. (2006) *Quiet Leadership: Help People Think Better—Don't Tell Them What to Do!* New York: HarperCollins.

Rock, D. and Ringleb, A.H. (2013) *Handbook of NeuroLeadership*. New York: CreateSpace Independent Publishing Platform.

Rock, D. and Schwartz, J. (2006) The neuroscience of leadership. *strategy+business magazine*, 43, 71–81.

Rogers, C.R. (1959) A theory of therapy, personality and interpersonal relationships, as developed in a client-centered framework. In S. Koch (Ed.), *Psychology: A Study of a Science*, vol. 3. Toronto: McGraw-Hill.

Rosenbaum. B.L. (1991) Leading today's technical professional. *Training and Development*, 45(10), 55–66,.

Rosenberg, S. (1997) Multiplicity of selves. In R.D. Ashmore and L. Jussim (Eds.), *Self and Identity: Fundamental Issues*. New York: Oxford University Press.

Rost, J.C. (1991) *Leadership for the Twenty-First Century*. New York: Praeger.

Rothstein, H.R., Schmidt, F.L., Erwin, F.W., Owens, W.A., and Sparks, C.R. (1990) Biographical data in employment selection: Can validities be made generalizable? *Journal of Applied Psychology*, 75, 175–184.

Rowland, D. (2016) Why leadership development isn't developing leaders. *Harvard Business Review*, October 14, 2016. Available at: https://hbr.org (accessed February 2, 2020).

Rowlinson, S., Ho, K.K., and Ph-Hung, Y. (1993) Leadership style of construction managers in Hong Kong. *Construction Management and Economics*, 11, 455–465.

Ryff, C.D. and Keyes, C.L.M. (1995) The structure of psychological well-being revisited. *Journal of Personality and Social Psychology*, 69, 719–727.

Sarros, J.C. (1992) What leaders say they do: An Australian example. *Leadership and Organization Development Journal*, 13(5), 21–27.

Sarros, J., Cooper, B., Hartican, A.M., and Barker, C. (2006) *The Character of Leadership: What Works for Australian Leaders – Making It Work for You*. Milton, Queensland: John Wiley and Sons, Ltd.

Schaubroeck, J., Walumbwa, F.O., Ganster, D.C., and Kepes, S. (2007) Destructive leader traits and the neutralizing influence of an "enriched" job. *The Leadership Quarterly*, 18, 236–251.

Schlenker, B.R. (1985) Identity and self-identification. In B.R. Schlenker (Ed.), *The Self and Social Life* (pp. 65–99). New York: McGraw-Hill.

Schriesheim, C.A., House, R.J., and Kerr, S. (1976) Leader initiating structure: A reconciliation of discrepant research results and some empirical tests. *Organizational Behavior and Human Performance*, 15, 297–321.

Schriesheim, C.A. and Kerr, S. (1977) Theories and measures of leadership. In J.G. Hunt and L.L. Larson (Eds.), *Leadership: The Cutting Edge* (pp. 9–45). Carbondale, IL: Southern Illinois University Press.

Schriesheim, J., Von Glinow, M.A., and Kerr, S. (1977) Professionals in bureaucracies: A structural alternative. In P.C. Nystrom and W.H. Starbuck (Eds.), *Prescriptive Models of Organizations* (pp. 55–69). New York: North-Holland.

Schutz, W. (1967) *Joy: Expanding Human Awareness*. London: Souvenir Press.

Schyns, B. and Schilling, J. (2013) How bad are the effects of bad leaders? A meta-analysis of destructive leadership and its outcomes. *The Leadership Quarterly*, 24(1), 138–158.

Scully, J.A., Henry, P.S., Judy, D.O., Eugene, R.S., and Kenneth A.S. (1996) Tough times make tough bosses: A meso analysis of CEO leader behavior. *Irish Journal of Management*, 17(1), 71–102.

Seale, C. (1999) *The Quality of Qualitative Research*. London: Sage.

Seeman, M. (1960) *Social Status and Leadership*. Columbus, OH: Ohio State University Press.

Seiling, J. and Hinrichs, G. (2005) Mindfulness and constructive accountability as critical elements of effective sensemaking: A new imperative for leaders as sensemakers. *Organization Development Journal*, 23(3), 82–88.

Seligman, M.E.P. (1999) The president's address. *American Psychologist*, 54, 559–562.

Seligman, M.E.P. (2002) Positive psychology, positive prevention, and positive therapy. In C. R. Snyder and S. J. Lopez (Eds.), *Handbook of Positive Psychology* (pp. 3–9). New York: Oxford University Press.

Seligman, M.E.P. and Csikszentmihalyi, M. (2000) Positive psychology: An introduction. *American Psychologist*, 55(1), 5–14.

Selmer, J. (2001) Expatriate selection: Back to basics ? *International Journal of Human Resource Management*, 12(8), 1219.

Seymour, D. and Elhaleem, T.A. (1991) 'Horses for courses': Effective leadership in construction. *International Journal of Project Management*, 9(4), 228–232.

Shakeel, F., Kruyen, P.M., and Van Thiel, S. (2019) Ethical leadership as process: A conceptual proposition. *Public Integrity*, 21(6), 613–624.

Shamir, B., Dayan-Horesh, H., and Adler, D. (2004) *Leading by biography: Towards a life-story approach to the study of leadership*. Paper presented at the Gallup Conference on Authentic Leadership, Omaha, USA.

Shamir, B., Dayan-Horesh, H., and Adler, D. (2005) Leading by biography: Towards a life-story approach to the study of leadership. *Leadership*, 1(1), 13–29.

Shamir, B., and Eilam, G. (2005) What's your story?: A life-stories approach to authentic leadership development. *The Leadership Quarterly*, 16(3): 395–417.

Shamir, B., House, R.J., and Arthur, M.B. (1993) The motivational effects of charismatic leadership: A self-concept theory. *Organization Science*, 4, 1–17.

Shamir, B. and Howell, J. (1999) Organizational and contextual influences on the emergence and effectiveness of charismatic leadership. *The Leadership Quarterly*, 10(2), 257–283.

Sheffield, P. (2020) The future is already here – civil engineering at the cutting edge. *Civil Engineering*, 173(1), 3–9.

Shelton, C. and Darling, J.R. (2001) The quantum skills model in management: A new paradigm to enhance effective leadership. *Leadership and Organization Development Journal*, 22(6): 264–273.

Shenhar, A.J. (2004) Strategic Project Leadership: Toward a strategic approach to project management. *R&D Management*, 34, 569–578,

Shnabel, N. and Nadler, A. (2008) A needs-based model of reconciliation: Satisfying the differential emotional needs of victim and perpetrator as a key to promoting reconciliation. *Journal of Personality and Social Psychology*, 94, 116–132.

Shultz, T.R. and Lepper, M.R. (1996) Cognitive dissonance reduction as constraint satisfaction. *Psychological Review*, 103, 219–240.

Sidani, Y.M. and Rowe, W.G. (2018) A reconceptualization of authentic leadership: Leader legitimation via follower-centered assessment of the moral dimension. *The Leadership Quarterly*, 29(6), 623–636.

Simon, L., Greenberg, J., and Brehm, J. (1995) Trivialization: The forgotten mode of dissonance reduction. *Journal of Personality and Social Psychology*, 68(2), 247–260.

Simonton, D.K. (1999) Significant samples: The psychological study of eminent individuals. *Psychological Methods*, 4, 425–451.

Sims, H.P., Jr., and Lorenzi, P. (1992) *The New Leadership Paradigm: Social Learning and Cognition in Organizations*. Newbury Park, CA: Sage.

Sims, H.P. and Manz, C.C. (1980) *Categories of observed leader communication behaviors*. Paper presented at Southwest Academy of Management Conference, March 1980, San Antonio, Texas.

Singapore Construction 21 Committee (1999) *Construction 21: Re-inventing Construction*, Singapore: Ministry of Manpower and Ministry of National Development.

Sjoberg, M., Wallenius, C., and Larsson, G. (2006) Leadership in complex, stressful rescue operations: A qualitative study. *Disaster Prevention and Management*, 15(4), 576–584.

Skipper, C.O. and Bell, L.C. (2008) Leadership development and succession planning. *Leadership and Management in Engineering*, 8(2), 77–84.

Slevin, D.P. and Pinto, J.K. (1991) Project leadership: Understanding and consciously choosing your style. *Project Management Journal*, 22(1), 39–47.

Smircich, L. and Morgan, G. (1982) Leadership: The management of meaning. *Journal of Applied Behavioral Science*, 18(3), 257–273.

Smith, G.R. (1999) Project leadership: Why project management alone doesn't work. *Hospital Material Management Quarterly*, 21(1), 88–92.

Smith, P.B., Peterson, M.F., and Schwartz, S.H. (2002) Cultural values, sources of guidance, and their relevance to managerial behavior: A 47-nation study. *Journal of Cross-Cultural Psychology*, 33, 188–208.

Songer, A., Chinowsky, P., and Butler, C. (2006) *Emotional intelligence and leadership behavior in Construction executives*. Paper presented at 2nd Specialty Conference on Leadership and Management in Construction, May 4–6, 2006, Grand Bahama Island, Bahamas.

Songer, A., Chinowsky, P., and Butler, C. (2008) Leadership behaviour in construction executives: Challenges for the next generation. *Construction Information Quarterly*, 10(2), 59–66.

Sooklal, L. (1991) The leader as a broker of dreams. *Human Relations*, 44, 833–850.

Sosik, J.J. (2005) The role of personal values in the charismatic leadership of corporate managers: A model and preliminary field study. *The Leadership Quarterly*, 16, 221–244.

Sosik, J.J., Avolio, B.J., and Jung, D.I. (2002) Beneath the mask: Examining the relationship of self-presentation attributes and impression management of charismatic leadership. *The Leadership Quarterly*, 13, 217–242.

Sparrowe, R.N. (2005) Authenticity and the narrative self. *The Leadership Quarterly*, 16(3), 419–439.

Spears, L. (1995) Introduction: Servant-leadership and the Greenleaf legacy. In L. Spears (Ed.), *Reflections on Leadership* (pp. 1–16). New York: John Wiley & Sons, Inc.

Spillane, J.P., Halverson, R., and Diamond, J.B. (2004) Towards a theory of leadership practice: A distributed perspective. *Journal of Curriculum Studies*, 36, 3–34.

Spisak, B.R., Grabo, A.E., Arvey, R.D., and van Vugt, M. (2014) The age of exploration and exploitation: Younger-looking leaders endorsed for change and older-looking leaders endorsed for stability. *The Leadership Quarterly*, 25(5), 805–816.

Stander, F.W., De Beer, L.T. and Stander, M.W. (2015) Authentic leadership as a source of optimism, trust in the organisation and work engagement in the public health care sector: Original research. *SA Journal of Human Resource Management*, 13(1): 1–12.

Stansbury, C. and Stansbury, N. (2018) Preventing corruption in the construction sector. In C. Egbu and G. Ofori (Eds.), *Proceedings of the International Conference on Professionalism and Ethics in Construction* (pp. 30–38).

Stansbury, N. (2005) Exposing the foundations of corruption in construction: Global Corruption Report. Transparency International. Available at: www.transparency.org

Staub, E., Pearlman, L., Gubin, A., and Hagengimana, A. (2005) Healing, reconciliation, forgiving and the prevention of violence after genocide or mass killing: An intervention and its experimental evaluation in Rwanda. *Journal of Social and Clinical Psychology*, 24(3), 297–334.

Steffens, N.K., Mols, F., Haslam, S.A., and Okimoto, T.G. (2016) True to what we stand for: Championing collective interests as a path to authentic leadership. *The Leadership Quarterly*, 27(5), 726–744.

Sternberg, R.J. (2002) Smart people are not stupid, but they sure can be foolish: The imbalance theory of foolishness. In R.J. Sternberg (Ed.), *Why Smart People Can Be So Stupid* (pp. 232–242). New Haven, CT: Yale University Press.

Sternberg, R.J. (2007) A systems model of leadership: WICS. *American Psychologist*, 62(1), 34–42.

Stewart, R. (1982) A model for understanding managerial jobs and behaviour. *Academy of Management Review*, 7, 7–13.

Stogdill, R.M. (1948) Personal factors associated with leadership: A survey of the literature. *Journal of Psychology*, 25, 35–71.

Stogdill, R.M. (1974) *Handbook of Leadership: A Survey of Theory and Research*. New York: Free Press.

Storey, J., Quintas, P., Taylor, P., and Fowle, W. (2002) Flexible employment contracts and their implications for product and process innovation. *The International Journal of Human Resource Management*, 13, 1–18.

Stovel, M. and Bontis, N. (2002) Voluntary turnover: Knowledge management friend or foe? *Journal of Intellectual Capital*, 3(3), 303–322.

Strang, K.D. (2007) Examining effective technology project leadership traits and behaviours. *Computers in Human Behavior*, 23, 424–462.

Strauss, A. and Corbin, J. (1990) *Basics of Qualitative Research: Grounded Theory Procedures and Techniques*. Newbury Park, CA: Sage.

Strauss, A. and Corbin, J. (1998) *Basics of Qualitative Research: Techniques and Procedures for Developing Grounded Theory* (2nd ed.). Newbury Park, CA: Sage.

Sulloway, F.J. (1996) *Born to Rebel: Birth-Order, Family Dynamics, and Creative Lives*. New York: Pantheon Books.

Takahashi, K., Ishikawa, J., and Kanai, T. (2012) Qualitative and quantitative studies of leadership in multinational settings: Meta-analytic and cross-cultural reviews. *Journal of World Business*, 47(4), 530–538.

Tannenbaum, R. and Schmidt, W.R. (1958) How to choose a leadership pattern. *Harvard Business Review*, 36, 95–102.

Taylor, F.W. (1911) *The Principles of Scientific Management*. New York: WW Norton.

Taylor, S.S. (1999) Making sense of revolutionary change: Differences in members' stories. *Journal of Organizational Change Management*, 12(6), 524–539.

Tedeschi, J.T. and Lindskold, S. (1976) *Social Psychology: Interdependence, Interaction, and Influence*. New York: Wiley,

Tennant, S. and Fernie. S. (2014) Theory to practice: A typology of supply chain management in construction. *International Journal of Construction Management*, 14, 56–66.

Tepper, J.B. (2000) Consequences of abusive supervision. *The Academy of Management Journal*, 43(2), 178–190.

Terry, R.W. (1993) *Authentic Leadership: Courage in Action*. San Francisco: Jossey-Bass.

Tesser, A. (2002) Constructing a niche for the self: A bio-social, PDP approach to understanding lives. *Self and Identity*, 1, 185–191.

The Diversity Practice (2007) Different women, different places. Report. Cheam, UK: The Diversity Practice.

The Economic Times (2019) Skill India: CREDAI says it will train 13,000 construction workers in 2019, July 2. Available at: https://economictimes.indiatimes.com/jobs/skill-india-credai-says-will-tra in-13000-construction-workers-in-2019-20/articleshow/70046013.cms?from=mdr

The Economist Intelligence Unit (2017) *The Critical Role of Infrastructure for the Sustainable Development Goals*. London: EIU.

The Economist Intelligence Unit (2019) EIU Democracy Index 2019. www.eiu.com/public/topical_report.aspx?campaignid=democracyindex2019

Thiry, M. (2004) How can the benefits of PM Training Programs be improved? *International Journal of Project Management*, 22(1), 13–18.

Thite, M. (2000) Leadership styles in information technology projects. *International Journal of Project Management*, 18(2), 235–241,

Thomas, J. and Mengel, T. (2008) Preparing project managers to deal with complexity–advanced project management education. *International Journal of Project Management*, 26(3), 304–315.

Thompson, W.R. (1992) Systemic leadership and growth waves in the long run. *International Studies Quarterly*, 36(1), 25–48.

Thoms, P. and Pinto, J.K. (1999) Project leadership: A question of timing. *Project Management Journal*, 30(1), 19–26,.

Tichy, N.M. and Devanna, M.A. (1986) *The Transformational Leader*. New York: John Wiley & Sons, Inc.

Toor, S.R. (2006) *Leadership flashback: An antecedental perspective to authentic leadership development*. Paper presented at Second Biennial Gallup Leadership Institute Summit, October 7–9, 2006, Washington, DC, USA.

Toor, S.R. (2008) *Management from an Islamic perspective: An historic opportunity, a timely challenge*. Paper presented at COMSATS International Conference on Management (CICM),January 2–3, 2008, Lahore, Pakistan.

Toor, S.R. and Ofori, G. (2006a) *In quest of leadership in construction industry: New arenas, new challenges!* Paper presented at Joint International Conference on Construction Culture, Innovation, and Management (CCIM), November 26–29, 2006, Dubai, UAE.

Toor, S.R. and Ofori, G. (2006b) An antecedental model for leadership development. In R. Pietroforte, E. De Angelis, and F. Polverino (Eds.), *Proceedings of the Joint CIB W65/W55/W86 Symposium, Construction in 21st Century: Local and Global Challenges*.

Toor, S.R. and Ofori, G. (2006c) Leadership motives: Moving one step further. In W. Kanok-Nukulchai, W. Munasinghe, and N. Anwar (Eds.), *Proceedings of the Tenth East-Asia Pacific Structural Engineering and Construction (EASEC-10)* (pp. 111–118).

Toor, S.R. and Ofori, G. (2006d) *Role of parents in developing leaders*. Paper presented at 7th Association of Pacific Rim Universities (APRU) Doctoral Students Conference, July 17–21, 2006, National University of Singapore, Singapore.

Toor, S.R. and Ofori, G. (2007) *Leadership research in the construction industry: A review of empirical work and possible future directions.* Paper presented at Construction Management and Economics, 25th Anniversary Conference on 'Past, Present and Future', July 16–18, 2007, University of Reading, Reading, United Kingdom.

Toor, S.R. and Ofori, G. (2008a) Leadership in the construction industry: Agenda for authentic leadership development. *International Journal of Project Management*, doi:10.1016/j.ijproman.2007.09.010.

Toor, S.R. and Ofori, G. (2008b) Taking leadership research into future: A review of empirical studies and new directions for research. *Engineering, Construction and Architectural Management*, 15(4), 352–371.

Toor, S.R. and Ofori, G. (2008c) Grounded theory as the most appropriate methodology for leadership research in the construction industry. In R. Haigh and D. Amaratunga (Eds.), *Proceedings of CIB International Conference on Building Education and Research: 'Building Resilience'* (pp. 1816–1831).

Toor, S.R. and Ofori, G. (2008d) Tipping points that inspire leadership: An exploratory study of emergent project leaders. *Engineering, Construction and Architectural Management*, 15(3), 312–329.

Toor, S.R. and Ofori, G. (2008e) Developing construction professionals of 21st century: A renewed vision for leadership. *ASCE Journal of Professional Issues in Engineering, Education, and Practice*, 134(3), 279–286.

Toor, S.R. and Ofori, G. (2008f) *Vigilant corporate leadership: A key to veritable performance of construction-related organizations in Singapore.* Paper presented at Specialty Conference on Leadership and Management in Construction, October 16–19, 2008, South Lake, Tahoe, USA.

Toor, S.R. and Ofori, G. (2008g) *Role of psychological capital (PsyCap) in leadership effectiveness.* Paper presented at CIB Joint International Symposium: Transformation through Construction, November 17–19, 2008, Dubai, UAE.

Toor, S.R. and Ofori, G. (2009a) Ethical leadership: Examining the relationships with full range leadership model, employee outcomes, and organizational culture. *Journal of Business Ethics*, 90, 533. https://doi.org/10.1007/s10551-009-0059-3.

Toor, S.R. and Ofori, G., (2009b) Authenticity and its influence on psychological well-being and contingent self-esteem of leaders in Singapore construction sector. *Construction Management and Economics*, 27(3), 299–313.

Toor, S.R., Ofori, G., and Arain, F.M. (2007) Authentic leadership style and its implications in project management. *Business Review*, 2(1), 31–55.

Toor, S.R. and Ogunlana, S.O. (2005) *What is crucial for success: Investigating the critical success factors and key performance indicators for mega construction projects.* Paper presented at SPMI Annual Symposium, October 12, 2005, Singapore.

Toor, S.R. and Ogunlana, S.O. (2006) *Successful project leadership: Understanding the personality traits and organizational factors.* Paper presented at CIB-W107, International Symposium, Construction in Developing Economies: New Issues and Challenges, January 18–20, 2006, Chile, Santiago.

Toor, S.R. and Ogunlana, S.O. (2008) Leadership skills and competencies for cross-cultural construction projects. *International Journal of Human Resources Development and Management*, 8(3), 192–215.

Toor, S.R. and Ogunlana, S.O. (2009) Ineffective leadership: Investigating negative attributes of project leaders and organizational neutralizers. *Engineering, Construction and Architectural Management*, 16(3).

Tugade, M.M. and Fredrickson, B.L. (2004) Resilient individuals use positive emotions to bounce back from negative emotional experiences. *Journal of Personality and Social Psychology*, 86, 320–333.

Tulving, D. (2005) Episodic memory and autonoesis: Uniquely human? In H.S. Terrace and J. Metcalfe (Eds.), *The Missing Link in Cognition: Origins of Self-Reflective Consciousness* (pp. 3–56). Oxford: Oxford University Press.

Transparency International (2011) Bribepayers' index. Berlin: Transparency International.

Trauth, E.M. (2001) *Qualitative Research in IS: Issues and Trends.* Hershey, PA: Idea Publishing.

Treviño, L.K. and Brown, M.E. (2004) Managing to be ethical: Debunking five business ethics myths. *Academy of Management Executive*, 18, 69–81.

Treviño, L.K., Brown, M., and Hartman, L.P. (2003) A qualitative investigation of perceived executive ethical leadership: Perceptions from inside and outside the executive suite. *Human Relations*, 55, 5–37.

Treviño, L.K., Hartman, L.P., and Brown, M. (2000) Moral person and moral manager: How executives develop a reputation for ethical leadership. *California Management Review*, 42(4), 128–142.

Uhl-Bien, M. (2006) Relational leadership theory: Exploring the social processes of leadership and organizing. *The Leadership Quarterly*, 17(6), 654–676.

UK Green Building Council (2016) *Health and Wellbeing in Homes*. London: UK Green Building Council.

UK Green Building Council (2019) *Circular Economy Guidance for Construction Clients: How to practically apply circular economy principles at the project brief stage*. London: UK Green Building Council.

United Nations (2017) *New Urban Agenda*. New York: United Nations.

United Nations (2020) *The Sustainable Development Goals Report 2020*. New York: United Nations.

University of Cambridge Institute for Sustainability Leadership (2017) *Global Definitions of Leadership and Theories of Leadership Development: Literature Review*. Cambridge: University of Cambridge.

van de Vliert, E. (2006) Autocratic leadership around the globe: Do climate and wealth drive leadership culture? *Journal of Cross-Cultural Psychology*, 37, 42–59.

van Dierendonck, D. (2011) Servant leadership: A review and synthesis. *Journal of Management*, 37(4), 1228–1261.

van Eijnatten, F.M., Putnik, G.D., and Sluga, A. (2007) Chaordic system thinking for novelty in contemporary manufacturing. *Manufacturing Systems: Proceedings of the CIRP Seminars*, 56(1), 447–450.

van Knippenberg, D.L. and Hogg, M.A. (2003) A social identity model of leadership effectiveness in organizations. *Research in Organizational Behavior*, 25, 243–295. doi:10.1016/S0191-3085(03)25006-1.

van Vugt, M. and von Rueden, C.R. (2020) From genes to minds to cultures: Evolutionary approaches to leadership. *The Leadership Quarterly*, 31. https://doi.org/10.1016/j.leaqua.2020.101404.

van Wart, M. (2005) *Dynamics of Leadership in Public Service: Theory and Practice*. Armonk, NY: M.E. Sharpe.

van Wart, M. (2014) Contemporary varieties of ethical leadership in organizations. *International Journal of Business Administration*, 5(5), 27. doi:10.5430/ijba.v5n5p27.

Varella, P., Javidan, M., and Waldman, D. (2005) The differential effects of socialized and personalized leadership on group social capital. In W.L. Gardner, B.J. Avolio, and F.O. Walumbwa (Eds.) *Authentic Leadership Theory and Practice: Origins, Effects and Development* (pp. 107–137). Oxford: Elsevier JAI.

Vee, C. and Skitmore, M. (2003) Professional ethics in the construction industry. *Engineering, Construction and Architectural Management*, 10(2), 117–127.

von Rueden, C. and van Vugt, M. (2015) Leadership in small-scale societies: Some implications for theory, research, and practice. *The Leadership Quarterly*, 26(6), 978–990. Vroom, V.H. and Jago, A.G. (1988) *The New Leadership: Managing Participation in Organizations*. Englewood Cliffs, NJ: Prentice-Hall.

Vroom, V.H. and Jago, A.G. (2007) The role of situation in leadership. *American Psychologist*, 62, 17–24.

Vroom, V.H. and Yetton, P.W. (1973) *Leadership and Decision-Making*. Pittsburgh, PA: University of Pittsburgh Press.

Waldman, D.A. and Siegel, D. (2008) Defining the socially responsible leader: Theoretical and practitioner letters. *The Leadership Quarterly*, 19, 117–131.

Walker, D., Myrick, F. (2006) Grounded theory: An exploration of process and procedure. *Qualitative Health Research*, 16(4), 547–559.

Walumbwa, F.O., Avolio, B.J., Gardner, W.L., Wernsing, T.S., and Peterson, S.J. (2008) Authentic leadership: Development and validation of a theory-based measure. *Journal of Management*, 34(1), 89–126. http://dx.doi.org/10.1177/0149206307308913.

Wamelink, H. and Kahkonen, K. (2016) Roadmap 'Organisation and Management of Construction', CIB W065. Available at: www.cibworld.nl/app/attach/tLFKJEfc/20152501/b9463c64b0ec2868042 06c5dfef413c7/Hans_Wamelink.pdf

Watermeyer, R. (2018) Client guide for improving infrastructure project outcomes. Johannesburg: Engineers Against Poverty and University of the Witwatersrand.

Watson, L.A. and Girard, F.M. (2004) Establishing integrity and avoiding methodological misunderstanding. *Qualitative Health Research*, 14(6), 875–881.

Watts, S. (2017) Operation Car Wash: Is this the biggest corruption scandal in history? *The Guardian*, 1 June. Available at: www.theguardian.com/world/2017/jun/01/brazil-operation-car-wash-is-this-the-biggest-corruption-scandal-in-history

Weber, M. (1968) *Max Weber on Charisma and Institution Building*. Chicago: The University of Chicago Press.

Weichun, Z., Chew, I.R.K., and Spangler, W.D. (2005) CEO transformational leadership and organizational outcomes: The mediating role of human-capital-enhancing human resource management. *The Leadership Quarterly*, 16, 39–52.

Weick, K. (1993) The collapse of sensemaking in organizations: The Mann Gulch disaster. *Administrative Science Quarterly*, 38, 628–652.

Weick, K.E. (1995) *Sensemaking in Organizations*. Thousand Oaks, CA: Sage.

Weick, K.E. (2001) *Making Sense of the Organization*. Oxford: Blackwell,

Weick, K.E., Sutcliffe, K.M., and Obstfeld, D. (2005) Organizing and the process of. sensemaking. *Organization Science*, 16(4), 409–421.

Wenger, E. and Snyder, W. (2000) Communities of practice: The organizational frontier. *Harvard Business Review*, January–February, 139–145.

Whittington, J.L.Kageler, W., Pitts, T., and Goodwin, V. (2005) Legacy leadership: The leadership wisdom of the apostle Paul. *The Leadership Quarterly*: Special Edition on Spirituality in Leadership, 16(4).

Wilber, K. (1996) *A Brief History of Everything*. Dublin: Newleaf.

Winch, G. M. (2014) Three domains of project organising. *International Journal of Project Management*, 32(5), 721–731.

Woetzel, J., Garemo, N., Mischke, J., Kamra, P., and Palter, R. (2017) Bridging infrastructure gaps: Has the world made progress? Available at: www.mckinsey.com/industries/capital-projects-and-infrastructure/our-insights/bridging-infrastructure-gaps-has-the-world-made-progress#section%201

Wong, J., Wong, P.N., and Li, H. (2007) An investigation of leadership styles and relationship cultures of Chinese and expatriate managers in multinational construction companies in Hong Kong. *Construction Management and Economics*, 25, 95–106.

Woolley, L., Caza, A., and Levy, L. (2011) Authentic leadership and follower development: Psychological capital, positive work climate, and gender. *Journal of Leadership and Organizational Studies*, 18(4), 438–448.

World Economic Forum (2014) *Global Leadership Index: Survey on the Global Agenda*. Geneva: World Economic Forum.

World Economic Forum (2015) *The Global Competitiveness Report 2015–2016*. Geneva: World Economic Forum.

World Economic Forum (2016a) *The Global Risks Report 2016*. Geneva: World Economic Forum.

World Economic Forum (2016b) *Shaping the Future of Construction: A Breakthrough in Mindset and Technology*. Geneva: World Economic Forum. Available at: www3.weforum.org/docs/WEF_Shaping_the_Future_of_Construction_full_report__.pdf

World Economic Forum (n.d.) *Future of Construction – Challenges*. Available at: https://futureofconstruction.org/challenge/.

World Economic Forum and Boston Consulting Group (2018) *An Action Plan to Solve the Industry's Talent Gap*. Geneva: World Economic Forum.

World Green Building Council Europe Regional Network (2019) A sustainable built environment at the heart of Europe's future. Available at: https://worldgbc.org/sites/default/files/WorldGBC%20European%20Advocacy%20Manifesto%20June%202019_0.pdf

World Public Opinion (2019) World poll finds global leadership vacuum. Available at: http://worldpublicopinion.net/wp-content/uploads/2019/12/WPO_Leaders_Jun08_packet.pdf

Wurf, E. and Markus, H. (1991) Possible selves and the psychology of personal growth. In D.J. Ozer, J.M. Healy, and A.J. Stewart (Eds.), *Perspectives in Personality: Self and Emotion* (pp. 39–62). Greenwich, CT: JAI Press.

Wyer, R.S. and Srull, T.K. (1989) *Memory and Cognition in Its Social Context*. Hillsdale, NJ: Erlbaum.

Young, N.W. Jr. and Bernstein, H.M. (2006) *Key Trends in the Construction Industry – 2006: SmartMarket Report*. New York: McGraw Hill.

Yukl, G. (1970) Leader LPC scores: Attitude dimensions and behavioral correlates. *Journal of Social Psychology*, 80, 107–212.

Yukl, G. (1994) *Leadership in Organizations*. Englewood Cliffs, NJ: Prentice-Hall.

Yukl, G. (1999) An evaluation of conceptual weaknesses in transformational and charismatic leadership theories. *The Leadership Quarterly*, 10(2), 285–306.

Yukl, G. (2002) *Leadership in Organizations* (5th ed.). Upper Saddle River, NJ: Prentice-Hall

Yukl, G. (2006) *Leadership in Organizations* (6th ed.). Upper Saddle River, NJ: Pearson Education.

Zaccaro, S.J. (2007) Trait-based perspectives of leadership. *American Psychologist*, 62, 6–16.

Zaccaro, S.J. and Horn, Z.N.J. (2003) Leadership theory and practice: Fostering an effective symbiosis. *The Leadership Quarterly*, 14(6), 769–806.

Zaccaro, S.J., Kemp, C.F., and Bader, P. (2004) Leader traits and attributes. In J. Ankonakis, A.T. Cianciolo, and R.J. Sternberg (Eds.), *Nature of Leadership*. Thousand Oaks, CA: Sage.

Zaccaro, S.J., Rittman, A.L., and Marks, M.A. (2001) Team leadership. *The Leadership Quarterly*, 12, 451–483.

Zacharatos, A., Barling, J., and Kelloway, E.K. (2000) Development and effects of transformational leadership in adolescents. *The Leadership Quarterly*, 11(2), 211–226.

Zaghloul, R.S. (2005) Risk allocation in contracts: How to improve the process. Unpublished doctoral thesis. University of Calgary, Calgary, AB. doi:10.11575/PRISM/17534.

Zaleznik, A. (1977) Managers and leaders: Are they different? *Harvard Business Review*, 55(3), 67–78.

Zanzi, A. and O'Neill, R. (2001) Sanctioned versus non-sanctioned political tactics. *Journal of Managerial Issues*, 13(2), 243–262.

Zhou, Z., Irizarry, J., Li, Q. and Wu, W. (2015) Using grounded theory methodology to explore the information of precursors based on subway construction incidents. *Journal of Management in Engineering*, 31(2). doi:10.1061/(ASCE)ME.1943-5479.0000597.

Zitzman, L. (2020) Women in construction: The state of the industry in 2020. *BigRentz*. Available at: www.bigrentz.com/blog/women-construction

Zillman, C. (2019) The Fortune 500 has more female CEOs than ever before. *Fortune*, May 16, 2019. Available at: https://fortune.com/2019/05/16/fortune-500-female-ceos/

Zyphur, M. J., Narayanan, J., Koh, G., and Koh, D. (2009) Testosterone-status mismatch lowers collective efficacy in groups: Evidence from a slope-as-predictor multilevel structural equation model. *Organizational Behavior and Human Decision Processes*, 110, 70–77.

Index

Printed in the United States
By Bookmasters